Light, Water, Hydrogen
The Solar Generation of Hydrogen by Water Photoelectrolysis

Craig A. Grimes • Oomman K. Varghese • Sudhir Ranjan

Light, Water, Hydrogen
The Solar Generation of Hydrogen by Water
Photoelectrolysis

 Springer

Craig A. Grimes
Pennsylvania State University
Department of Electrical Engineering
Department of Materials Science & Engineering
217 Materials Research Lab.
University Park, PA 16802

Oomman K. Varghese
Pennsylvania State University
Materials Research Institute
208 Materials Research Lab.
University Park, PA 16802

Sudhir Ranjan
Pennsylvania State University
Materials Research Institute
208 Materials Research Lab.
University Park, PA 16802

ISBN 978-0-387-33198-0   e-ISBN 978-0-387-6828-9

Library of Congress Control Number: 2007933414

Printed on acid-free paper.

9 8 7 6 5 4 3 2 1

springer.com

TP
359
.H8
G75
2008

Foreword

In addition to domestic animals the earliest records of mankind indicate that slavery, until the use of coal became widespread, has always been a significant aspect, or part, of nearly every society. Consider for example ancient Attica (Greece), in which 115,000 out of a total population of 315,000 were slaves [1]. For the lucky rulers slaves represented power, Joule/second or Watt. On a steady state basis a healthy adult generates about 100 Watts, or 100 J/s, while a highly conditioned endurance athlete can generate about 300 W for perhaps an hour. Today we obtain our energy from fossil fuels, that magical brew of latent-heat chemistry that allows us to run the world without having to rely on people or domestic animal power. We owe much if not all of modern civilization to fossil fuels, no more than stored solar energy, which provide the 40-plus Terawatts that annually powers the $\approx$ 7,000,000,000 people on this planet, with our fossil fuel burn rate growing to accommodate the annual increase of some additional 100,000,000 or so souls.

The foundation of modern society is a pile (lake) of priceless, irreplaceable fossil fuel that, by any measure of the energy you get and what you pay, is all intents free, and being virtually free we have and continue to burn our way through it as fast as we possibly can. It is the tragedy of the (fossil fuel) commons. Take away fossil fuels with their life giving energy and for all intents and purposes you are back in the 16th century, with an impact that should be obvious. A gallon of gasoline has an energy equivalent of 121.8 MJ, a remarkable number that is entirely sufficient to explain the modern politics of the Middle East and the vast military presence there of the United States. Converted to 100 W people power, a single gallon of gasoline is equivalent in energy to the fulltime dedication of 14 people for 24-hours. From that perspective gasoline at $100/gallon can be considered a rare bargain.

Since fossil fuels are so remarkably energy dense (a single tank of gas can move your fully laden inefficient car hundreds of miles), essentially free (try walking instead), and relatively safe to use its good that they will continue to be freely available to us for the imaginable future, that is the next few years. A couple of decade's worth of oil remains, at least for the fortunate rulers with the strongest armies. Depending upon the expert, the earth has 80 to 300 years or so of coal, and about 50 years worth of natural gas, with the supply lifetimes dependent upon a variety of factors such as the desire of rich countries to turn their coal into gasoline.

However experiencing fossil fuel depletion, today, would be a great thing from the perspective of keeping modern civilization intact, as otherwise we will simply cook ourselves, having set the house on fire and then finding there are no exits, no way out from planet earth. In a collective effort we have succeeded in depositing vast amounts of CO_2 in the atmosphere, today reaching the highest concentrations seen on this planet in, at least, the last 500,000 years. As we keep on doing what we are doing (burning fossil fuels) to provide the more than 40 Terawatts of energy we use every year earth's atmospheric CO_2 levels will reach heights never, to our knowledge, seen outside of planet Venus. Since atmospheric CO_2 lasts a long time in the atmosphere, and acts as an insulator trapping heat, the logical outcome is a very hot earth for a very long time. Not an ice age, but a steam age. It will be a Saurian hot house for future generations with, finally, the oceans receiving the vast amount of atmospheric CO_2 and subsequently becoming acidic. So we can say to future generations, "Sorry for the mess," and "Good Luck."

To summarize the issue, approximately 7 billion people on the planet the support of whom is virtually all based upon fossil fuels which are: Point {1} rapidly being depleted, and Point {2} when burned in such vast quantities, after a modest time lag, appears likely to make life on earth un-tenable but for a substantially reduced population. Point {3}, there is no backup plan, no Energy Plan B on how we might even begin to provide the vast amounts of energy used by humans on this planet once fossil fuels are either depleted or their use made politically unacceptable due to environmental consequences. In essence modern society is being bet on a Faith-based Energy Policy, that is to say let us hope for a miracle. Point {4}, the discussion here is not of another crisis du jour, of which most of us are familiar and can ignore without consequence, rather an increasingly inescapable grinding reality of our future that was first foretold in 1949 by M. K. Hubbert [2]; since 1949 the story has stayed the same but the ending grown increasingly unpleasant.

Perhaps the above noted points are not to be considered a problem. Nature defines good as that which survives, and bad as that which goes under; if we have some intrinsic weakness as a species that allows us to cook ourselves while feasting at the fossil fuel table maybe that's just how it is, and we shouldn't dwell upon it. On one point all historians agree, civilizations begin, flourish, decline, and disappear. "They failed as a species since they were unable to look beyond their immediate gratification," an observer might one day write. Another could pencil "their civilization declined through failure of its intellectual and political leaders to meet the challenges of change. Alas."

Let us consider where the path we are on leads. Notably it appears one that includes the potential for massive wars over the remaining fossil fuel supplies. Whatever the pretences, the real points of interest for the Middle East deserts is the oil that lies beneath them, an interest the United States at least does not take lightly.[1] Without a viable Energy Plan B as the energy-noose

---

[1] U. S. Military presence in the Middle East: BAHRAIN: Navy 5th Fleet headquarters - 1,200 sailors; Joint Venture HSV-X1 – 50 troops. DJIBOUTI: Camp Lemonier – 1,300 U.S. troops. HORN OF AFRICA: Elements of Combined Joint Task Force Horn of Africa, 10th Mountain Division, 478th Civil Affairs Division; about 1,300 troops. MEDITERRANEAN: Harry S. Truman Carrier Battle Group/Carrier Air Wing 3 (Marine Fighter-Attack Squadron 115) – 7,610 sailors and Marines; Theodore Roosevelt Carrier Battle Group/Carrier Air Wing 8 – 7,445 sailors; 26th Marine Expeditionary Unit trains aboard Iwo Jima Amphibious Ready Group. PERSIAN GULF: Amphibious Task Force East – 5,000 sailors; Amphibious Task Force West – 4,080 sailors; Tarawa Amphibious Ready Group w/15th Marine Expeditionary Unit – 1,700 sailors, 2,200 Marines; 2nd Marine Expeditionary Brigade returns to ships of Amphibious Task Force East; Echo Company, Battalion Landing Team 2nd Battalion (with Nassau Amphibious Ready Group); Coast Guard cutters (2) and patrol boats (4) - 690 Coast Guardsmen. RED SEA: Attack submarine Boise – 112 sailors; Attack submarine Toledo – 112 sailors; Attack submarine San Juan – 112 sailors. TURKEY: Elements of 1st Infantry Division – 2,000 soldiers; Incirlik Air Base – F-15 and F-16 aircraft, 4,000 airmen. DIEGO GARCIA: AF 20th Bomb Squadron; 917th Bomb Wing Air Force Reserve. EGYPT: 1st Battalion, 180th Infantry Regiment, Oklahoma National Guard - 865 soldiers. GULF OF ADEN:

tightens pressures on governments to obtain more of a precious, vitally needed dwindling resource will grow more intense, resulting in ever-larger military establishments and appropriations, with the freedoms of democracy disappearing to the discipline of arms. History indicates time and again such statements to be fully accurate. We compete peacefully when there is enough to go around, but let the mouths out-run the food and it is us-versus-them, and violent. Yet for all that, as climate changes go non-linear a few more wars will probably be the least of our troubles.

We submit that it would be nice to pass on a civilized heritage to our children. If this is to be accomplished over the next generation it will require many creative individuals with initiative and clarity of mind, and resources, to rise to meet this challenge of energy. How do we power a planet of 10 billion souls (2050 estimated population) without cooking ourselves by the release of more carbon? If we accept this is a problem, what are our options? Our apparent default option is to do nothing, choose the *Everything is OK* option and ultimately fight other countries over the dwindling pile of fossil fuels until they are gone, or we run out of water, or food. Although this is not actually a solution, let's ask if it is cost

---

Command ship Mount Whitney - 700 sailors, 400 troops. IRAQ: 82nd Airborne; 3rd Infantry Division; 4th Infantry Division; 101st Airborne Division; 173rd Airborne Brigade; V Corps; 1st Armored Division - 250,000 soldiers. I Marine Expeditionary Force. KUWAIT: Elements of the 101st Airborne Division - about 20,000 soldiers; Elements of 3rd Infantry Division - 13,500 soldiers; 325th Airborne Infantry Regiment, 82nd Airborne Division – 4,000 soldiers; Other Army elements – 10,800 soldiers; Army reservists – 5,299 soldiers; Elements of 293rd Infantry, Indiana National Guard - 600 soldiers; 190th Fighter Squadron, Idaho National Guard - 200 soldiers; Elements of I Marine Expeditionary Force – 45,000 Marines; Regimental Combat Team I – 6,000-7,000 Marines; 15th Marine Expeditionary Unit – 2,200 Marines; A-10 and F-16 aircraft; 2nd Marine Expeditionary Brigade – 6,000 to 7,000 Marines; 1042nd Medical Company, Oregon National Guard - 18 soldiers. OMAN: B-1B bombers and AC-130 gunships. QATAR: Al Udeid Air Base – F-15 and F-16 fighters, KC-135s and KC-10s, 3,500 airmen; Camp As Sayliyah - Central Command battle command; 205th Area Support Medical Battalion. SAUDI ARABIA: Prince Sultan Air Base – 4,500 US military personnel, un-disclosed number of F-15 and F-16 fighters; 1042nd Medical Company, Oregon National Guard - 10 soldiers. UNITED ARAB EMIRATES: Al Dhatra Air Base – reconnaissance aircraft, 500 airmen. From http://www.militarycity.com/map/

effective? While we do not mean to imply the current war in Iraq has or had anything to do with its significant oil reserves, solely as a point of reference the war in Iraq currently (Summer, 2007) has cost the U.S. approximately $500,000,000,000, and many, many lives. Multiplying this effort across the half-dozen remaining oil rich countries, against several opponents, indicates costs that would be difficult to long sustain.

One can consider other energy options. For example, to supply 40 to 60 Terawatts of energy via nuclear fission is possible, it could be done. However it necessitates increasing by almost a factor of x500 the number of nuclear power plants ever built. The consequence of such demand is that we would soon deplete earth's uranium supplies. Breeder reactors are an un-stable possibility, like mixing matches, children, and gasoline. Depending upon ones viewpoint fusion remains either a to be hoped for miracle, or an expensive civil-works project.

At the end of the energy discussions there realistically appears to be only one solution: inexpensive yet efficient means of harvesting solar energy. Energy politics currently in vogue look to solve this energy-civilization conundrum by growing plants that we can subsequently turn into automobile fuel. While this is an excellent scenario for buying the votes of farmers, unfortunately the solar to fuel conversion efficiencies of plants are quite low, on an annual basis approximately 0.1%. Furthermore the crops of plants we plan to use on a massive scale as fuel for the ≈ 1 billion cars on the planet are already used as either food, or the organic matter that soil is made of which sustains agriculture. Additionally global warming has kicked-in enough to create severe droughts across wide swaths of land that used to support large harvests, and it looks to get significantly worse. No doubt the rich will be able to drive biofuel based cars while the poor lack food to eat, providing more grist for the philosophers. We certainly can and need to do more with the harvesting of wind energy, however there are not a lot of suitable high-wind activity locations. Furthermore global warming models indicate we are headed for tremendous droughts punctuated by tremendous storms. The point being, expensive wind turbines don't last long in a tornado nor Category 5 hurricane.[2]

Direct, inexpensive hence widespread conversion of solar energy into electrical energy, with efficient means to store it, is our best and realistically only scenario for avoiding the issues summarized above. While hydrogen is not an ideal fuel (save for its associated CO_2 nothing is as good a fuel as petroleum), it appears to be the best foreseeable option. In combination with a suitable semiconductor, the combination of water and light results in hydrogen and oxygen via water splitting; this is the general topic of our book. This vision begs the rhetorical question of 'How hard is this to do, and how are we going to do it?' The answer is it is hard, and we are not going to do it unless we at least try. However the fact that some major league baseball players have annual salaries roughly equal to what the United States spends on supporting research towards new solar energy technologies is not an encouraging sign. Actually we do not mean to pick on major league baseball players, the reality is that more money is spent on almost *anything* one can think of than on developing solar energy technologies. It is as if our political leaders want our future societies to fail, as if we are trapped in an eternal present.

Unless we are willing to bet modern civilization on a miracle, which we currently are, it will take a substantial investment in solar energy technologies to get us out of the mess we are in, well beyond the several tens of millions of dollars now annually invested in the field. It will require nano to km scale engineering at the highest levels, with the best and brightest minds the world has. It will take a scale of investment similar to what the U.S. has spent on the latest Iraq war, which might be considered the second of the recent major oil wars. We submit that everyone, from top to bottom on this planet would be vastly better off if we invested two billion dollars per week on developing low cost solar energy technologies today, then spend vastly, vastly larger amounts fighting an end-game over the dwindling supplies of fossil fuel energy that currently keep

---

[2] Note: If you would like to see the hard numbers on the global energy perspective presented in a cogent, concise manner we strongly suggest to all a web-broadcast talk by Professor Nate Lewis of Caltech at http://nsl.caltech.edu/energy.html

our economies functioning, and keeps modern civilization modern. We had best get started.

Craig A. Grimes
Oomman K. Varghese
Sudhir Ranjan

The Pennsylvania State University
University Park, Pennsylvania 16802

1.   Gomme AW (1933) The population of Athens in the fifth and fourth centuries B.C. Oxford Press, London pp. 20-26.
2   Hubbert MK (1949) Energy from fossil fuels. Science 109:103-109.

Preface

The solar production of hydrogen by water photoelectrolysis is an open field in which chemists, electrical engineers, material scientists, physicists, and others practice together. This book is intended as a modern text on the subject for an advanced (undergraduate and up) reader that leads, hopefully, to a point from which the current literature in much of the field can be read with a critical understanding and appreciation. Semiconductors are intrinsic to and hence implicit in water photoelectrolysis, yet the astute reader will have already noticed 'semiconductor' is missing from the title; we mean no disrespect to semiconductors, rather we prefer short and hopefully catchy titles.

The authors have tried to distill the subject, and the literature, to its problem solving essence. That is to say, the authors are aware that more importantly than another book on this topic we need problem solutions leading to affordable, commercial-scale hydrogen production by water photoelectrolysis. To that end the authors have tried to bring the discussion to the key problem-solving points of the field.

The sequence of chapters is chosen to give the reader an appreciation of the field and technical problems that currently limit it. While Chapter 1 is the Introduction proper, Chapter 2 is dedicated to an introductory overview of other, at least conceptually renewable means of hydrogen generation including water electrolysis, thermochemical water splitting, and water biophotolysis. An understanding of water photoelectrolysis systems cannot be achieved without understanding the charge-transfer characteristics of semiconductors. Hence the core of the text associated with the title begins with Chapter 3, by examining foundational concepts of semiconductor theory, then describing their application to water photoelectrolysis with immersion of the semiconductor in an electrolyte coupled with illumination for electron-hole pair generation. Since various publications give an assortment of reports on photoconversion efficiency, several of which have drawn considerable response if not ire, Chapter 3 also includes a discussion of photoconversion efficiency determination.

Chapters 4 through 7 consider the properties and application of semiconductor electrodes for water photoelectrolysis with respect to photocorrosion, band edge matching, light absorption and charge transfer properties.   Chapter 4 examines the use of oxide semiconductor electrodes as a function of composition and crystallinity for water photoelectrolysis in photoelectrochemical cells (anode, cathode), with Chapter 5 considering the effects of the material architecture, or geometry, on the resultant photoelectrolysis properties of TiO_2. In Chapter 6 water splitting by suspended oxide nanoparticles is considered.   The application of non-oxide semiconductors to water splitting is the topic of Chapter 7, including both photoelectrochemical cells and suspended solutions. Particular interest is paid to schemes by which the low bandgap materials, ideally suited for collecting solar spectrum energy, can be protected from photocorrosion to enable their practical application. Chapter 8 ties back to the topic of Electrolysis presented in Chapter 2, with the source of the electrical energy driving the reaction obtained from semiconductor solar cells.

As noted, the water photoelectrolysis field is inherently interdisciplinary, variously attracting and mixing the work of chemists, electrical engineers, material scientists, and physicists. Each field has its own nuances of description, which makes a book such as this a mix of units, terminology, and symbols. We have done our best to concisely describe the subject matter in way a generic scientist, regardless of the discipline in which they hope to or have received their Ph.D., could understand.   Certainly the successful researcher in the field must be able to appreciate and integrate concepts and principles that cut across traditional disciplinary boundaries.

Selection of subject matter for the book was, of course, a most difficult task. It is noted that at one time a desk of one of the authors was loaded with three thousand journal papers on the subject (!). Simply listing the papers relevant to the field in a legible font size would make a book at least half the width of that which you are holding. Ultimately though decisions have to be made regarding what to include, and what not, and of those works included what were the key points and key outcomes, and where does the kernel of knowledge in the work lead us? While we have done our best within personal limitations we must acknowledge the feeling, in reviewing and distilling the literature, that we perhaps missed more than we caught.

Finally we note that there are many fields where the output of a researcher doesn't have to be correct, that is to say there is no end product that needs to work or be validated. In contrast the focus of photoelectrolysis research is to obtain a material system that is affordable and does work. The combination of this affordable semiconductor material architecture with light, and water, to make hydrogen will validate the field as something more than a scientific curiosity or laboratory demonstration, rather something that can save a significant fraction of the world. To that end we hope this text contributes meaningful insight to the issues, spanning a variety of fields, helping the researcher to take the next step to our desired objective.

Craig A. Grimes
Oomman K. Varghese
Sudhir Ranjan

The Pennsylvania State University
May 2007

Acknowledgment

Partial support of this work by the Department of Energy under grant DE-FG02-06ER15772 is gratefully acknowledged.

Contents

Chapter 1

FROM HYDROCARBONS TO HYDROGEN: TOWARDS A SUSTAINABLE FUTURE

1.1 Introduction

The backbone of civilization is formed by our energy system, which facilitates the advancement of technology and, in turn, a higher standard of living. Energy is a fundamental component of productivity, as important as raw materials, capital and labor. The combination of human energy amplified by non-human energy is the basis of the high standard of living in the industrialized nations. The discovery and use of new fuels of higher energy content has seen a corresponding increase in the advancement of technology and world human population. The transition from wood to coal fueled the industrial revolution in the 18th and 19th centuries and from coal to petroleum products, a shift from solid to liquid state fuel, assisted the unprecedented technological development that revolutionized the living standards in the latter half of the 20th century. Energy dense fossil fuels including coal, petroleum - the most coveted fuel that has ever been discovered, shale oil, tar sands and natural gas are mineral organic compounds embedded in the earth's crust. According to biogenic theory, these were formed as a result of the transformation of biological residue under high temperature and pressure for millions of years. In addition to their combustion for energy, fossil fuels are used as raw materials for making plastics, paints, fibers, rubber, drugs, cloth, chemicals, drugs, and lubricants.

Sustainable development is a strategic goal of modern society reflecting contemporary demand for economic, social, political and environmental development. Access to affordable and reliable energy drawn from environmentally acceptable sources of supply is an important feature of sustainable development. According to the 1987 Brundtland Report from the United Nations [1] **Sustainable development** is a process of developing land, cities, business, communities, etc. that "meets the needs of the present without compromising the ability of future generations to meet their own needs." The report continues: *When a kilogram of fuel is burned, it makes 3.2 kilograms of carbon dioxide and 1.0 kg of water. The extraction and combustion of fossil fuels is a major threat to the environment because of land damage, smog, acid rain and changes in the composition of the atmosphere. Environmental damage and atmospheric changes may soon alter the weather and climate patterns of the earth resulting in grave problems of all its inhabitants. Hydrocarbons have extraordinary value as the source of chemicals used to produce goods and other essentials for living beings.*

Unfortunately our current societies and hence populations are not sustainable. In fact they are in peril, and it behooves us to immediately consider the situation and act. The problem arises in that our societies are now largely based on the use of fossil fuels. Consider:

1.1.1 Problem Number One: It's Going Fast

Through many tens of millions of years of biogenic-abiogenic processes Nature has stored vast amounts of excess solar energy in the form of mineral organic compounds. These substances include coal, petroleum, shale oil, tar sands and natural gas. When commercial oil drilling started in Pennsylvania in 1859, with a daily production of 15-20 barrels per day, the world had a supply of petroleum equal to approximately 1.8 trillion barrels. As of today we are left with roughly 0.9 trillion barrels of petroleum, and petroleum world demand is approximately 105 million barrels per day. Coupled with increasing world population, 10 billion by 2050, and world modernization the projected global energy demand is expected to double by 2050. One can of course do the linear calculations to estimate how much 'time' we have left, but it is only part of the problem. The oil we have consumed to date was the oil relatively easy to obtain, that is to say large oil lakes at or relatively near the surface. With the easily obtained oil gone or accounted for, the energy input for new oil extraction continues to increase, in some instances the energy required to obtain a barrel of oil represents 50% of the energy inherent in the oil itself. Hence rather than a complete depletion of the remaining oil, a halt in oil extraction becomes a question of economics when the energy spent on obtaining oil becomes significantly higher than the energy output from the fuel. Considering the increasing economical and technological difficulties in obtaining the remaining petroleum, any realistic predictions of its duration would not go beyond several decades.

The total natural gas reserve, estimated on well lifetimes obtained several decades ago, is about 1.4×10^{14} m^3 with production at about 2.4×10^{12} m^3 per year. From past to present, however, well lifetimes continue to significantly shorten [2]. Apparently lateral migration of the natural gas gave the early wells, in comparison to modern wells, greatly extended lifetimes; it is upon the lifetimes of these early wells that our total reserves are estimated. The coal reserve is estimated at 9.1×10^{11} tons and annual consumption is about 5×10^9 tons. Although it is estimated that the world has about 250 years of 'harvestable free' energy (coal for 250 years, natural gas for 60 years, nuclear fuel for another 200 years at the current use rate) of economically recoverable reserves, the devil is always in the details. As one source is depleted another will suffer higher rates of consumption with, in turn, more rapid depletion. There are also the unanticipated losses, such as coal mine fires, unanticipated oil well fires due to wars, or sabotage of oil pipelines. Certainly of the available reserves 100% is not recoverable. Tar sands, oil shale and gas hydrates cannot long support the energy status quo, particularly as converting them into a useful form will rapidly consume available water resources. Both fossil and fissile energy sources are finite and, on our current projected course, will soon be exhausted. Considering the scale of the problem, i.e. a world population of 10 billion by 2050, energy conservation can only modestly delay the day societies find themselves, as they say, between a rock and a hard place. The survival of modern society, as we know it, faces collapse unless alternate routes for energy production are established. Ideally, any new energy system that replaces the existing one should possess the desirable qualities of the present oil based system, that is high energy density, relatively safe handling, portable, and at the same time it should not have any harmful effects, such as threatening the existence of life on the planet.

1.1.2 Problem Number Two: The Steam Age

There is a penalty associated with hydrocarbon fuels that even their staunchest supporter has to admit [3], and that is environmental pollution. In 1896 Arrhenius [4] put forth the concern that carbon dioxide from fossil fuel burning could raise the infrared opacity of the atmosphere enough to warm earth [5]. Numerous studies have shown that the level of carbon dioxide in earth's atmosphere has shown a periodic increase and decrease since its formation. Directly coupled with the variations in carbon dioxide concentration are variations in earth's average temperature. Carbon dioxide is a greenhouse gas trapping long wavelength radiation (heat) emitted by earth.   Hence more CO_2 in the atmosphere and the temperature increases, less CO_2 and the temperature decreases. Earth's baseline temperature, and that which we are comfortable with, appears to be finely balanced. Consider how even the modest tilting of the earth is enough to incur dramatic temperature variations. Tilt the earth a small amount one direction and it is 90°F in Pennsylvania, tilt the earth a similar modest amount in the other direction and it is 10°F in Pennsylvania (twenty years ago the balance was ≈ 75°F to -10°F, but things are quite warmer here now). The point being, small variations in this fine balance can result in significant consequences. The global average temperature change between the last ice age and the climate of today is about 1°C. Computer models of the temperature change associated with our global consumption of fossil fuels ranges from a few, to several, to 10-12°C. However it is difficult, if not fundamentally impossible, for linear models to accurate predict non-linear systems, and earth's temperature definitely behaves in a non-linear fashion. This can and probably will be to our great surprise.

Approximately 6 billion tons of carbon dioxide are produced every year by human activity, with 80% from the burning of the fossil fuels. Coal produces 430 grams of carbon dioxide/kWh, while natural gas produces 190 g of CO_2/kWh. Current CO_2 levels are easily measured, and future levels accurately predicted by simple arithmetic: burning so many tons of coal releases so many tons of CO_2. The released CO_2 passes into the atmosphere of known volume, allowing concentrations to be accurately calculated. Since measurements began, all studies have shown an unprecedented continuous increase in carbon dioxide levels, particularly during the latter half of the 20th century as the global burn rate significantly increased with modernization and increased population. At the beginning of the industrial revolution in the 18th century the atmospheric carbon dioxide level was 270 ppm, a value that had held steady for millions of years. It increased to 370 ppm within the 20th century, reaching 383 ppm in 2007. Climate models suggest that levels of 550 ppm would lead to a magnitude of warming equal to that of the cooling seen in the last ice age. So rather than an ice age, it appears we may soon have a '*steam age.*' Like the ice age one can suggest that humans will survive, but one can also anticipate some substantial and unpleasant disruptions.

From 1990 to 2004 world CO_2 emissions increased by 24.4% [6]. However if anything approaching the current situation continues (≈ 1 billion gasoline powered vehicles) we will easily and soon surpass 550 ppm well into the upper 600 ppm levels. At that point anything can happen such as, for example, heat induced thawing of the permafrost tundra driving global warming to even higher levels. While there was great excitement in the press that astronomers recently discovered an earth-like planet 20 light-years away, we note there is no way for us to move there; the Warp-drive of Star Trek space craft is not reality but a fictional TV show. While it may seem trite to say it, we have only one planet to live on, and we would be wise to take care of it.

While the following issues may pale in comparison to global warming, we also note that there is particulate, or soot, pollution association with fossil fuel consumption that we both see, and breathe. As evidenced by Black Lung Disease, these particulates are not good for us. There are other invisible pollutants, including cadmium, mercury, arsenic, nitrogen oxides, sulfur, and beryllium vapors that are all, to varying degrees, emitted with the combustion of fossil fuels, particularly coal [7-13]. For example, coal-burning power plants account for over 40 percent of all domestic human-caused mercury emissions [14], with regional variability depending upon where the coal-burning power plants are located and prevailing winds. Sulfur of course is converted to sulfur dioxide, one of the major contributors to acid rain. Nitrogen oxides and sulfur dioxide, together with the decomposition products of hydrocarbon burning (and/or unburned hydrocarbons) react with atmospheric materials creating smog, a significant problem particularly in major cities. Nitrogen oxides and sulfur dioxide react with oxygen and water to form nitric acid and sulfuric acid respectively. The cost of the environmental and health damage caused by the burning of fossil fuels is commonly ignored, as it is a non-specific person that suffers.

1.1.3 Problem Number Three: The Oil Wars

Modern society is habituated to a high degree of mobility, fast communication, and daily comfort, all of which require considerable energy input. As supply now exceeds demand, large energy users are able to readily meet energy requirements by importation. In the United States today approximately 66% of the oil consumed is imported, 66% percent of this is spent on transportation, and of this 66% is used to enable cars and light trucks. The status-quo will likely change dramatically when demand exceeds supply, something economists refer to as Inelastic Demand where prices continues to rise without a corresponding reduction in demand. If prices rise to a degree that users are not able to afford, the choice is give up the lifestyles societies have come to expect, or use other means of obtaining the needed energy. The government of the United States has deemed freely available Middle Eastern oil to be of the utmost importance, and backs this up with a vast (and expensive) military presence across the Middle East [15]. Two key points to keep in mind are: (1) nations always have, are currently, and will continue to fight wars over needed resources, and (2) societies will not happily go back to the life style of the preindustrialized world.

1.1.4 What Can Be Done?

The world's most authoritative body of climate scientists conclude that a 55-85% reduction in greenhouse gas emissions are necessary to stabilize atmospheric concentrations [16]. To eliminate anthropogenic CO_2 emissions we require non-carbon-emitting energy sources. We cannot continue to burn fossil fuels and somehow sequester the produced CO_2 efficiently enough to address global warming, as the processes of concentrating and burying or transforming the CO_2 are themselves energy intensive. Storage of CO_2 is only a temporary option, and should a CO_2 sequestration reservoir break and the CO_2 make it to the surface, it would effectively displace O_2 because CO_2 is heavier than air. Certainly CO_2 upgraded to reusable hydrocarbon resources, such as methanol and hydrogen, would benefit the environment and humanity.

It is imperative to find a suitable and sustainable energy alternative to the present energy system. We note that hydrogen appears a promising, useful candidate to replace hydrocarbon fuels, and that the sun gives an energy of 3×10^{24} joules/ year to the earth which is 10,000 times our present global energy consumption. Furthermore, the world is blessed with a vast abundance of desert lands (19.2×10^6 km^2) [17] that, one imagines, would be ideally suited for capturing solar energy in a sustainable fashion for the use of society. The topic of using solar energy to generate hydrogen in a sustainable fashion, via water splitting, is the topic of the book.

1.2 Hydrogen: A Historical Perspective

British scientist Henry Cavendish (1731-1810) demonstrated to the Royal Society of London in 1766 that there were different types of air: 'fixed air' or carbon dioxide and 'inflammable air' or hydrogen. Cavendish evolved hydrogen gas by reacting zinc metal with hydrochloric acid. He proved that hydrogen is much lighter than air and was the first to produce water by combining hydrogen and oxygen with the help of an electric spark in the late 1770s.

In 1783, Jacques Alexander Chales, a French Scientist, launched the first hydrogen balloon flight. Known as "Charliere" the unmanned balloon flew to an altitude of 3 km. Three months later, he himself flew in his first manned hydrogen balloon. In 1785 Lavoisier repeated Cavendish's experiments and proved that hydrogen and oxygen were the basic elements of water. Named by Lavoisier in 1788 Hydrogen is from two Greek words, hydro meaning water and genes meaning 'born of'.

In 1800 William Nicholson and Sir Anthony Carlisle discovered electrolysis and initiated the science of electrochemistry. In their experiments they employed a voltaic pile to liberate oxygen and hydrogen from water. They discovered that the amount of oxygen and hydrogen liberated by the current was proportional to the amount of current used.

In 1838 Swiss chemist Christian Friedrich Schoenbein discovered that hydrogen and oxygen can be combined to produce water and electric current- the fuel cell effect. Sir William Robert Grove was an English scientist and judge who demonstrated in 1845 Schoenbein's discovery on a practical scale by creating a 'gas battery' and earned the platitude 'father of the fuel cell'. This led to his development of the 'gaseous voltaic battery," the forerunner of the modern fuel cell. The Grove cell as it came to be called, used porous platinum electrodes and sulfuric acid as the electrolyte. In 1780, water gas was produced for the first time, a mixture of 50% hydrogen, 40% CO plus CO_2 and nitrogen. After that (early 1800's to mid 1900's) followed town gas, consisting of 50% hydrogen, 30% methane, 6% CO and traces of other gases. The use of these gases containing a high percentage of hydrogen only ceased in the 1960s when natural gas became abundant.

Konstantin Tsiolkovsky first proposed hydrogen-fueled rocket propulsion for space flights in the late 1890s. In 1911 Carl Bosch directed the development for ammonia and fertilizer to be manufactured from hydrogen and nitrogen gases, leading to the manufacturing of synthetic fertilizers. During the 1920s Rudolf Erren converted the internal combustion engines of trucks, buses, and submarines to use hydrogen or hydrogen mixtures. J.B.S. Haldane produced hydrogen by using wind-generated electricity in 1923. Rudolph A. Erren, a developer of hydrogen fueled motor vehicles, demonstrated their use in fleet service during the 1930s. In 1950 Akira Mitsui was successful in biologically producing hydrogen using special types of algae and microorganisms. In 1959 Francis T.

Bacon of Cambridge University made the first practical hydrogen-air fuel cell. The 5kW system powered a welding machine. He named it the 'Bacon cell'. Hydrogen fuel cells, based on Bacon's design have been used to generate on-board electricity, heat and water for astronauts aboard the Apollo spacecraft and all subsequent space shuttle missions. In the 1960s several scientists proposed the use of solar energy to split water into hydrogen and oxygen, which would then be recombined in fuel cells. The 20th century saw hydrogen being effectively used for manufacturing ammonia, methanol, gasoline, heating oil, fertilizers, glass, refined metals, vitamins, cosmetics, semiconductor circuits, soaps, lubricants, cleaners, margarine, peanut butter and rocket fuel.

The year 1970 marked the first use of the phrase 'hydrogen economy' though the idea was expressed in a book 'Conduction mechanisms and energy conversion' by E. Justi in 1965 [18]. During a brainstorming session at the General Motors Technical Laboratory, Warren, MI, electrochemist J.O'M Bockris commented, "We should be living in a hydrogen society", to which Neal Triner exclaimed: "There will be a hydrogen economy". The premise of a hydrogen economy is that hydrogen is the energy carrier, readily transported to where needed.   The late 20th and early 21st centuries saw many industries worldwide begin producing hydrogen, hydrogen-powered vehicles, hydrogen fuel cells and other hydrogen products. Iceland is committed to be the first hydrogen economy by 2030.

In 1990 the world's first solar-powered hydrogen production plant became operational, at Solar-Wasserstoff-Bayern, a research and testing facility in southern Germany. In 1994 Daimler Benz demonstrated its first NECA I (New Electric CAR) fuel cell vehicle at a press conference in Ulm, Germany. In 1999, Europe's first hydrogen fueling stations were opened in the German cities of Hamburg and Munich. In 2000 Ballard Power systems presented the world's first production-ready PEM fuel cell for automotive applications at the Detroit Auto Show. In 2004 the world's first fuel cell-powered submarine, German Navy, underwent deepwater trials.

The future (?): Water replaces fossil fuels as the primary resource of hydrogen. Hydrogen is distributed via national networks of hydrogen transport pipelines and fueling stations. Hydrogen energy and fuel cell power, clean, abundant, reliable, and affordable, become an integral part of all sectors of the world economy.

1.3 Renewable Energy and Hydrogen

Hydrogen is an energy carrier and not an energy source; it is not readily available in nature but energy from an energy source can be used to produce it. Renewable energy sources are inexhaustible, replenishing themselves and allowing energy harvesting at a rate decided by nature. It is, obviously, imperative that hydrogen should come from renewable energy sources. Solar radiation (photons), wind, ocean currents, tides and waves are examples of renewable energy. The main attractions of renewable energy are that it is abundant, intrinsically clean and will last as long as the planet exists. The principal renewable energy source is the sun. Solar energy is presently converted into electricity and heat using devices such as solar cells or black absorbers (solar thermal energy). A solar cell can produce modest amounts of electricity even when it is cloudy by the conversion of the diffuse light to power. Calculations show an area of 100 x 100 miles square would yield an energy equivalent to that used annually in the entire United States, of 1.019×10^{20} Joules per year (in 1999). The energy received by 48 contiguous states of USA is 1.766×10^{23} Joules per year (assuming 8.38 Joules/cm^2/min). The supply of energy from the Sun to the Earth is gigantic: 3×10^{24} joules a year, or approximately 10,000 times more than the global population currently consumes. Covering 0.1% of the Earth's surface with solar cells with an efficiency of 10% would satisfy our present energy requirements. This remains an enormous challenge, and the challenge is made greater in that we lack a serious effort to achieve it. Since our energy is now predominately provided by fossil fuels, we note that fabrication of solar cells comes with a CO_2 emission price tag. For a silicon solar cell, it appears three years of (non carbon emitting) power generation are required to compensate for the CO_2 emitted during fabrication [19]. As manufacturing processes improve, or new solar cells technologies requiring less energy investment arise, the time required for CO_2 payback can be expected to further decrease.

Of course a variety of renewable energy sources exist. Certain methods of biomass utilization are environmentally friendly and renewable, such as biomass fermentation to produce hydrogen. Fuels such as soy methyl ester (biodiesel from soybeans) can be distributed using our current infrastructure, and used in conventional diesel vehicles. Wind energy represents the nearest term cost-competitive renewable energy source. Produced by the solar heating of earth, wind as an energy resource, proportional to the cube of its velocity, presents a dual-use technology: the land can still be used for farming, ranching and forestry, and the collection of solar energy. Hydroenergy has been used throughout the world for centuries but, however, has little room for future growth. Energy from the ocean is appealing but difficult to achieve. For example, the temperature difference between the surface and deep-ocean can be used for thermally driven power generation. However biofouling, corrosion, and storms make the prospect challenging. While tides have been exploited since the 18th century for power generation the small change in potential energy associated with the tides, corresponding to a height difference of 1 m to 2 m, is difficult to harvest. Small scale harvesting of wave energy appears to be marginally feasible.

Geothermal energy is obtained by extracting heat from water or rocks deep underground. A practical geothermal site requires that there be high temperature rocks and/or water within approximately 300 m of the surface. San Jose, CA produces 850 megawatts from geothermal sources. There are places where the holes are drilled 1500 m deep and 300°C superheated water pumped to the surface for electricity generation. Unlike solar energy geothermal energy is not intermittent and hence reliable for, typically, several years. However, heat can and typically is extracted from the rocks much more rapidly than it is restored from the ambient. Therefore the geothermal plants have limited lifetimes, requiring the periodic drilling of new holes (with their associated costs) for continued operation.

The availability of energy from renewable sources varies in space and time. The energy may not be available per the need of users. Hence a medium, or a buffer system, is required to store the energy. A renewable source can be considered useful if it can deliver energy in electrical or chemical energy form. Energy storage technologies include H_2, batteries, flywheels, superconducting magnetic energy storage technologies (SMES), advanced electrochemical capacitors, pumped hydro and compressed gas. As renewable sources are localized, typically requiring transportation of energy to places in need, the energy carrier should be portable. Hydrogen is considered the most appropriate medium where renewable energy can be effectively stored and transported.

Whereas stationary users can use fuels in any physical state, solid, liquid or gas the transportation sector must use a fuel that can be quickly and easily loaded into the vehicle storage tank. As a result, the selection of the future chemical fuel is strongly driven by the needs of the transportation sector. Except for fossil fuels, none of the primary energy sources (thermonuclear energy, nuclear breeders, solar energy, hydropower, geothermal energy, ocean currents, tides and waves are primary sources of energy) can be directly used as a fuel and hence they must be used to manufacture a fuel or fuels, as well as to generate electricity. For useful purposes these primary sources should be converted to the energy carrier needed by the customers. There are many candidates for this including natural gas, methanol, ethanol and hydrogen. The criteria for selecting a future transportation fuel are versatility, efficiency, environmental compatibility, safety, and economics. Also it is necessary to find an energy vector that will enable us to replace the market currently supplied by hydrocarbons, without making any great alterations. This new vector must act as a bridge between the new sources of primary renewable energies and the various sectors of consumption. It must also be economical to produce, easy to transmit and store, renewable and non-polluting and, if possible, more efficient than current fuels. Hydrogen has all the necessary qualities, however as yet we have no scale means of its renewable generation.

1.4 The Energy Carriers: Hydrogen or Electricity?

Renewable energy sources such as wind and sunlight are intermittent, necessitating a buffer medium for the uninterrupted supply of energy to users. Both hydrogen and electricity need an energy source for production, but are useful as intermediates to convert and store energy from renewable or non-renewable sources. Electricity, and to some extent hydrogen have already become a part of modern society, hence these energy carriers are intrinsic to an energy system utilizing renewable sources. Any realistic energy system should consider the utilization of both hydrogen and electricity as they have complimentary functions. This will enable a complete replacement of the fossil fuel systems. For sustainable development both should be produced from renewable sources rather than from the fossil fuels. The new energy system must be able to supply electricity and chemical fuel in a proper balance to best serve the customer.

Hydrogen is an ideal energy carrier because: (1) it can be produced from and converted into electricity at relatively high efficiencies; (2) its raw materials for production is water; (3) it is a renewable fuel; (4) it can be stored in gaseous, liquid or metal hydride form; (5) it can be transported over large distances through pipelines or via tankers; (6) it can be converted into other forms of energy in more ways and more efficiently than any other fuel; (7) it is environmentally compatible since its production, storage, transportation and end use do not produce pollutants, green house gases or any other harmful effects on the environment (except in some cases the production of nitrous oxides). Unfortunately hydrogen currently suffers from high production costs, but of course everything is expensive in comparison to free (thanks Nature!), the cost basis that fossil fuels have been provided to mankind.

The major drawback of electricity is that it is difficult to store. In contrast chemical fuels can be easily stored for long periods, with the gas placed in a container of sufficient strength to resist the gas pressure and prevent leakage. Chemical fuels have very desirable characteristics for the transport of energy being portable, readily carried on trucks, trains, planes and boats. The continent wide high-pressure natural gas lines in the USA serve as storage without any need of local storage. In a large-scale distribution system, the pipelines provide a large volume continent-spanning storage reservoir in addition to their primary function as a conduit. Energy losses in the pipeline are due to leakage and friction, with intermediate pumping stations used to compensate for the losses. Because of the transient nature of electricity, none of the techniques used for the storage of chemical fuels are of use in storing electricity. Similarly electrical lines also suffer from transmission losses due to heat production and arcs at high voltage. The best electrical transmission system provides $\approx 80\%$ of their generator output to the customer.

The common storage container for electrical energy is the battery. In single use batteries chemical energy is converted into electrical energy and cannot be recharged from the grid. In multiple use storage batteries, the electrical energy input is converted into an active chemical within the battery. The chemicals are stored within the battery for later regeneration of electricity. One of the biggest problems of batteries is their tendency to accept less total charge each time they are recharged. The lead-acid batteries store modest amounts of energy, with a battery lifetime of 2 to 5 years. The lead used in the batteries is toxic, and care is needed for disposal. Nickel-cadmium batteries store comparatively more energy but are far more expensive than lead-acid batteries, and Cd is toxic. Sodium-sulfur batteries can handle more power and recharge cycles without significant loss of capacity but they should be operated at 350°C. Very long heating and cooling times are a shortcoming, and if damaged the hot sulfur and molten sodium burns. A rechargeable battery must store all the chemicals involved in the reaction within the confines of the battery case. This requirement, to carry all the reactants at all times, is the primary reason that batteries cannot store as much energy per unit mass as can chemical fuels that generate energy by reacting with air.

The best of all possible chemical battery cells, a beryllium-air cell, would store energy at a rate of 24.5 megajoules per kg of reactants. Gasoline stores 45 megajoules per kg and hydrogen stores about 120 megajoules per kg. Storing electricity in the magnetic fields of superconducting coils is an option, if suitable room temperature super conductors are discovered and one is not worried of electrical shorts. The transportation of energy through discrete bulk shipment is not possible with electricity. Energy for use by moving vehicles is not possible by electricity except where direct electrical lines can be used, e.g. trains or city buses. The use of electricity in vehicles is not going to be practical using present technology because of lack of proper storage batteries and the lack of fast recharging systems.

Hydrogen can be utilized directly for internal combustion engines, as well as utilized as a chemical feed stock for almost all the things that are presently being manufactured using fossil fuels. The production of hydrogen from renewable sources can be used for the intermittent availability of the energy from these sources, with the stored hydrogen later used as needed.   An approximate comparison of hydrogen-electricity total losses: Transmission line loss is 8%; pipeline energy consumption is 12%. Hydrogen requires additional conversion steps, with electrolysis consuming 10-15% of the original electricity. Re-converting H_2 to electricity takes 30-40% of remaining energy. Storing hydrogen, then using this hydrogen to generate electricity provides only 45-55% of the original energy compared to 92% if transmitted directly as electricity. Hence wind energy sent as electricity provides roughly twice the end use benefits as wind energy delivered as H_2.

1.5 Hydrogen as a Chemical Fuel

Although any flammable material can theoretically be used as a fuel, only a few are realistically practical. Some of them have low energy content, and/or produce toxic gases and other polluting vapors on combustion, e.g. the pyrophoretic combustion of iron particles. In contrast, the combustion of hydrogen in pure oxygen results in only heat and water.

The criteria for a potential chemical fuel include large energy per unit mass; that it be a gas or a liquid for ease in transportation; non toxic, and provide only gaseous products when burned in air; environmentally benign; be made of readily available chemical elements to ensure abundant supply; be made from elements available in most locations; easily manufactured by a low cost process; and be easy to use in existing power/energy generation equipment to aid the transition from one primary energy source to another. Hydrogen satisfies all these stated criteria. Hydrogen releases 142 MJ/kg on combustion. Of all fuels it produces the least pollution when ignited in air, residual traces of NO_x. By careful control of hydrogen/air mixture ratio and in some cases the use of chemical reaction accelerators (catalysis) the output of nitrogen oxides can be reduced to low and safe levels. Automobiles, buses, trucks etc. currently running on natural gas can readily be run on hydrogen.

Hydrogen is the most abundant element in the universe, representing 90% of the universe by weight, and the third most abundant in our planet. Hydrogen is the lightest element with a density of 0.08988 g/liter at STP. Hydrogen is colorless, odorless, tasteless, and non-toxic under normal conditions. Liquid hydrogen has a density of 0.07 g/cc compared 0.75 g/cc for gasoline. It is the lightest gas, providing it with some advantages as well as disadvantages. The advantage is that H_2 stores $\approx$3x the energy per unit mass as gasoline, and $\approx$7x the energy per unit mass as coal. The disadvantage is that it needs about 4 times the volume for a given amount of gasoline energy. While hydrogen releases about 2.45 times more energy per unit weight than methane, its light weight results in volumetric energy content only 30% of methane. Due to its low density, the H_2 handling and storage equipment of today is large and bulky. The low temperature of liquid hydrogen, 20 K, necessitates the use of superinsulated storage vessels. These vessels must be equipped with appropriate safety features and equipment to accommodate the slight and continuous boil-off of liquid hydrogen. Hydrogen has a wide flammability region of 4% to 75% in oxygen; self-ignition occurs at 576°C.

Current major industrial uses of hydrogen include the production of ammonia, hydrotreating of petroleum feedstocks, hydrogenation of petrochemicals, and methanol synthesis. Due to biological processes and decomposition of water by ultraviolet radiation hydrogen is always present in the air, at about 4 ppm. Water, which covers 70% of earths' surface (a figure increasing each day!), represents a vast source of hydrogen. When hydrogen is burned in pure oxygen a flame temperature of 2700°C is measured. Hydrogen burns with a pale-blue, almost invisible flame, making hydrogen fires difficult to see. The energy content for 1 kg (424 standard cubic feet) of hydrogen, while forming water by reacting with oxygen, is 141,600 kJ = 134,200 Btu = 39.3 kWh.

Hydrogen represents an energy storage medium of high specific energy (energy to weight ratio), high specific power (power to weight ratio). Hydrogen can be stored months or years without issue. Hydrogen can be piped and transported like other gaseous and liquid fuels. Assuming economies of scale it should be inexpensive facilitating widespread adoption, and clean, i.e. can be used without creating pollution.

1.6 The Hydrogen Economy

The hydrogen economy has become a synonym for a sustainable energy system, referring to a scenario in which the whole energy system of a country is based upon hydrogen and its counterpart, electricity. For a hydrogen economy to be successful the hydrogen has to be generated from renewable sources, preferably using water photoelectrolysis where water and solar light are the sources for hydrogen generation. Hydrogen would replace fossil fuels and be used as a fuel, or energy carrier, as well as chemical feedstock. Factors surrounding the production, storage and infrastructure will determine the feasibility of a hydrogen economy. In theory hydrogen and electricity can satisfy all the energy needs of society, forming an energy system that would be permanent and independent. In a hydrogen economy, energy transport to end users, depending on distance and overall economics, would either be in the form of electricity or in the form of hydrogen. Some of the hydrogen would be used to generate electricity (via fuel cells) depending on demand, geographical location, or time of the day. In a hydrogen economy hydrogen and electricity would be produced in large quantities from available energy sources and used in every application where fossil fuels are being used today.

Recent interest in hydrogen as a clean fuel has surged as concerns over the costs of fossil fuels to the economy, environment and national security have become paramount. The ultimate hydrogen economy is expected to have: i) low cost production of hydrogen using either renewable or inexhaustible primary nuclear energy sources; ii) oxidation of hydrogen in such designs as fuel cells in fuel cell hybrid vehicles with high conversion efficiencies not capped by the Carnot limit as for conventional engine cycles; and iii) safe operation in spite of the wide flammability limits of hydrogen-air mixtures, 4% to 74%, and the associated ease of ignition.

About 45 million metric tons of hydrogen are produced globally each year from fossil fuel. Approximately half of this hydrogen goes to making ammonia (NH_3), which is a major component of fertilizer and a familiar ingredient in household cleaners. Approximately 37% of the hydrogen is used in refineries for chemical processes such as removing sulfur from gasoline and converting heavy hydrocarbons into gasoline and diesel fuel. The remaining hydrogen is used for applications including edible fat hydrogenation, methanol production, float glass production, generator cooling, weather balloons and rockets.

1.7 Hydrogen Production [20]

Hydrogen can be produced by numerous methods, through a variety of chemical reactions many of which have been known for centuries. However most of these reactions raise severe safety and environmental issues, while the availability of raw materials is a critical problem. Hydrogen evolving examples include the reaction of aluminum with lime water (calcium hydroxide) used for making hydrogen filled balloons, and the (violent) reaction of sodium with water to form sodium hydroxide and hydrogen. The emergence of the hydrogen economy will depend upon the development of safe, efficient, widely available and environmentally responsible means for generating hydrogen. Since fossil fuel sources are currently abundant almost all hydrogen is produced from these sources. However, the aim of any realistic sustained energy system based on hydrogen necessitates the fabrication of hydrogen from renewable sources. In this regard hydrogen from water and sunlight is the ultimate choice. We briefly introduce various hydrogen production technologies below.

1.7.1 Hydrogen from Fossil Fuels

About 96% of the worlds' hydrogen is currently produced from natural gas, oil or coal (48% from natural gas, 30% from oil- most of which is consumed in refineries, 18% from coal, and the remaining 4% via water electrolysis). In general the purity of hydrogen from fossil fuels is about 98% although purity in excess of 99.99% can be achieved with advanced purification techniques. Natural gas is the cleanest of all fossil fuels as it has methane as the major component. Methane has the highest hydrogen to carbon ratio among hydrocarbons, hence the carbon and related compounds generated as by-products are minimal. Most of the today's hydrogen for the United States (about 95%) is produced via steam-methane reforming (SMR).

The SMR process consists of two steps. The first is the reformation process in which methane mixed with steam is passed over a catalyst bed at high temperature and pressure to form a mixture of hydrogen and carbon monoxide (reaction 1.1), called syngas. The second step is the shift reaction in which carbon monoxide from the first stage reacts with additional steam to release carbon dioxide and more hydrogen (reaction 1.2).

$$CH_4 + H_2O \rightarrow CO + 3H_2 \quad \Delta H = 206.1 kJmol^{-1} \quad \text{at } 398K \quad (1.1)$$
$$CO + H_2O \rightarrow CO_2 + H_2 \quad \Delta H = -41.1 \ kJmol^{-1} \quad \text{at } 298K \quad (1.2)$$

The overall reaction can be written as

$$CH_4 + 2H_2O + \text{thermal energy} \rightarrow 4H_2 + CO_2$$
$$\Delta H = 165 \ kJmol^{-1} \text{ at } 298K \quad (1.3)$$

Reaction 1.3 is endothermic, driven by methane combustion. Reaction 1.1 is performed at 700-900°C and pressures between 1.5 to 3 Mpa. Reaction 1.2 is performed in one or two stages. The high temperature shift reaction occurs at temperatures between 340°C and 450°C, and the low temperature reaction between 190°C and 250°C. Nickel or nickel based alloys (nickel supported by alumina, iron-nickel etc.) are generally used as catalysts. Prior to the reformation process the feed gases are subjected to a desulphurization process to avoid sulphur poisoning of the catalyst. The product stream consisting of hydrogen and carbon dioxide is cooled and then carbon dioxide is removed using CO_2 sorbants. The energy conversion process efficiency, fossil fuel to hydrogen fuel, is about 65 to 85%. Using the latest techniques to remove carbon dioxide it can produce hydrogen of about 99.99% purity. SMR is not suitable for hydrocarbons heavier than naphtha.

Partial oxidation can be used for feedstocks ranging from methane, to heavy oil, to coal. Additionally with the shift reaction (1.2), the process takes place via the following two reactions:

$$C_nH_m + n/2\ O_2 \rightarrow nCO + m/2\ H_2 \text{ (exothermic)} \qquad (1.4)$$
$$C_nH_m + nH_2O \rightarrow nCO + (n+m/2)\ H_2 \text{ (endothermic)} \qquad (1.5)$$

The exothermic first stage is syngas generation, where the hydrocarbon is oxidized with oxygen. The additional energy needed for reaction (1.5) is supplied by burning additional hydrocarbons. The generated carbon monoxide is converted to carbon dioxide in the shift reaction (reaction 1.2). The reactions are performed at temperatures in excess of 1150°C for heavy hydrocarbons and at about 600°C for fractions lighter than naphtha. The pressures range from 3 to 12 Mpa. The energy conversion efficiency of the process is about 50%.

Autothermal reforming is a combined form of partial oxidation and SMR. This makes use of heat from the exothermic reaction to sustain the SMR reactions without external heating. In this case, oxygen is fed into a combustion zone where partial combustion of methane takes place. The hot gas mixture is then fed to the SMR zone. Thus the heat energy needed for SMR is created internally. Thermal cracking of methane is another method useful for hydrogen production. Here, the methane is split at high temperatures using a methane-air flame to produce hydrogen and carbon.

Another process used for hydrogen generation involving fossil fuels is coal gasification. The coal undergoes partial oxidation at very high temperatures and pressure (~5 Mpa) with the help of oxygen and steam producing a mixture of hydrogen, carbon monoxide, carbon dioxide, methane and other compounds. At temperatures above about 1000°C at 1 bar, mainly hydrogen and CO remain. The general process can be represented by the following reactions

$$C + \frac{1}{2}\ O_2 \rightarrow CO \quad \text{(exothermic)} \qquad (1.6)$$
$$C + H_2O \rightarrow CO + H_2 \quad \text{(endothermic)} \qquad (1.7)$$

Production of 1kg of H_2 by steam reforming of methane yields 10.6621 kg of CO_2, 146.3 grams of methane and traces of benzene, CO, oxides of nitrogen (NO and NO_2), nitrous oxide (N_2O), non-methane hydrocarbons, particulates and oxides of sulfur (SO_x and SO_2). The US hydrogen industry produces 11 million tons of hydrogen a year. In so doing, it consumes 5% of the U.S. natural gas usage and releases 74 million tons of CO_2. A typical SMR hydrogen plant with the capacity of one million m^3 of hydrogen per day produces 0.3-0.4 million standard cubic meters of CO_2 per day, which is normally vented into the atmosphere. The amount of CO_2 is doubled if coal gasification is used.

Modification of the conventional steam methane reforming (SMR) process to incorporate an adsorbent in the reformer to remove CO_2 from the product stream may offer a number of advantages over conventional processes. Upsetting the reaction equilibrium in this way drives the reaction to produce additional hydrogen at lower temperatures than conventional SMR reactors. Although still in the research stage, the cost of hydrogen from this modified process is expected to be 25%-30% lower, primarily because of reduced capital and operating costs.

1.7.2 Hydrogen from Biomass [21]

Biomass is considered an abundant renewable resource, predicated of course upon suitable water and soil nutrients- a point commonly ignored. Biomass contributes about 12% of today's world energy supply. Herbaceous energy crops, woody energy crops, industrial crops, agricultural crops and aquatic crops, crop and animal waste, forestry waste and residues and industrial and municipal wastes can be used for energy production. Hydrogen can be produced from biomass using thermochemical and biological processes. Pyrolysis and gasification are very feasible thermochemical routes for hydrogen production whereas biophotolysis, biological gas shift reaction and fermentation are highly promising biological processes.

Hydrogen production via biomass pyrolysis involves heating the biomass rapidly to high temperatures in the absence of oxygen. Products like hydrogen, methane, carbon monoxide, carbon dioxide, tar, oils, carbon and other solid, liquid and gaseous products depending upon the organic nature of the biomass, are produced as a result of pyrolysis. Methane and other hydrocarbons can also be converted into hydrogen using appropriate methods mentioned in the previous section. The temperature used for pyrolysis ranges from 400 to 600°C and pressure ranges from 0.1 to 0.5 MPa.

Another promising route for hydrogen production via biomass is gasification. Biomass gasification is done in the presence of oxygen and steam. The conversion process can be represented as

$$Biomass + thermal\ energy + steam \rightarrow H_2 + CO + CO_2 + CH_4 + hydrocarbons + ash$$

Methane, hydrocarbons and carbon monoxide can be converted into hydrogen using SMR, partial oxidation and gas shift reactions respectively. Gasification is generally used for biomass with a moisture content less than 35%.

Biological methods for hydrogen production are still at the research level. Hydrogen production via biophotolysis is discussed in detail in Chapter 2. Some photoheterotrophic bacteria survive in the dark via water-gas shift reaction (reaction 1.2). An example of such bacterium is *Rhodospirillum rubrum*. These bacteria generate ATP (adenosine triphosphate) using carbon monoxide by coupling the oxidation of CO to the reduction of H^+ to H_2. Under anaerobic conditions, CO induces the synthesis of several proteins, including CO dehydrogenase, Fe-S protein and CO-tolerant hydrogenase. Electrons produced from CO oxidation are conveyed via the Fe-S protein to the hydrogenase for hydrogen production. Photosynthetic bacteria such as *Rhodobacter spaeroides* can generate hydrogen through photofermentation of organic wastes. A more promising process is dark fermentation of carbohydrate-rich substrates by anaerobic bacteria and green algae. For example, glucose and sucrose from sugar refinery waste water can be fermented to produce hydrogen. Biomass glucose gives a maximum yield of four mol of H_2 per mol of glucose with acetic acid as the by-product.

$$C_6H_{12}O_6 \rightarrow 2CH_3COOH + 2CO_2 + 4H_2$$

The yield reduces to half with butyrate as the fermentation end product.

$$C_6H_{12}O_6 \rightarrow CH_3CH_2CH_2COOH + 2CO_2 + 2H_2$$

Although carbon dioxide is released in most of the biomass-to-hydrogen conversion processes, the net CO_2 emission is zero due to the photosynthesis of green plants.

1.7.3 Hydrogen from Water Using Renewable Energy

In discussions on the hydrogen economy many justifiably quote from the 1875 book written by Jules Verne, *The Mysterious Island* (Part II, Chapter 11). The lead character Cyrus Smith, a man of great capabilities, is quoted as saying: "Yes, my friends, I believe that water will one day serve as our fuel, that the hydrogen and oxygen which compose it, used alone or together, will supply an inexhaustible source of heat and light, burning with an intensity that coal cannot equal. One day, in the place of coal, the coal bunkers of steamers and the tenders of locomotives will be loaded with these two compressed gases which will burn in furnaces with an enormous heating power... I believe that when the coal mines have been exhausted, we will both heat and be heated with water. Water is the coal of the future." To which a thoughtful friend replied; "That I should like to see."

So too would many of us. Water is the most abundant source of hydrogen on the planet. However, hydrogen is strongly bonded to oxygen and a minimum energy of 1.229 eV (at 25°C, 1 bar) is required for splitting water into hydrogen and oxygen. Although this energy can be supplied via renewable or non-renewable sources, hydrogen from water using renewable energy sources is the only permanent solution, and one that can be obtained peacefully as well.

Water splitting can be accomplished via techniques such as electrolysis, photoelectrolysis, thermochemical, and biophotolysis. Electrolysis, thermochemical, and biophotolysis techniques are discussed in Chapter 2. Chapters 3-7 are dedicated to the discussion of photoelectrolysis. Chapter 8 discusses water electrolysis using energy derived from solar cells. Although each technique has its own merits and demerits and will have a role in hydrogen economy, solar energy coupled with electrolysis and solar water photoelectrolysis appear to be the most logical solution pathways.

1.8 Hydrogen and Transportation

Transportation systems, today based on petroleum fuels, are the lifeblood of the modern world. The status quo however will not last for more than a few years. A permanent solution is to form a stable, environmentally friendly transportation sector based on renewable sources of hydrogen and/or electricity.

As far as road transportation is concerned, the present efforts are oriented in two directions: to make hydrogen based internal combustion vehicles and to make hydrogen fuel cell-based electrically driven vehicles. Liquid H_2 carrying vehicles with internal combustion engines offer the following advantages: Total elimination of adding of carbon dioxide to the atmosphere, near total elimination of harmful smog producing pollution (unburned hydrocarbons, CO and sulfur compounds), great reduction in the production of nitrogen oxides, and improved cold starting performance. The advantages of hydrogen fuel cell vehicles are: Termination of the addition of carbon dioxide to the atmosphere, total elimination of harmful smog producing pollution, total elimination of nitrogen oxides, great reduction in vehicle noise, approximately a doubling of overall vehicle energy efficiency and improved cold starting performance. Unfortunately their introduction into the commercial market faces the eternal dilemma of the 'chicken and the egg': Customers will not purchase hydrogen-based vehicles unless adequate fueling is available; manufacturers will not produce vehicles that people will not buy; fuel providers will not install hydrogen stations for vehicles that do not exist.

The power output of hydrogen internal combustion IC engines can be comparable to or even higher than that of gasoline engines, while the high diffusivity of hydrogen rapidly creates a uniform fuel-air mixture. The energy needed to ignite hydrogen at stoichiometric air/fuel ratio is almost one order of magnitude less than that needed for gasoline, 0.017 mJ for hydrogen vs 0.24 mJ for gasoline. It has a wide flammability range, 4% to 75% for hydrogen vs 1.4% to 7.4% for gasoline, and the engine can work in wide a range of H_2 fuel/air ratios from stoichiometric to very lean mixtures. It has a high auto-ignition temperature (576°C) allowing use of high compression ratios. Yet the low density of hydrogen makes the volumetric energy density of the air/fuel mixture low. The stoichiometric air to hydrogen ratio by volume is 2.4:1. Therefore, in stoichiometric conditions, hydrogen will occupy 29.6% of the cylinder volume whereas gasoline occupies only about 2%.

Many of the air pollutants currently emitted from an automobile are carbon compounds that include unburned hydrocarbons, carbon monoxide, aldehydes, ketones and carbon dioxide. In contrast hydrogen combustion cannot produce any carbon compounds. However, due to the presence of nitrogen in the combustion zone of a hydrogen IC engine, at high temperatures NO_x can be produced. If the ratio of hydrogen to oxygen is adjusted to provide a slight excess of hydrogen, nitrogen oxides are not produced. This is a result of the much greater affinity of hydrogen for oxygen, than nitrogen for oxygen. The mixture ratio for complete reaction is two volumes of hydrogen to one volume of oxygen. A 1% excess of hydrogen is sufficient to suppress the formation of the nitrogen oxides. Hydrogen combustion will produce no acid and no solid carbon nor lead residues, hence well-designed hydrogen engines should require little maintenance and have extended lifetimes.

The low volumetric energy density of hydrogen poses an intrinsic limit to the range of a vehicle. Compressed hydrogen cylinders kept in the vehicle will not provide a driving range comparable to that of gasoline driven engines. While liquid hydrogen is a better option in terms of energy density, and hence range, as the liquefying temperature of hydrogen is 20K advanced techniques providing high insulation are required to avoid losses and safety hazards. However, hydrogen vehicles will not have cold start problems as present in current vehicles in winter due to the low boiling point of hydrogen. Car manufacturers such as BMW, e.g. the BMW 750 hL, have already developed internal combustion engines using hydrogen.

Progress in the commercialization of fuel-cell based vehicles are, not surprisingly, dependent on the development of efficient and economically viable hydrogen fuel cells. As of today fuel cells are expensive, and the technology is not yet mature enough for wide spread use in vehicles. However due to the current interest in a hydrogen based energy system efforts are being made in various countries to develop efficient, reliable and economically viable fuel cells.

Fuel cells operate in a manner reverse to that of electrolysis, discussed in Chapter 2, combining fuel to make electricity. The basic design consists of two electrodes separated by an electrolyte. The oldest type of fuel cell is the alkaline fuel cell where an alkaline electrolyte like potassium hydroxide is used. The hydrogen enters through the anode compartment and oxygen through the cathode compartment. The hydrogen is ionized by the catalytic activity of the anode material and electrons are released into the external circuit. The protons react with the hydroxyl ions in the electrolyte to form water. The reaction can be written as:

$$H_2 + 2OH^- \rightarrow 2H_2O + 2\ e^-$$

The cathodic reaction supplies the hydroxyl ions for the reaction at the anode by consuming electrons from the external circuit per the reaction

$$O_2 + 2H_2O + 4e^- \rightarrow 4OH^-$$

The overall reaction is

$$2H_2 + O_2 \rightarrow 2H_2O$$

The theoretical cell voltage is 1.229V, while in application cell voltages are typically a bit more than half the theoretical value. To achieve larger voltages several cells are stacked, that is connected in series. Alkaline fuel cells operate at relatively low temperatures, 70°C – 120°C, however these cells require both pure hydrogen and pure oxygen for operation. The theoretical maximum efficiency is 83%; in practice, alkaline fuel cell efficiencies are about 60%. A critical advantage of a fuel cell is that it is not a heat engine and thus does not require ever-higher temperatures, and as a consequence expensive temperature-resistant materials, for energy-efficient operation.

Other promising types of fuel cells utilizing hydrogen are proton exchange membrane fuel cells (PEMFC), phosphoric acid fuel cells (PAFC), molten carbonate fuel cells (MCFC) and solid oxide fuel cells (SOFC). The PEM fuel cells use a proton conducting solid polymer membrane as electrolyte. The operating temperature ranges from 60°C to 120°C, with an efficiency of 40 –50%. The PAFC uses liquid phosphoric acid as the proton conducting electrolyte, usually held in a silicon carbide matrix. The operating temperature ranges from 150°C to 210°C. MCFC uses an alkaline mixture like lithium and potassium carbonate as the electrolyte. It operates at temperatures about 650°C where the alkali carbonates form a high conductivity molten salt. These fuel cells need a supply of carbon dioxide along with oxygen at the cathode. The efficiency is about 55 to 65%. SOFC uses ion conducting ceramic electrolyte like zirconia (oxygen ion conductor). The operating temperature is about 1000°C and efficiency can be 50 to 70%.

In contrast to batteries, fuel cells cannot be discharged: like an IC engine, as long as fuel is available a fuel cell can continue to provide power. When energy efficiencies of the fuel cell automobile are taken into account, including that of the electric motor and transmission, the total system efficiency will be about 50% to 70%. Hence the fuel cell automobile will have about four times the energy efficiency of current automobiles, and possibly twice that of the new generation of hybrid automobiles. The fuel cell is most efficient at small loads; the efficiency decreases as the load increases and it is least efficient at full load. Thus the fuel cell operates efficiently under the most common driving conditions of partial load. The large efficiency reduces the amount of fuel that has to be carried on the vehicle. Since the amount of hydrogen needed is small, it is feasible to use simple, low cost, high-pressure gas storage for the hydrogen. With increased efficiency less hydrogen has to be stored on the vehicle for its operation, in turn reducing the cost of vehicle operation. Fuel cell driven vehicles do not require crankcase oil as required for lubrication of IC engines.

Besides high efficiency, fuel cell driven engines offer noiseless and clean operation. Nitrogen oxides, the only pollutants emitted by hydrogen IC engines, are not produced by a fuel cell. Fuel cells have been effectively used in space programs where they convert hydrogen and oxygen fuel into electricity needed for space ship operation, with the by-product pure water used by the crew.

1.9 Environmental Effects of Hydrogen

The burning of hydrogen in pure oxygen produces only water. However, if hydrogen is burned in air oxides of nitrogen (NO_x) can be formed known to be the cause of urban smog and acid rain. However in comparison to the burning of fossil fuels the amount of these oxides generated by burning hydrogen are essentially nil.

The atmospheric hydrogen concentration is about 4 ppm. Hydrogen added to the air, by its use as a fuel, will enter without making any measurable change in the long-term concentration. If hydrogen is used as an aviation fuel large amounts of water vapor will be emitted into the lower stratosphere, possibly resulting in stratosphere cooling [22]. However, water vapor is by far a less noxious pollutant than the by products of aviation kerosene.

1.10 Hydrogen Storage

Due to its low energy per unit volume, hydrogen is generally stored as a compressed gas or liquid for practical applications. Hydrogen becomes liquid at 20K; a volume of liquid hydrogen weighs only 10% as much as the same volume of gasoline. Obviously handling problems are severe at 20K. In order to serve as a practical fuel for transportation hydrogen must be highly compressed to minimize fuel storage volume. Typical hydrogen storage pressures are 2000 to 5000 psi. Existing hydrogen generation technologies typically produce hydrogen gas at atmospheric pressure to 375 psi. Mechanical compression is needed to raise gas pressure to levels needed for almost all applications. Unfortunately, this compression process carries with it an energy penalty, requiring capital expenditure for the compression equipment, and incurs ongoing operation and maintenance costs. Two alternatives to pressurized gaseous hydrogen storage are available: liquefied hydrogen and metal hydrides. However, liquefying hydrogen calls for expensive, elaborate equipment to cool the gas to −253°C, and all this takes energy. Metal hydrides are expensive, heavy and have a limited lifespan, showing a decay in energy storage capacity after repeated cycling. Research is needed on approaches for hydrogen storage using the reversible formation of safely transportable chemical compounds.

1.11 Hydrogen Safety

The risks with hydrogen are of the same order of magnitude as gasoline or natural gas; all can be used safely with a proper understanding of their properties. Safety precautions for use of hydrogen in the home and in industrial applications are similar to those required for natural gas. The safety advantages of hydrogen are its great speed of dissipation, low flame luminosity, lack of toxic gas production on combustion, and total lack of toxicity of the hydrogen. Hydrogen rapidly diffuses making the build up of flammable concentrations more difficult. Hydrogen can be safely used at homes with no CO release. Hydrogen, like natural gas, can leak and cause an explosive hazard, but there is no danger from poison gas. Hydrogen fires will flash and float away due to its buoyancy, while in contrast gasoline fires persist and produce toxic carbon monoxide. Natural gas can produce carbon monoxide when it burns. Hydrogen is somewhat harder to ignite with heat than is natural gas, but is more easily ignited with electric sparks. The safety advantages of H_2 are partly offset by the wide combustion mixture range of hydrogen (4 to 75 vol %) compared to natural gas (5 to 15 vol%).

Hydrogen is commonly viewed as a dangerous gas due to its wide flammability range and low ignition energy. The topic of hydrogen safety often brings memories of the Hindenberg tragedy: in 1937 the hydrogen-filled (7 million cubic feet) zeppelin Hindenberg burst into flames while landing at Lakehurst, NJ with a death toll of 35 people out of 97 onboard. While hydrogen is commonly viewed as the culprit, studies conducted by a 1997 NASA team showed that the fire resulted from a weather-related static electricity discharge that ignited a coating applied to the zeppelin, which contained nitrocellulose (smokeless gunpowder) and powdered aluminum (rocket propellant). The hydrogen floated upwards, burning safely above the passengers and did not cause a single death.

References

1. UN General Assembly document (1987) A/42/427
2. Campbell CJ (1997) The Coming Oil Crisis. Multi-Science Publishing Company & Petroconsultants S.A., Essex, United Kingdom
3. Bush Orders First Federal Regulation of Greenhouse Gases. WASHINGTON, DC, May 14, 2007 (ENS) - After resisting the regulation of greenhouse gases since he took office in 2001, President George W. Bush today signed an Executive Order directing four federal agencies to develop regulations limiting greenhouse gas emissions from new mobile sources. Greenhouse gases, such as carbon dioxide emitted by the combustion of fossil fuels, contribute to global climate change. The President directed the U.S. Environmental Protection Agency, EPA, the Department of Transportation, the Department of Energy, and the Department of Agriculture to work together "to protect the environment with respect to greenhouse gas emissions from motor vehicles, nonroad vehicles, and nonroad engines, in a manner consistent with sound science, analysis of benefits and costs, public safety, and economic growth," the Executive Order states
4. Arrhenius S (1896) On the influence of carbonic acid in the air upon the temperature of the ground. Philosophical Magazine 41:237-276
5. Hoffert MI, Caldeira K, Benford G, Criswell, DR, Green C, Herzog H, Jain AK, Kheshgi HS, Lackner KS, Lewis JS, Lightfoot HD, Manheimer W, Mankins JC, Mauel ME, Perkins LJ, Schlesinger ME, Volk T, Wigley TML (2002) Advanced technology paths to global climate stability: Energy for a greenhouse planet. Science 298:981-987
6. Jefferson M (2006) Sustainable energy development: performance and prospects. Renewable energy 31:571-582
7. Linak WP, Wendt JOL (1994) Trace metal transformation mechanisms during coal combustion. Fuel Processing Technology 39:173-198

8.   Wong CSC, Duzgoren-Aydin NS, Aydin A, Wong MH (2006) Sources and trends of environmental mercury emissions in Asia. Science of the Total Environment 368:649-662

9.   Sun L, Yin X, Liu X, Zhu R, Xie Z, Wang Y (2006) A 2000-year record of mercury and ancient civilizations in seal hairs from King George Island, West Antarctica. Science of the Total Environment 368:236-247

10.   Muir DCG, Shearer RG, Van Oostdam J, Donaldson SG, Furgal C (2005) Contaminants in Canadian arctic biota and implications for human health: Conclusions and knowledge gaps. Science of the Total Environment 352: 539-546

11.   Wang D, He L, Wei S, Feng X (2006) Estimation of mercury emissions from different sources to atmosphere in Chongqing, China. Sci. Total Environment 366:722-728

12.   Pacyna EG, Pacyna JM, Pirrone N (2001) European emissions of atmospheric mercury from anthropogenic sources in 1995. Atmospheric Environment 35:2987-2996

13.   Sunderland EM, Chmura GL (2000) The history of mercury emissions from fuel combustion in Maritime Canada. Environmental Pollution 110:297-306

14.   http://www.epa.gov/mercury/about.htm

15.   http://www.militarycity.com/map/   BAHRAIN: Navy 5th Fleet headquarters - 1,200 sailors; Joint Venture HSV-X1 – 50 troops. DJIBOUTI: Camp Lemonier – 1,300 U.S. troops. HORN OF AFRICA: Elements of Combined Joint Task Force Horn of Africa, 10th Mountain Division, 478th Civil Affairs Division; about 1,300 troops. MEDITERRANEAN: Harry S. Truman Carrier Battle Group/Carrier Air Wing 3 (Marine Fighter-Attack Squadron 115) – 7,610 sailors and Marines; Theodore Roosevelt Carrier Battle Group/Carrier Air Wing 8 – 7,445 sailors; 26th Marine Expeditionary Unit trains aboard Iwo Jima Amphibious Ready Group. PERSIAN GULF: Amphibious Task Force East – 5,000 sailors; Amphibious Task Force West – 4,080 sailors; Tarawa Amphibious Ready Group w/15th Marine Expeditionary Unit – 1,700 sailors, 2,200 Marines; 2nd Marine Expeditionary Brigade returns to ships of

Amphibious Task Force East; Echo Company, Battalion Landing Team 2nd Battalion (with Nassau Amphibious Ready Group); Coast Guard cutters (2) and patrol boats (4) - 690 Coast Guardsmen. RED SEA: Attack submarine Boise – 112 sailors; Attack submarine Toledo – 112 sailors; Attack submarine San Juan – 112 sailors. TURKEY: Elements of 1st Infantry Division – 2,000 soldiers; Incirlik Air Base – F-15 and F-16 aircraft, 4,000 airmen. DIEGO GARCIA: AF 20th Bomb Squadron; 917th Bomb Wing Air Force Reserve. EGYPT: 1st Battalion, 180th Infantry Regiment, Oklahoma National Guard - 865 soldiers. GULF OF ADEN: Command ship Mount Whitney - 700 sailors, 400 troops. IRAQ: 82nd Airborne; 3rd Infantry Division; 4th Infantry Division; 101st Airborne Division; 173rd Airborne Brigade; V Corps; 1st Armored Division - 250,000 soldiers. I Marine Expeditionary Force. KUWAIT: Elements of the 101st Airborne Division - about 20,000 soldiers; Elements of 3rd Infantry Division - 13,500 soldiers; 325th Airborne Infantry Regiment, 82nd Airborne Division – 4,000 soldiers; Other Army elements – 10,800 soldiers; Army reservists – 5,299 soldiers; Elements of 293rd Infantry, Indiana National Guard - 600 soldiers; 190th Fighter Squadron, Idaho National Guard - 200 soldiers; Elements of I Marine Expeditionary Force – 45,000 Marines; Regimental Combat Team I – 6,000-7,000 Marines; 15th Marine Expeditionary Unit – 2,200 Marines; A-10 and F-16 aircraft; 2nd Marine Expeditionary Brigade – 6,000 to 7,000 Marines; 1042nd Medical Company, Oregon National Guard - 18 soldiers. OMAN: B-1B bombers and AC-130 gunships. QATAR: Al Udeid Air Base – F-15 and F-16 fighters, KC-135s and KC-10s, 3,500 airmen; Camp As Sayliyah - Central Command battle command; 205th Area Support Medical Battalion, Missouri National Guard – 50 soldiers. SAUDI ARABIA: Prince Sultan Air Base – 4,500 US military personnel, un-disclosed number of F-15 and F-16 fighters; 1042nd Medical Company, Oregon National Guard - 10 soldiers. UNITED ARAB EMIRATES: Al Dhatra Air Base – reconnaissance aircraft, 500 airmen

16.   http://www.ilea.org/articles/CEF.html
17.   Desert land areas are given in square miles. The list total is 7.4 million square miles. African deserts include: Sahara at 3.5 million; Arabian 1 million; Kalahari 220,000; Namib 13,000. Australian deserts include: Australian Desert 120,000; Great Sandy 150,000; Great Victoria 250,000; Simpson and Sturt Stony 56,000. North American deserts include: Mojave 54,000; Sonoran 120,000; Chihuahuan 175,000; Great Basin 190,000; Colorado Plateau 130,000. Asia deserts include: Thar 175,000; Taklamakan 105,000; Gobi 500,000; Kara-Kum 135,000; Kyzyl-Kum 115,000; Iranian 100,000. South American deserts include: Atacama 54,000; Patagonian 260,000
18.   Bockris JOM (2002) The origin of ideas on a hydrogen economy and its solution to the decay of the environment. International Journal of Hydrogen energy 27:731-740
19.   Turner JA (1999) A realizable renewable energy future. Science 285:687-689
20.   Yürüm Y Ed. (1995) Hydrogen energy system: production and utilization of hydrogen and future aspects. Kluwer academic publishers, Boston
21.   Ni M, Leung DYC, Leung MKH, Sumathy K (2006) An overview of hydrogen production from biomass. Fuel processing technology 87:461-472
22.   Tromp TK, Shia RL, Allen M, Eiler JM, Yung YL (2003) Potential environmental impact of a hydrogen economy on the stratosphere. Science 300:1740-1742

Chapter 2

HYDROGEN GENERATION BY WATER SPLITTING

2.1 Introduction

While today hydrogen is predominately generated from the processing of fossil fuels, which as a by-product results in the release of CO_2, this does us no good within the scheme of trying to prevent catastrophic global warming, nor any good within the less-immediate but still real issue of fossil fuel depletion. This chapter briefly discusses avenues to hydrogen production (other than photoelectrolysis) by splitting of the water molecule to generate hydrogen and oxygen that do not, necessarily, result in the emission of climate-altering gases.

The 1973 oil embargo motivated significant research efforts in hydrogen production [1-4] that were subsequently abandoned when the fossil fuel flood-gates were re-opened. Recently emerging concerns over energy security, and the maintenance of our global ecosystem has rekindled the search for the processes that are efficient, economical and practical for large-scale production of hydrogen; herein we consider some possibilities.

2.2 Hydrogen Production By Water Electrolysis

Electrolysis of water to generate hydrogen and oxygen has a history of more than 200 years [5]. It is the simplest of all the water-splitting techniques and comparatively efficient. The net reaction is

$$H_2O \text{ (liquid/vapor)} + \text{electrical energy} \rightarrow H_2 \text{ (g)} + \tfrac{1}{2} O_2 \text{ (g)} \quad (2.2.1)$$

Oxygen, a highly useful gas, is the only by-product. Water electrolyzers today satisfy approximately 3.9% of the world's hydrogen demand [6]. Electrolyzers are used to produce hydrogen at levels ranging from a few cm^3/min to several thousand m^3/hour, with units ranging in physical size from portable to essentially immovable [4,7,8]. Electrolyzers generally use electricity from the power grid, which in turn is predominately powered by the combustion of coal, resulting in expensive and eco-hostile operation [9]. For sustained development the electrolyzers will need to be run using electrical energy derived from renewable sources such as windmills or solar cells.

At standard ambient temperature and pressure, the change in Gibb's free energy ΔG for the water splitting reaction (2.2.1) is positive and hence the reaction is non-spontaneous; for the reaction to occur the electrical energy equivalent to the change in the Gibb's free energy ΔG of the reaction must be supplied. In an electrolytic cell the electrical energy is supplied by applying a potential difference between two electrodes placed in an electrolyte. Electrical energy to chemical energy conversion takes place at the electrode-solution interface through charge transfer reactions. A potential difference V applied between the electrodes can be used to do maximum work of nFV, where F is the Faraday constant and n is the number of moles of electrons involved in the reaction. If V_{rev} is the minimum voltage needed to drive the water splitting reaction,

$$\Delta G = nFV_{rev} \qquad (2.2.2)$$

This is the maximum amount of useful work that can be derived from the system on driving the reaction in the opposite direction. Thus, V_{rev} corresponds to the reversible work and is consequently called the thermodynamic reversible potential. At 25°C and 1 bar, the ΔG for the water-splitting reaction is 237.178 kJ/mol [10]. Therefore,

$$V_{rev} = \Delta G/nF = 1.229V \qquad (2.2.3)$$

The reaction at this voltage is endothermic and hence at isothermal conditions heat energy ($=T\Delta S$, where is S is entropy and T is the absolute temperature) must be absorbed from the surrounding environment for the increase in entropy associated with water dissociation. On operating an electrolytic cell above V_{rev} heat is generated due to the losses in the cell and a fraction of this provides the additional energy needed for the entropy change.

When energy exactly equal to the enthalpy $\Delta H = \Delta G + T\Delta S$ (=285.83 kJ/mol at 1 bar and 25°C) for water splitting is supplied, no heat is absorbed or evolved by the system [10]. The voltage corresponding to this condition, the thermoneutral voltage V_{tn} is given by

$$V_{tn} = \Delta H/nF = 1.482 \ V \qquad (2.2.4)$$

At V_{tn} the electrolysis generates enough heat to compensate for $T\Delta S$. When the cell is operated above $V_{tn,}$ the reaction becomes exothermic and heat must be removed from the electrolytic cell for isothermal operation. However energy losses, associated with reaction kinetics as well as charge transport through electrical leads and the electrolyte, necessitate electrolyzer operation in this voltage regime. In the case of practical devices the operating voltage can be expressed as

$$V_{op} = V_{rev} + \eta_a + |\eta_c| + \eta_\Omega \qquad (2.2.5)$$

where η_a and η_c are the anodic and cathodic overpotentials associated with the reaction kinetics (electrode polarization effects) and η_Ω is the Ohmic overpotential that arises due to the resistive losses in the cell [7].

The overvoltage represents the voltage in excess of the thermodynamic voltage, required to overcome the losses in the cell and obtain the desired output in terms of current density or amount of hydrogen from the cell. Slow electrode/electrolyte reactions and large resistive losses result in high overpotentials. The overpotentials should be reduced to increase the optimum current density. The electrode, both anodic and cathodic, overpotential arises as a result of several polarization effects. These include low activity of the electrodes in the electrolyte (known as activation overpotential) leading to slow-charge transfer processes and ion deficiency at the electrode surface for charge transfer to take place due to the poor mass transport through diffusion (diffusion overpotential), migration or convection [7,11,12]. The electrode overpotential increases logarithmically with current density as given by the Tafel relation [13]

$$\eta = a + b \log (j) \qquad (2.2.6)$$

where j is the current density and 'a' and 'b' are characteristic constants for the electrode system. The constant 'a' provides information about the electrocatalytic activity of the electrodes, and 'b' the slope of the Tafel plot (η Vs log j) carries information regarding the electrode reaction mechanisms.   The electrode overpotential can be minimized by selecting electrode materials with high electrocatalytic activity, and maximum real to apparent surface area [14]. The Ohmic overvoltage is a consequence of the resistive losses in the cell that occur mainly at the electrodes, electrical lead wires, metal-metal joints, and inside the electrolyte. Reduction in the electrode separation and electrolyte-resistance lowers the Ohmic overvoltage [4,7,15]. Bubble formation by the product gases is also a source of activation and Ohmic overpotentials [15-18]. An increase in operating temperature helps to reduce both activation and Ohmic overvoltages as it enhances the reaction rate and reduces the electrolyte resistance. Overvoltage minimization is essential for high efficiency operation of the electrolytic cells.

The efficiency of water electrolysis is defined as ratio of the energy content of hydrogen (the energy that can be recovered by reoxidation of the hydrogen and oxygen to water) to the electrical energy supplied to the electrolyzer [19,20]. In terms of voltage, the efficiency can be expressed as

$$\varepsilon = V_{tn}/V_{op} \qquad (2.2.7)$$

Ideally, a cell operating at V_{tn} can produce hydrogen at 100% thermal efficiency. The energy in excess of ΔG need not be supplied in the form of electrical energy, hence a voltage as low as V_{rev} can also be used to split water if the system is allowed to absorb heat from its surroundings. The operation is called allothermal operation. The cell is about 120% voltage efficient when the operating voltage is V_{rev}. This means that the fuel value of the hydrogen produced will be 120% of the heating value of the electrical energy input at this operating condition [21]. When a cell is operated above V_{tn} heat generated inside the cell due to losses supply the extra energy needed for driving the water splitting reaction, a process called autothermal operation. The cells should be operated at low voltages and high current densities to achieve high efficiencies and high hydrogen production rates. Practical efficiencies lie in the range 50 to 90% [22].

The cell voltage V_{op} is largely decided by the operating temperature and pressure. The thermodynamic voltage for water splitting (V_{rev}) reduces with an increase in operating temperature as given by the relation [23,24] for °K.

$$V^0_{rev, T} = 1.5184 - 1.5421 * 10^{-3} * T + 9.523 * 10^{-5} * T * \ln T \quad (2.2.8)$$
$$+ 9.84 * 10^{-8} * T^2$$

V_{rev} takes a value of 1.18V at 80°C. The over-voltages are lowered considerably at elevated temperatures due to the increased conductivity of the electrolyte and higher electrode activities. The reduction in thermodynamic voltage as well as over-voltages at elevated temperatures lower the operating voltage of the electrolytic cell, thereby increasing the water splitting efficiency with resultant energy savings. When temperature and pressure effects are considered it is accurate to use V_{HHV}, which is the higher heating value voltage, in place of V_{tn} for efficiency calculation [24]. V_{HHV} corresponds to the heat content of the dry product gases with respect to the liquid water at 25°C. Therefore,

$$\varepsilon = V_{HHV}/V_{op} \quad (2.2.9)$$

The absolute temperature dependence of V_{HHV} can be given by the relation [7,24].

$$V^0_{HHV,T} = 1.4146 + 2.205 * 10^{-4} * T + 1.0 * 10^{-8} T^2 \quad (2.2.10)$$

V_{HHV} increases slightly with temperature taking a value 1.494 V at 80°C.

The reversible voltage V_{rev} is related to the operating pressure as

$$V_{rev,T,P} = V^0_{rev,T} + (RT/2F) \ln [(P-P_w)^{1.5}/(P_w/P^0_w)] \qquad (2.2.11)$$

where R is the universal gas constant, F is Faraday constant, P is operating pressure of the electrolyzer and P_w and P^0_w are, respectively, the partial pressures of water vapor over the electrolyte and over pure water [23]. An increase in pressure raises the reversible voltage at the rate of ~ 43 mV for every ten-fold increase in pressure where it has only a negligible effect on V_{HHV} [4,18,23-25]. The over-voltages are considerably reduced at higher pressures and the savings in energy due to this is considered higher than the extra energy needed to overcome the theoretical voltage. Noted advantages of operation at higher pressures are: (1) higher optimum current density compared to the lower pressure cells operating at the same voltage; (2) enabling of higher operating temperatures; (3) suppression of bubble formation at high current densities; (4) elimination of equipment and its energy used for pressurizing the hydrogen as needed for practical applications; (5) prevention of water loss by evaporation from cells operating above 80°C [15,25]. However a recent analysis shows higher overall electrical energy consumption in pressurized electrolyzers compared to those operating at atmospheric pressure [18]. Furthermore problems such as corrosion, hydrogen embrittlement and other instabilities in the cell components are worse in pressurized electrolyzers. A source of inefficiency and purity loss in high-pressure electrolysis is the recombination of hydrogen and oxygen dissolved in the electrolyte that crosses over, respectively, into the anode and cathode compartments of the cell.

For hydrogen production from water, pure water (pH=7.0) is seldom used as an electrolyte. Water is a poor ionic conductor and hence it presents a high Ohmic overpotential. For the water splitting reaction to proceed at a realistically acceptable cell voltage the conductivity of the water is necessarily increased by the addition of acids or alkalis. Aqueous acidic and alkaline media offer high ionic (hydrogen and hydroxyl) concentrations and mobilities and therefore possess low electrical resistance. Basic electrolytes are generally preferred since corrosion problems are severe with acidic electrolytes. Based on the type of electrolytes used electrolyzers are generally classified as alkaline, solid polymer electrolyte (SPE), and solid oxide electrolyte (SOE) which we now consider further.

Alkaline Electrolyzers

As the name indicates alkaline electrolyzers use high pH electrolytes like aqueous sodium hydroxide or potassium hydroxide. This is the oldest, most developed and most widely used method of water electrolysis. Hydrogen evolution takes place at the cathode, and oxygen evolution takes place at the anode. The cathodic reaction can be represented by the following steps [26,27]

$$H_2O + M + e^- \rightarrow M\text{-}H + OH^- \quad (2.2.12)$$
$$H_2O + M\text{-}H + e^- \rightarrow H_2 + M + OH^- \quad (2.2.13)$$
$$2M\text{-}H \rightarrow H_2 + 2M \quad (2.2.14)$$

M-H represents the hydrogen adsorbed at the active sites on the electrode surface. The hydrogen is first adsorbed onto the active sites on the cathode surface (reaction 2.2.12). Under the catalytic action of the electrode surface, the adsorbed hydrogen atoms combine together to form hydrogen molecules (reaction 2.2.13 and/or reaction 2.2.14). These molecules accumulate until a bubble forms which breaks away and rises to the electrolyte surface. The hydroxyl ions are discharged at the anode leading to oxygen evolution. The process can be represented as [28]

$$OH^- + M \rightarrow M\text{-}OH + e^- \quad (2.2.15)$$
$$M\text{-}OH + OH^- \rightarrow M\text{-}O + H_2O + e^- \quad (2.2.16)$$
$$2M\text{-}O \rightarrow O_2 + 2M \quad (2.2.17)$$

An alkaline electrolyzer consists mainly of a cell frame, electrolyte, anode, cathode and a separator [29]. A two-cell configuration is shown in **Figure 2.1.** The cell frame is generally made of stainless steel. The electrolyte should provide high ion conductivity, should not suffer chemical decomposition on the application of voltage so as to restrict the process to splitting of water molecules, and be capable of withstanding pH changes during the process as a result of the rapid changes in hydrogen ion

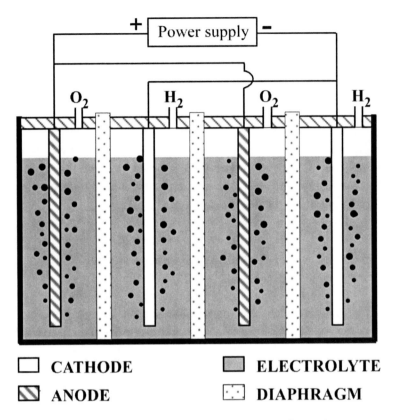

CATHODE ☐   ELECTROLYTE ▨

ANODE ▨   DIAPHRAGM ⬚

Fig. 2.1: Two-cell conventional alkaline electrolyzer configuration.

concentration at the electrodes. As acidic electrolytes cause corrosion problems, strong bases such as NaOH or KOH are commonly used as electrolytes [4,7]. The conductivity of the aqueous KOH is maximum when its concentration is near 28% and hence most of the electrolyzers use a concentration of 25% to 35% KOH in water [30]. The conductivity of this electrolyte decreases with increasing temperature reaching a maximum at approximately 150°C [7]. The electrolyte is circulated through cells, heat exchangers and filters to maintain a constant operating temperature, reduce polarization effects due to bubble formation and concentration gradients, and remove suspended solids and possible corrosion products. The electrolyte concentration is maintained by addition of high purity water as needed depending upon the hydrogen production rate. A schematic of an electrolyzer unit is given in **Fig. 2.2**. The evolved hydrogen and oxygen vapor also contain that of the electrolyte, hence the vapors are initially passed through electrolyte-gas separators for purification; the hydrogen gas is then dried by passing through a condenser.

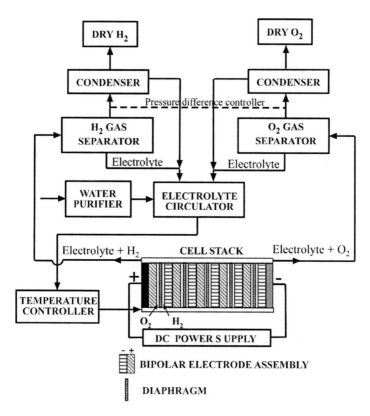

Fig. 2.2: Schematics of hydrogen production using an alkaline electrolyzer unit with bipolar electrode geometry.

The electrodes are made of electrocatalytic materials to ensure low overvoltages by facilitating rapid electrode-electrolyte charge transfer, and nucleation of the gas bubbles at the electrode surface as well as their self-detachment from the electrode surface at the operating cell voltage. To reduce the overvoltages the electrode should provide an interface between catalyst and the electrolyte of large surface area. Hence various electrode geometries have been used, including finned bodies, screens, perforated plates and flat plates with electrochemically roughened surfaces [7]. The elements from group 8 (Fe, Ru, Os), 9 (Co, Rh, Ir) and 10 (Ni, Pd, Pt) and their alloys are conventionally used as electrode materials [31]. Noble metals like platinum and iridium have high catalytic activity with high corrosion resistance, however these materials are expensive and their use is limited to electrolysis using acidic media. Nickel, cobalt and iron (stainless steel) have low overpotentials in aqueous caustic soda or potash solutions and hence are used widely as cathodes for alkaline electrolysis [27,32-37]. Ni has high corrosion stability and hence both Ni electrodes and Ni coated steel electrode supports are common in alkaline electrolyzers, especially in those using elevated temperatures (above 80°C) and high concentration electrolytes [14]. Raney nickel is a porous material with high catalytic activity that shows low hydrogen overpotentials in alkaline solutions [7,38]. Advanced electroyzers prefer Raney nickel as the cathode material and perovskite oxides like $LaCoO_3$ and $LaNiO_3$ as anode materials [39].

Porous diaphragm-like separators are used to divide the anode and cathode regions to prevent mixing of the evolved hydrogen and oxygen gases while allowing for passage of the electrolyte solution. The pores of a separator must remain full of liquid so that gas cannot penetrate them, as well as allow passage of current without an appreciable resistance. Use of a separator allows adjacent placement of the electrodes thus reducing the Ohmic overpotential. A separator has pore sizes smaller than the diameter of the smallest gas bubble, for hydrogen this is around 10 μm, and to keep the electrical resistance low a porosity greater than 50%. Commonly used separator materials, which must be corrosion resistant and structurally stable, are asbestos [$Mg_3Si_2O_5 (OH)_4$], woven cloth, polymer materials like polytetrafluoroethylene (PTFE), oxide ceramic materials like NiO, and cermets like $Ni-BaTiO_3$ [21,30,40].

Water electrolyzer units typically consist of several cells or electrodes arranged in two basic configurations, tank type operated in unipolar configuration, or filter press type operated in bipolar configuration. The most common configuration, see **Fig. 2.1**, is the unipolar tank type where each electrode has only one polarity and all the electrodes of the same polarity are connected in parallel. The anodes and cathodes are alternately connected, with the electrodes and separators kept immersed within an electrolyte containing tank. The unipolar tank cells require relatively simple construction, few parts, low cost, and are easy to maintain. However, they are also bulky and cannot be operated at high temperatures or pressures; the unipolar tank type cells normally operate at temperatures between 60°C and 90°C (higher temperatures result in rapid electrolyte evaporation) and near ambient pressure [30]. The relatively long current paths manifest in the leads and electrodes result in unwanted energy losses.

In bipolar filter press type configuration, see **Fig. 2.2**, the cells are connected in series with each cell consisting of a separator diaphragm pressed on either side by electrodes. An electrically conducting solid separator plate joins the electrodes in adjacent cells and serves as the partition between the hydrogen cavity of one cell and the oxygen cavity of the other. The electrode at one side of the separator plate acts as a cathode of one cell, whereas that on the other side acts as the anode of the adjacent cell.   Advanced electrolyzers have a zero-gap geometry [20] with the anode and cathode directly formed on opposite sides of a porous diaphragm. The bipolar filter-type electrolyzers are relatively compact, and in comparison to tank type cells capable of operating at higher temperatures and pressures with lower Ohmic losses. The bipolar filter type cells can operate at temperatures as high as 150°C and pressures up to about 30 bar [23]. Consequently in comparison to tank type cells, filter-type electrolyzers can be operated at lower cell voltages with higher current densities. However, they need precise construction and they are more expensive and difficult to maintain.

Alkaline electrolyzers are used all over the world for electrolytic hydrogen production, delivering hydrogen with a nominal purity of 99.8%. Conventional cells operate at voltages 1.8-2.2 V with current densities below 0.4 A/cm^2; advanced electrolyzers operate at relatively lower voltages (as low as 1.6 V) and higher current densities (up to about 2 A/cm^2) [21,30]. Conventional tank-type electrolyzers have efficiencies in the range of 60-80%, while modern zero-gap electrolyzers can be up to 90% efficient.   Hydrogen production rates in normal alkaline electrolyzers are between 0.01 and 10 m^3/h, whereas some large-scale units can reach 10 to 100 m^3/h.   However, electrolyzers suitable for large-scale hydrogen production necessarily have large-scale electrical power requirements, and there are environmental concerns related to the chemicals used in the electrolytes. Within the global energy context and the need to reduce CO_2 emissions, the most logical approach for hydrogen generation by water electrolysis appears to be the adjacent coupling of either windmills or photovoltaic grids (see Chapter 8) to the electrolyzer.

Solid Polymer Electrolyzers

The development of solid polymer electrolyte (SPE) water electrolyzers is coupled to the invention of proton exchange membrane (PEM) fuel cells [9,41,42]. In the mid 1950s researchers at the General Electric Corporation (GE) developed fuel cells using a sulfonated polystyrene electrolyte [41,42]. In 1966 fuel cells employing a much superior membrane, DuPont's Nafion, were developed and used for NASA space projects. In 1973 GE developed SPE water electrolyzers using proton exchange membrane technology that were initially used for oxygen generation in nuclear submarines [41]. SPE water electrolyzers have now become an industrially viable, well-accepted technology. These electrolyzers are compact and ecologically clean; in comparison to alkaline electrolyzers, SPE electrolyzers are able to operate at lower cell voltages, higher current densities, as well as higher pressures and temperatures.   SPE electrolyzer efficiencies can reach near 100%, with normal device operation the 80-90% range [30,43,44]. An additional major advantage of SPE technology is that it can generate very high purity (>99.999%) hydrogen.

    SPE electrolyzers, see **Fig. 2.3**, have a device config-uration similar to that of zero-gap bipolar filter press type alkaline electrolyzers, see **Fig. 2.2**, but a proton conducting perfluorinated polymer membrane like Nafion (also known as perfluorosulfonic acid) having side chains terminated in sulphonate ion exchange groups, serves simultaneously as electrolyte and the separator [43]. Highly pure water circulated through the cell is split into hydrogen and oxygen with the help of electrocatalysts on the membrane surface.   The membrane is normally a 150 − 300 μm thick sheet that is impermeable to water and product gases. It possesses poor electronic conductivity but high proton conductivity when saturated with water. The sulfonic acid groups ($-SO_3H$) incorporated in the membrane become hydrated when exposed to water and then dissociate ($-SO^-_{3aq} + H^+_{aq}$) facilitating proton conduction [45]. On the anode, when in contact with water, hydroxy bonds are formed at the active surface sites, X, on the membrane with release of protons [46].

$$X + H_2O \rightarrow X\text{-}OH_{ads} + H^+ + e^- \qquad (2.2.18)$$

The hydroxy bonds then break releasing more protons.

$$X\text{-}OH_{ads} \rightarrow X\text{-}O + H^+ + e^- \qquad (2.2.19)$$

The protons, or more correctly hydrated protons ($H^+.nH_2O$), migrate through the membrane towards the cathode by hopping from one fixed sulfonic group to another; at the cathode they collect electrons and form hydrogen gas. The X-O bonds at the anode side then break and release oxygen gas, leaving the active sites free for another adsorption-desorption process. In addition to Nafion, materials like polyetheretherketone, polytetrafluoroethylene (PTFE), polyethylene, polybenzimidazole, polyimide, polyphosphazene and methyl methacrylate are also used to fabricate membranes for electrolysis after appropriate chemical modifications [33,47,48]. Methyl methacrylate monomer is an OH^- conductor instead of a proton conductor and hence the use of noble metal catalysts can be avoided.

The structure of a SPE cell is shown in **Fig. 2.3**. The basic unit of a SPE electrolyzer is an electrode membrane electrode (EME) structure that consists of the polymer membrane coated on either side with layers (typically several microns thick) of suitable catalyst materials acting as electrodes [43,49,50], with an electrolyzer module consisting of several such cells connected in series. The polymer membrane is highly acidic and hence acid resistant materials must be used in the structure fabrication; noble metals like Pt, Ir, Rh, Ru or their oxides or alloys are generally used as electrode materials. Generally Pt and other noble metal alloys are used as cathodes, and Ir, IrO_2, Rh, Pt, Rh-Pt, Pt-Ru etc. are used as anodes [43,46]. The EME is pressed from either side by porous, gas permeable plates that provide support to the EME and ensure homogeneous current distribution across the membrane surface. These porous plates provide exit paths for the gas generated at the electrodes. Titanium and carbon are the commonly used materials at the anode side whereas carbon is the preferred material at the cathode side. The bipolar plates are in direct physical contact (see **Fig. 2.3**) with the porous structures that separate adjacent cells while allowing the current to pass from one cell to the next. The cavity between the bipolar plate and current collector allows the distribution of feed water and the collection of product gases. The plate material must have high electronic conductivity, negligible permeability to product gases, and high corrosion resistance. Graphite is generally used for making the bipolar plates.

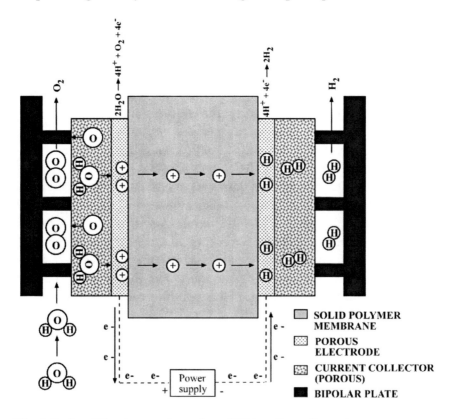

Fig. 2.3: A solid polymer electrolyte (SPE) cell configuration.

In comparison to alkaline electrolyzers, SPE electrolyzers are more efficient, reliable, and safer. Since SPE electrolyzers use solid electrolyte, there is no risk of corrosive chemical leaks nor issues of unwanted gas cross-over. As the cells in SPE electrolyzers are connected in series high voltage dc supplies can be used which are less expensive compared to the low voltage high current dc supplies. Currently these cells generally operate at temperatures of 80-150°C and pressures of about 30 bar [30], with high pressure ($\approx$ 135 bar) SPE electrolyzers being developed. Cell voltages range from 1.4 V to 2 V, with current densities up to about 2 A/cm^2. These cells can intrinsically adjust to variations in electrical power hence are well suited for operation using power from inherently intermittent solar cells or wind mills. However, SPE electrolyzers are comparatively expensive due to the high cost of the polymer membranes and noble metal electrodes, as well as requiring very high purity water. Other design nuances are that precise control of the differential pressure across the anodic and cathodic compartments is necessary for membrane stability. In thin membranes hydrogen diffusion to the anode can adversely affect device efficiency [23].

Solid Oxide Electrolyzers

Solid oxide electrolyte (SOE) cells were first developed in the early 1970s [5]. These cells operate at very high temperatures, near 1000°C, exploiting the useful effects of high temperatures on the kinetic and thermodynamic parameters controlling electrolytic water splitting to generate hydrogen from steam. The total energy demand for water splitting ΔH is lower in the vapor phase (241.8 kJ/mol) than in the liquid phase (285.83 kJ/mol) [10,12]. The thermoneutral voltage is around 1.287 V at 800°C [51]. As discussed at the beginning of this section, high temperatures lower the thermodynamic reversible potential for water splitting as well as the activation overpotentials. Thus the electrical energy requirements are much lower for water electrolysis in vapor phase at high temperatures. The extra energy needed to drive the reaction is supplied as heat, which is more efficient to obtain and hence cheaper than electricity. **Figure 2.4** shows the demand of electrical and heat energies with electrolysis temperature. It can be seen from **Fig. 2.4** that the ratio of ΔG (electrical energy demand) to ΔH (total energy) is about 93% at 100°C and about 70% at 1000°C, which shows that in high temperature electrolysis about 30% of the energy involved can be supplied as heat [52]. The heat requirements of the cell can be met either internally through Ohmic heating or externally using a heat source depending upon whether the operation is allothermal or autothermal. The reduced electrical energy requirements enables the SOE cells to operate at lower voltages (about 0.95 to 1.33 V) than those required for other types of electrolyzers [30].

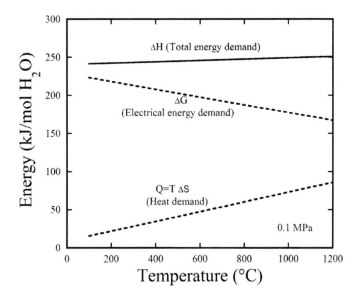

Fig. 2.4: Requirement of electrical and heat energies for steam electrolysis.

SOE cells utilize solid ceramic electrolytes (e.g. yttria stabilized zirconia) that are good oxygen ion (O^{2-}) conductors at very high temperatures in the range of 1000°C [8]. The operating temperature is decided by the ionic conductivity of the electrolyte. The feed gas, steam mixed with hydrogen, is passed through the cathode compartment. At the cathode side, the reaction is

$$2H_2O + 4e^- \rightarrow 2H_2 + 2O^{2-} \tag{2.2.20}$$

As the electrolyte is impermeable to hydrogen gas, the O^{2-} ions migrate through the electrolyte towards the anode under the action of the electric field, where they gain electrons and form oxygen gas.

$$2O^{2-} \rightarrow O_2 + 4e^-$$ (2.2.21)

The electrolyte is impermeable to water vapor and product gases, thus it effectively separates hydrogen and oxygen.

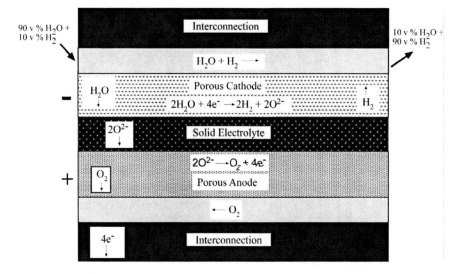

Fig. 2.5: A solid oxide electrolyte (SOE) electrolyzer configuration (planar geometry).

SOE cells are fabricated in planar, see **Fig. 2.5**, as well as tubular geometries [51-55], with the voltage and current losses minimized in the planar geometry. In both planar and tubular cases the electrolyte is pressed between porous electrodes. The cells are attached in series using interconnecting elements that serve as electrical conductors as well as current distributors. The cells with planer geometries are simply stacked to form the electrolyzer unit. In the tubular geometry each cell is fabricated in the shape of a short circular tube, which are then connected along their lengths forming a long tube; the electrolyzer module consists of a stack of several such tubes. The feed gas is passed through the inner surface of the tube and oxygen is evolved through the outer surface. Although a number of solid oxide electrolytes have been tested, yttria stabilized zirconia is the favorite electrolyte material of today [56]. Cermets such as Ni-ZrO_2 and Pt-ZrO_2 are commonly used as cathode materials, and conducting perovskites like $LaNiO_3$, $LaMnO_3$, and $LaCoO_3$ used as anodes [57]. Materials like Ferritic steel and doped $LaCrO_3$ are used to make the interconnections, and $CaZrO_3$ used for insulation of the assembly.

We note that SOE cell technology is still in the developmental stage. Severe materials problems related to the high temperature operation, typically 700-1000°C, need to be solved. Low life times of the materials used for cell construction, intermixing of adjacent phases and engineering problems related to thermal cycling and gas sealing are some of the factors that prevent the technology from successful commercialization [8]. However, this technology has some unique advantages that make it attractive in comparison to other technologies [58]. Compared to other types of electrolyzers the power consumption is less; in comparison to advanced alkaline electrolyzers [59] SOE cells require at least 10% less energy. The efficiency of high temperature electrolysis can essentially be 100% [30]. Furthermore SOE electrolyzers are relatively eco-friendly, not requiring any corrosive electrolytes for operation [58].

2.3 Hydrogen Production by Thermochemical Water-Splitting

Thermochemical processes utilize thermal energy, either directly or through different chemical reactions, to carry out water splitting for hydrogen generation. The net reaction is

$$H_2O + \text{thermal energy} \rightarrow H_2 + \tfrac{1}{2}O_2 \qquad (2.3.1)$$

A portion of the supplied thermal energy is expelled with the products. The overall water splitting process, see **Fig. 2.6**, in general requires both heat and work input. In **Fig. 2.6** q_r represents the heat energy supplied at high temperature T_r, q_0 represents the heat rejected at lower temperature T_0, and W_i is the useful work input, if any, for the process. The enclosed region contains only water, or water and materials involved in different reactions. Water is supplied into the enclosed region where it is split using thermal energy, releasing hydrogen and oxygen as the by-products.

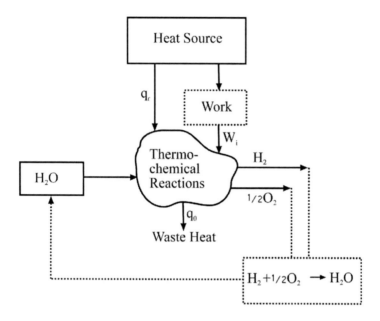

Fig. 2.6: General representation of the thermochemical water splitting process

As the water splitting process is cyclic (reversible behavior shown by dotted lines in **Fig. 2.6**) it has limitations imposed by second law of thermodynamics [60]. Hence the operating temperatures T_r and T_0 are crucial in determining the thermal efficiency of the process. In a water splitting process using both heat and work inputs, the thermal efficiency in general is defined as [60,61]

$$\varepsilon = \Delta H_0 / (\Sigma_r \, q_r + \Sigma_i \, W_i / \xi_i) \tag{2.3.2}$$

Where q_r represents the heat input from different sources, W_i the useful work input and ξ_i denoting the heat to useful work conversion efficiency associated with it. Depending upon the process, work input W_i may be required for different process steps like driving the reaction, reaction-product separation, and mass transfer. ΔH_0 is the enthalpy change involved in water splitting; generally the higher heating value (HHV) of hydrogen is assigned to ΔH_0 in equation (2.3.2) [62].

Equation (2.3.2) shows that a reduction in work input W_i is necessary if, realistically, high operating efficiencies are to be obtained. If the work input is zero, i.e. if only thermal energy is supplied, using the first and second laws of thermodynamics it can be shown that [2,4,60,61]

$$\varepsilon \le (\Delta H_0/\Delta G_0) \, [(T_r - T_0)/T_r] \qquad (2.3.3)$$

Here $T_r - T_0/ T_r$ is the Carnot efficiency of a heat engine working between T_r and T_0 [60]. This relation gives the maximum attainable thermal (heat to heat conversion) efficiency of an ideal water splitting process. The efficiency of an ideal process reaches over 80% when T_r and T_0 are, respectively, 1200 K and 393 K [3]. This indicates that high efficiencies are possible if water splitting can be done at high temperatures. It should be noted that a parameter that is important for assessing the commercial viability of a thermochemical process is the exergy (useful work) efficiency [63,64]. The exergy efficiency is defined as the ratio of the Gibbs free energy of the water formation reaction to the total energy necessary to split water using a thermochemical cycle. Exergy analysis gives the information about the requirements and losses in each process component. However, in the following discussion 'efficiency' implies thermal efficiency.

Emil Collett, in 1924, hypothesized a two-step water splitting process involving mercury and mercury oxide that, while a remarkable idea, was later proved infeasible [4,65]. In 1960s the interest in utilizing nuclear waste heat for useful purposes like hydrogen production initiated research in practical thermochemical water splitting [1], as pioneered by Funk and Reinstrom [60]. The 1973 oil embargo motivated significant research efforts in thermochemical hydrogen production [1-4] that were subsequently abandoned due to lack of financial support when the immediate energy crisis passed.

Historically, the initial objective behind the use of thermochemical processes for water splitting was to produce hydrogen directly from the heat released by nuclear waste, rather than through a waste heat- electricity- electrolysis route in which the overall efficiency of hydrogen production is the product of the electrical energy production efficiency and the electricity-to-hydrogen conversion efficiency. The low thermal to electrical energy (useful work) conversion efficiency in the waste heat-electricity- electrolysis route reduces the overall efficiency of hydrogen production. For example, the overall efficiency of hydrogen production from an advanced electrolyzer working at an efficiency of 90% using the electricity produced from a nuclear reactor at an efficiency of 40% (present day feasible value) is 36%. However, the overall thermal to hydrogen conversion efficiency in the thermochemical water splitting process can be 50% or above [66,67]. A 50% efficient process can produce 10 tons of hydrogen a day from a 30 MW thermal energy source.

The maximum temperature of a high temperature nuclear reactor, as determined by safety considerations, is 1573 K [68]. Selection of a suitable reactor type for thermochemical hydrogen production depends upon whether the coolant used in the reactor can effectively transfer the heat to a thermochemical process. Studies show that high temperature gas cooled reactors (HTGR), molten-salt-cooled reactors (MSR), and heavy metal cooled reactors (HMR) are the most suitable candidates for thermochemical hydrogen production technology [69]. Of these helium-based HTGR are appropriate for use in processes operating in the range of 1100-1300 K. The environmental as well as safety concerns regarding the use of nuclear reactors are, of course, applicable to the nuclear-thermochemical technology.

Thermochemical water splitting, making use of concentrated solar radiation for supplying the necessary thermal energy to drive the reactions, is a promising technology. Very importantly this solar energy based approach is renewable, while very high temperatures can be achieved that far exceed those obtainable with nuclear reactors. This high temperature operation offers higher ideal thermal efficiency according to equation (2.3.3). The concentrators collect the solar radiation and focus it into the chamber where reactions take place. Depending upon the geometry of the concentrators, parabolic trough, tower, dish etc., different concentration ratios and hence temperatures can be achieved [70-72]. Temperatures up to about 3000 K can be obtained using concentration ratios in excess of 10000 suns [70]. Hence the technology is particularly useful for thermochemical cycles that require very high temperatures. However, solar concentrators require space, and the installations are in practice limited to the desert or sun-belt regions of the world (of which there are many) where there are minimum diurnal and seasonal variations. Different options have been considered to avoid shut down of thermochemical plants during cloudy days. One option is to store solar energy in a thermal storage system during sunshine hours while simultaneously using the solar energy directly for running the thermochemical plant [73].

Different strategies have evolved for thermochemical hydrogen production to effectively utilize the potentials of, in particular, nuclear and solar thermal energy sources. These strategies, which we discuss below, can be categorized depending upon the number of process steps involved and whether electrolysis is employed in a reaction.

One-step Thermochemical Process

In a process called direct thermolysis, at a high enough temperature thermal energy is sufficient to split water into hydrogen and oxygen. Only one reaction is involved in this process, that of equation (2.3.1). In **Fig. 2.6**, if the input water and output gas mixture are at the same temperature T_0, the minimum work input required to effect water splitting at temperature T_r can be written [60]

$$W_r = \Delta G = \Delta G_0 - \Delta S \, (T_r - T_0) \qquad (2.3.4)$$

where ΔG and ΔG_0 are, respectively, the free energy changes when the reaction is carried out at T_r and T_0. ΔS does not increase significantly with temperature [3]. Thus the useful work requirement reduces on increasing the reaction temperature T_r. At a pressure of 1 bar ΔG becomes zero at about 4300 K [1,3,70]; at this operating point 100% of the water molecules are split. Only partial dissociation of the water molecules is achieved at lower temperatures. At 2000 K and 1 bar pressure about 1% of water is split; this increases to about 9% at 2500 K, and 36% at about 3000 K [3]. The fraction dissociated at a particular temperature increases with a reduction in pressure. At 2500 K and 0.05 bar about 25% of the water is dissociated.

When water splitting occurs in the temperature range 2000 to 3000 K, 1 bar pressure, only H_2O, OH, H, O, H_2 and O_2 exist in significant concentrations (and in different molar ratios); the possibility of the presence of species such as HO_2, O_3 and H_2O_2 can be neglected [74-77]. To achieve any reasonable yield of H_2 using direct thermolysis the temperature should be around 3000 K or above. For example, at 2500 K and 1 bar, direct thermal water splitting can yield about 4% molecular hydrogen (H_2 species) [3]. Nevertheless, due to materials limitations at this high temperature the preferred operating temperature regime of today is 2000- 2500 K [74]. As the water splitting reaction is reversible, and hydrogen yield is low in the 2000- 2500 K temperature regime, hydrogen and oxygen should be separated in a timely manner before they can recombine to form water.

As noted, separation of the evolved hydrogen and oxygen is an important step in any thermochemical water splitting cycle. Rapid separation of the gases is required not only to avoid efficiency loss due to hydrogen-oxygen recombination, but also to avoid hydrogen and oxygen forming an explosive mixture. Among the different methods available for product separation, methods such as quenching and high temperature separation are appropriate for a one step process [74-76,78]. In the quenching method, the hot product gas is rapidly cooled to 450 K using low-temperature steam or inert gas [74], then further cooled to remove water vapor through condensation. The mixture of dry oxygen and hydrogen are then separated using non-porous solid membranes like palladium or palladium-silver alloys. Quenching is not an energy efficient process as it is not possible to retrieve the energy contained in the un-split water cooled along with product gas mixture.

In high temperature separation, hydrogen is separated from the hot product gas at the reaction temperature using porous membranes made of materials such as zirconia. The porous membranes separate the gases through, in general, either mass diffusion or molecular effusion [3,74,79]. In mass diffusion, the gases are separated by the difference in the diffusion coefficients of the component gases with respect to an auxiliary gas like water vapor [3]. In molecular effusion, hydrogen is separated from the reacting gas mixture by effusing through the porous ceramic membrane in the Knudsen flow regime. For this process to occur the pore size is kept less than the mean free path of hydrogen at the separation temperature and pressure. This process is efficient and hence recommended for one-step process. The hot unused steam rejected after the selective effusion of hydrogen can be further dissociated using high temperature electrolysis. Compared to other separation processes this hybrid process is considered more efficient and economical [74]. Instead of porous membranes, solid oxygen ion conductors like zirconia can also be used to remove oxygen from the hot gas by applying an electrical bias across the membrane.

The high temperatures needed for the one-step process can only be supplied using solar energy concentrators. The reaction is usually conducted in a cavity where the solar radiation is focused through a quartz window using dish type mirrors with large collection areas. Materials used for cavity construction need to be able to withstand extremely high temperatures and large temperature fluctuations; as of today graphite (in inert ambient) and zirconia are the preferred materials [75]. Immediately after formation hydrogen is separated from the mixture; the hot gases are then passed through heat exchangers for cooling and then compressed for storage and use. The theoretical efficiency of the process is about 64% at 2500 K and 1 bar [71]. Although the water splitting yield is higher at pressures below 1 bar the overall efficiency is considerably less due to the high work input needed to compress the hydrogen [80].

Direct thermal water splitting is an environmentally benign technology that, once the system is built, requires only solar energy and water. The process efficiency is high due to the high temperatures involved. However the high operating temperatures, i.e. 2000 K to 3000 K, pose severe challenges to the materials used for construction of reaction chamber, separators and heat exchangers. At the reaction temperature gas separation, and collection of the various hydrogen species, is difficult. Furthermore re-radiation from the reactor at these extreme temperatures lowers the absorption efficiency [81]. In the 2000 to 3000 K temperature range a significant amount of water entering the reaction chamber is heated without undergoing dissociation [3,81,82]. For achieving reasonably high practical efficiencies, methods to effectively recover high temperature heat from this unutilized steam need to be implemented. Hence at present the direct thermal splitting of water is not considered an economically viable technology that can be used for commercial hydrogen production.

Two-step Thermochemical Processes

If water splitting is performed through more than one step, let us say j reactions with the reactions occurring at different temperatures, then the minimum useful work requirement given in equation (2.3.4) can be written as [2,60,61]:

$$W_r = \Delta G = \Delta G_0 - \sum_{j=1}^{i=j} \Delta S_i (T_{r(i)} - T_0) \qquad (2.3.5)$$

Thus, by selecting reaction steps with appropriate ΔS and T_r values it is theoretically possible to reduce the work input to zero while minimizing the T_r of each reaction. That is to say, when more than one reaction step is involved water splitting can be done using thermal energy at much lower temperatures than that required for a single step process. The simplest example of this is, of course, a two-step process. Two-step processes involving metal oxide cycles are largely considered the most promising. The process, in general, consists of the following steps

Metal oxide $\rightarrow$ Reduced metal oxide + O_2 (endothermic)   (2.3.6)
Reduced metal oxide + $2H_2O \rightarrow$ metal oxide + $2H_2$ (2.3.7)

In the first step, a high valence metal oxide is reduced to a low valence metal oxide or a low valence metal oxide is reduced to the corresponding metal by the high temperature heat energy, releasing oxygen. The metal or oxygen deficient metal oxide then reacts with water at a lower temperature releasing hydrogen and regaining stoichiometry. The net reaction is the splitting of water into hydrogen and oxygen. The temperature range required for the first reaction is about 1700-3000 K. Hence, for practical applications the option is limited to the use of heat from concentrated solar radiation.

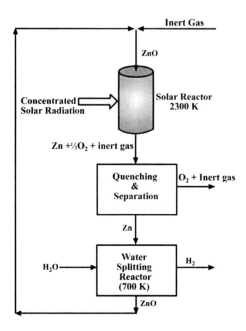

Fig. 2.7: Process flow schematic of a two-step Zn/ZnO cycle.

In 1977, Nakamura proposed a two-step cycle involving the Fe_3O_4/FeO redox pair [80]. Since then many metal oxides systems including Co_3O_4 /CoO, Mn_2O_3/ MnO, TiO_2/TiO_x, ZnO/Zn and different oxide ferrite systems like $ZnFe_2O_4/Zn/Fe_3O_4$ have been proposed by different groups [83-89]. Of these materials the ZnO/Zn pair is considered the most promising [70,88,90,91]. The ZnO/Zn cycle, see **Fig. 2.7**, consists of the following reaction steps

$$ZnO \rightarrow Zn + 0.5O_2 \qquad (\text{at } 2300 \text{ K})$$
$$Zn + H_2O \rightarrow ZnO + H_2 \qquad (\text{at } 700 \text{ K})$$

Zinc oxide in solid or fine particle form is kept in a reactor cavity that is subjected to irradiation from solar concentrators [92]. The dissociation products are zinc (vapor) and oxygen; for this first reaction $\Delta G=0$ at about 2235K [91]. The reactor is made of materials like inconel steel, zirconia, silicon carbide or graphite [68,89,92]. The graphite is used in special designs to avoid direct contact with chemical species [68]. The dissociation products are then cooled rapidly to separate zinc and oxygen, transporting the products using an inert carrier gas to a low temperature zone. In this zone oxygen escapes, and zinc exists in fine liquid or solid particulate form [90] that is subjected to hydrolysis by passing steam. Zinc reacts with water and forms zinc oxide while releasing hydrogen [68,90,91,93], and the remaining zinc oxide is now available to begin the next cycle. A Zn to ZnO conversion rate of 83% by hydroloysis was reported by Wegner and co-workers [90]. The cycle needs only water as a feed material, and produces hydrogen and oxygen as products. The theoretical thermal efficiency of this cycle is 53% [94], whereas the exergy efficiencies at solar concentration ratios of 5000 and 10000 suns without considering any heat recovery are 29% and 36% respectively [91].

Compared to a single step process, two-step processes need lower temperatures yet provide a higher hydrogen yield (ratio of amount of hydrogen actually produced to maximum amount of hydrogen that can be produced). The product gas separation is relatively easy in a two-step process as oxygen and hydrogen are produced in two separate stages. No corrosive chemical is involved in the process. For many two-step cycles involving a metal oxide, the high temperature reaction is carried out around 2000K, which can be achieved using solar concentration ratios between 1000 and 5000 suns. This simplifies the solar concentrator design and implementation. However, this technology is still in its infancy due to a number of practical problems. For example, high temperature re-radiation loss is a major concern, which needs to be minimized using appropriate materials and designs for the reactor windows [91]. Also, after dissociation of the oxide there is a tendency for metal/reduced metal oxide and oxygen to recombine when the metal/reduced metal oxide cools as it proceeds towards the lower temperature zone [68]; rapid cooling is required to prevent this. High temperature sintering of the metal oxide particles is another problem, which effectively deactivates the metal oxide for subsequent cyclic reactions. Solid metal deposits on the reactor walls also need to be removed periodically. Finally the temperature regime used in a two-step metal oxide process is still not low enough to enable large-scale hydrogen production. As in the case of a one-step process, finding durable construction materials for different components of the reaction chamber remains an un-solved challenge. As of today, the overall practical efficiency of this process is not high enough to compete with electrolysis.

Multi-step Thermochemical Processes

Multi-step cycles are designed in accordance with equation (2.3.5) in such a way that the maximum temperature required for any reaction in the cycle is limited to the temperature that can be supplied from a nuclear reactor [2,4,95], approximately 1500K. Furthermore all the chemicals involved in the reaction are confined to the enclosed region in **Figure 2.6** within a closed cycle, the operation of which is essential to avoid heat and material losses. With recycling of the chemicals only water is consumed, generating hydrogen and oxygen.

The suitability of a cycle for hydrogen production depends upon the overall thermal efficiency and operational feasibility. A highly endothermic reaction step is required in a cycle to achieve effective heat-to-chemical energy conversion. For efficient mass and momentum transfer a fluid based system is preferred [96] and, ultimately, for large-scale hydrogen production other factors such as environmental effects and cost effectiveness must also be considered.

The first multi-step process having a theoretical efficiency above 50% was proposed by Marchetti and Beni in 1970 [2]. This process became the *Mark-1* process of the European Commission's Joint Research Center (JRC), Ispra, Italy. A large number of cycles with reasonably high efficiencies were designed after this; Abanades and co-workers have recently made a compilation of the available thermochemical cycles that yielded about 280 cycles [67]. A large number of cycles are listed in references [2,4,67,95,97]. However only a few of these processes have exhibited potential for high overall efficiencies and technical feasibility. We discuss a few cycles that are presently under study and promising for practical implementation.

Sulfur-iodine Cycle

This cycle was developed by General Atomics Corporation in the mid 1970s [88,99]; the Japanese atomic energy research institute (JAERI) has also been actively involved in developing their version of this cycle [100-103]. The reaction steps involved in this cycle are given below.

$$H_2SO_4 \rightarrow SO_2 + H_2O + \tfrac{1}{2} O_2 \qquad \text{(at 1123 K)}$$
$$I_2 + SO_2 + 2H_2O \rightarrow 2HI + H_2SO_4 \qquad \text{(at 393 K)}$$
$$2HI \rightarrow I_2 + H_2 \qquad \text{(at 723 K)}$$

A typical sulfur-iodine (S-I) process flow diagram is given in **Fig. 2.8**. The reactions are carried out in three different compartments [102-104]. The sulfuric acid decomposition is a high temperature endothermic process. The second step is the Bunsen reaction (exothermic) where sulfuric and hydrioidic acids are produced from iodine and sulfur dioxide in aqueous solution. In a critical step the sulfuric acid and hydriodic acids are separated in liquid phase, then purified and concentrated. For facilitating effective separation an excess of molten iodine is used in this reaction to form two separate phases, a *light* phase of sulfuric acid and *heavy* phase containing hydrogen iodide and iodine [99,103]. In an endothermic process the hydriodic acid is decomposed at a lower temperature to hydrogen and iodine. The net reaction is splitting of water into hydrogen and oxygen.

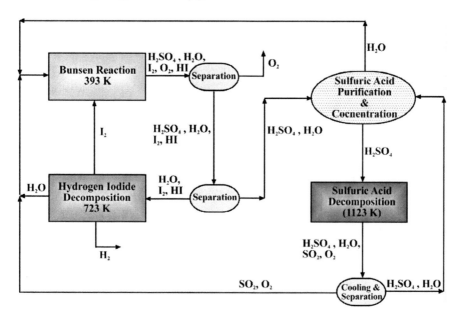

Fig. 2.8: Process flow schematics of a sulfur-iodine cycle.

The S-I cycle consists of reactions all taking place at moderate temperatures that can be achieved using nuclear waste heat, or concentrated solar rays sufficient to achieve temperatures comparable to its nuclear counterpart. According to the claims made by different groups regarding the efficiency of the S-I cycle, the values of efficiency lie in the range 42 to 56.8% showing this to be one of the most efficient cycles [68,100,105]. All reactions involved in the S-I cycle are fluid based, and there are no effluents. However there remain technical challenges.   The reactions involving decompositions of sulfuric acid and hydriodic acid create an aggressive chemical environment and hence chemically stable materials must be used in construction of the process chamber. After the Bunsen reaction, a complete removal of the sulfuric acid from the hydriodic acid phase is necessary to avoid sulfur formation in the HI separation and HI decomposition steps. Further, complete recovery of hydrogen from HI and I_2 from the last reaction is difficult, limiting the process efficiency. In the sulfuric acid decomposition step SO_3 can be formed along with SO_2 that needs to be promptly separated. To eliminate problems associated with SO_3 an operating temperature above 1400K is necessary [104].

The UT-3 cycle

The UT-3 cycle was named after the University of Tokyo where the process was developed in late 1970s. As of today the UT-3 cycle is considered to have the greatest potential for commercialization. UT-3 cycle is a four-step process involving the reactions [73,106-110]

$CaBr_2$ (s) + H_2O (g) → CaO (s) + 2HBr (g)   (at 1033 K)
CaO (s) + Br_2 (g) → $CaBr_2$ (s) + ½ O_2 (g)   (at 845 K)
Fe_3O_4 (s) + 8HBr (g) → $3FeBr_2$ (s) + $4H_2O$ (g)+Br_2 (g)   (at 493 K)
$3FeBr_2$ (s)+ $4H_2O$ (g)→ Fe_3O_4 (s)+ 6HBr (g) + H_2 (g)   (at 833 K)

The first two reactions ensure the formation of hydrobromic acid releasing oxygen, and the other two ensure the reduction of water releasing hydrogen. The reaction takes place in four separate reactors in isothermal or adiabatic conditions. The reactors are paired; one pair contains calcium compounds (reaction 1 and 2) and the other contains iron compounds (reactions 3 and 4). The complete cycle takes about an hour to complete. After each cycle the functions of the paired reactors are switched so that reactor 1 now carries CaO and reactor 2 contains $CaBr_2$ (similarly reactors 3 and 4), thus producing hydrogen and oxygen continuously without transferring solid products from one reactor to the other. The theoretical thermal efficiency of a UT-3 hydrogen production system is around 49%, and exergy efficiency is about 53% [73,107]. One limitation of the processes is that the solids used in the cycle, especially those working near their melting points, could suffer structural changes due to sintering effects leading to deterioration of process efficiency. Operational issues associated with alternately switching the reactors for exothermic and endothermic reactions have yet to be effectively addressed.

The Copper-Chlorine cycle

In comparison to the S-I and UT-3 cycles the Cu-Cl cycle is attractive due to the significantly lower reaction temperatures [69,111]. The cycle reactions consists of [66]:

$$2CuCl \rightarrow CuCl_2 + Cu \qquad \text{(at} < 373 \text{ K)}$$
$$2CuCl_2 + H_2O \rightarrow CuO. CuCl_2 + 2HCl \qquad \text{(at 573-648 K)}$$
$$CuO. CuCl_2 \rightarrow 2CuCl + \tfrac{1}{2} O_2 \qquad \text{(at 723-803 K)}$$
$$Cu + HCl \rightarrow CuCl + \tfrac{1}{2} H_2 \qquad \text{(at 698-723 K)}$$

This cycle uses solid reactants. Small dendritic copper particles are used to carry out the last reaction to make the transformation of all the solid copper to CuCl, thereby maximizing hydrogen yield. The reported efficiency of this cycle is 49% [66]. This low temperature cycle is believed to eliminate many of the engineering and materials issues associated with the other two previously discussed cycles, however this cycle is also in the initial stages of development [111]. The temperature ranges are such that lower temperature nuclear reactors, e.g. sodium-cooled fast reactors, could be used with this cycle [69]. A hybrid version of this cycle is under investigation in Argonne National Laboratory [66,112].

In general, the efficiency of a multi-step process reduces with increasing number of reaction steps due to thermal losses, and difficulties involved in product separation from each reaction. However if thermochemical technology is to be implemented to put the hydrogen in a hydrogen economy then multi-step processes appear as the favorite candidates for large-scale hydrogen production. Compared to one-step and two-step processes, in practice these cycles offer much better efficiencies (> 50%) and are believed to have comparatively low capital and operating costs. Hydrogen and oxygen are produced in separate reactions hence costs associated with separation are avoided. While in multi-step processes there is greater potential for the presence of hazardous impurities in the hydrogen and oxygen product streams, which may necessitate purification facilities to meet health and environmental standards, still it appears that volume scale hydrogen can be realistically produced at a purity in excess of 99.99%.

Hybrid Thermochemical Processes

Hybrid thermochemical processes use both heat and work inputs, with at least one of the reactions driven by electricity. The electricity driven, electrochemical reaction is selected in such a way that the overall work input is much less than that required for water electrolysis. An advantage of this technology is the possibility to establish two-step processes that require much lower temperatures compared to the pure two-step thermochemical cycles and hence greater simplicity in the chemical processes. Several hybrid cycles were developed during the 1970s and 1980s [2,113,114]. A hybrid cycle initially studied at Los Alamos Scientific Laboratory and developed by Westinghouse Corporation in 1975, called the sulfuric acid hybrid cycle or the Westinghouse sulfur process (WSP) is considered promising and still under study [65,97,114-118]. This process is named the Mark 11 cycle by the European commission's Joint Research Center (JRC), Ispra, Italy [T24]. The reactions involved are:

$$H_2SO_4 \rightarrow SO_2 + H_2O + \frac{1}{2} O_2 \qquad \text{(at 1100 K)}$$
$$2H_2O + SO_2 \rightarrow H_2SO_4 + H_2 \qquad \text{(electrolytic; at 363 K)}$$

The reversible potential for the sulfur dioxide electrolysis is only 0.17 V, less than 10% that of water electrolysis (minimum of 1.23V at 298K and 1 bar) [65,69]. However corrosion problems in the electrolysis step are severe due to the presence of high concentration (about 50%) sulfuric acid. The overall thermal efficiency of the process, considering both thermal and electrical energy input derived from the same heat source, is estimated as 48.8% [116]. However, in terms of economics and process complexity the hybrid cycles face tough competition from advanced water electrolyzers.

2.4 Hydrogen Production By Water Biophotolysis

Biophotolysis is the process of splitting water into hydrogen and oxygen through a series of biological activities utilizing solar radiation. The overall process can be represented as

$$H_2O + light\ energy \xrightarrow{\text{(biological activity)}} H_2 + \tfrac{1}{2} O_2 \qquad (2.4.1)$$

Biophotolysis is an attractive method for generating hydrogen as it is both renewable and environmentally friendly. As the whole process involves only plants, sunlight and water, all the components in the process can be recycled producing no hazardous wastes. Carbon dioxide is recycled in this process and hence there is no net accumulation of this greenhouse gas. It is considered as a method for harvesting hydrogen directly, rather than obtaining hydrogen by planting, harvesting and processing of biomass. In principle, biophotolysis requires lower initial investments compared to other hydrogen generation processes.

Biophotolysis has its root in photosynthesis. Photoautotrophic plant species, both micro and macro, generate carbohydrates needed for metabolism using solar energy and water through the photosynthesis process. This process of converting solar energy to chemical energy utilizes water as an electron donor. It consists of a series of visible light induced redox reactions leading to the reduction of carbon dioxide received from the atmosphere and formation of carbohydrates. Photosynthesis involves light dependent and light independent processes, with photosynthetic systems in plants and algae consisting of various complex components such as light-harvesting antennae, energy and electron-transfer systems, and redox centers.

The light dependent processes take place in the thylakoid membrane located in the chloroplast. The light harvesting takes place in two separate light absorbing membrane-integrated protein complexes each containing pigments tuned for light absorption in the visible region of the solar spectrum. These are denoted as Photosystem I and II (PS I and PS II) according to the chronological order in which these were discovered [119,120]. Each photosystem contains an array of light absorbing pigments called antenna pigments and a reaction center. The antenna consists of *chlorophyll a* and accessory pigments like *chlorophyll b, c* and *d, carotenes, xanthophylls* and *phycobiliproteins*. The presence and amount of the accessory pigments vary from species to species. The reaction center is a *chlorophyll a* protein complex consisting of an interacting pair of chlorophyll molecules. The reaction centers in PS I and PS II are denoted, respectively, as P_{700} and P_{680} according to the wavelength in the red region at which maximum light absorption occurs. P_{680} has two absorption peaks; one at 680 nm and the other around 430 nm [120]. The light absorption behavior of each accessory pigment differs from that of another. The overall absorption spectrum of the pigments shows lowest absorption in the green-yellow region. The process of light harvesting involves light absorption by antenna pigments, which get excited and transfer energy to successive elements in the array via resonance energy transfer and finally to the reaction center. The oxygen evolution via water splitting takes place at PS II; the electrons generated by this process pass through PS II and PS I to help in the synthesis of the reductant NADPH (reduced nicotinamide adenine dinucleotide), needed for carbon fixation [121].

The process of photosynthesis is generally explained by the modified Z-scheme [122-127]. P_{680} in PS II absorbs light and goes to an excited state. It releases an electron to a primary acceptor named pheophytin. P^+_{680} thus formed receives electrons successively from a manganese cluster that is a part of a water-oxidizing complex. The water splitting reaction,

$$2H_2O \rightarrow O_2 + 4H^+ + 4e^- \qquad (2.4.2)$$

is mediated by a redox-active tyrosine residue [127,128]. The electrons are transferred to the manganese ions and protons are released into the lumen. The electron transferred to pheophytin moves to plastoquinone Q_A, then to polypeptide bound plastoquinone Q_B and to a cytochrome b_6f complex containing the cytochrome f and b, iron-sulfur cluster and quinone. From this complex, the electron is transferred to plastocyanin (PC), see **Fig. 2.9**. The electron is now available for the processes in PS I. P_{700} in PS I also goes to the excited state by receiving light energy. It transfers an electron to the chlorophyll primary electron acceptor A_0 and receives an electron from plastocyanin. The electron transfer takes place from A_0 to the phylloquinone secondary electron acceptor A_1 [129,130], then to an iron-sulfur center and to Ferredoxin (Fd) [131]. Ferredoxin provides an electron for the reduction of $NADP^+$ (oxidized nicotinamide adenine dinucleotide) that combines with a proton in the stroma to form NADPH. The reaction is catalyzed by the flavoenzyme, ferredoxin-NADP oxidoreductase (FNR) [129,132]. A process called photophosphorylation takes place in parallel to the electron transport process across the thylakoid membrane. The electron transport leads to the creation of a proton gradient across the thylakoid membrane. The electrochemical energy from this proton gradient drives the synthesis of ATP (adenosine triphosphate) from ADP (adenosine diphosphate) enzyme and inorganic phosphate [133]. In some species a cyclic transport of electrons takes place in PS I that lead to the synthesis of ATP instead of NADPH, a process that does not require water oxidation [134].

The light independent reactions take place in the stroma with the help of ATP and NADPH. In a process called the Calvin-Benson cycle, or carbon fixation, carbon dioxide from the atmosphere is captured and converted into carbohydrates [135]. The reaction is catalyzed by the enzyme RuBisCO (ribulose-1,5-biphosphate carboxylase/oxygenase). The overall photosynthesis process can be represented as

$$H_2O + CO_2 + light \rightarrow CH_2O + O_2 \tag{2.4.3}$$

As discussed above, although water is split in green plants no molecular hydrogen evolution takes place. A study conducted by Jackon and Ellms in 1896 showed that hydrogen was evolved by a filamentous cyanobacterium *Anabaena* [136]. The basic research in the field started in 1920s with bacterial hydrogen production. Studies showed that both eucaryotic and prokaryotic types of microalgae can generate hydrogen under certain conditions. In 1942, Gaffron and Rubin, in a study using green algae *Scenedesmus Obliquus*, observed that if the organism was allowed to adapt to a dark anaerobic condition for a certain period it would produce hydrogen the rate of which becomes dramatically higher when illuminated after this adaptation period [137]. This discovery did not receive much attention till the 1970s, co-incident with the Arab oil embargo and correlated interest in developing hydrogen from renewable sources. The basic difference between photosynthesis and biophotolysis processes is that in photosynthesis ferredoxin reduces NADP to NADPH with the addition of a proton that is used by RUBisCO for carbohydrate preparation with the help of carbon dioxide, whereas in biophotolysis ferredoxin activates certain enzymes that produce molecular hydrogen [138,138]. The focus of biophotolysis research has generally been on microalgae, both green algae (eucaryote) and cyanobacteria or blue-green algae (prokaryote) that possess enzymes capable of acting as catalysts for the photo assisted production of hydrogen. Microalgae have the highest photosynthetic capability per unit volume [140], and grow much faster than higher-level plants and with minimum nutrition requirements.

The unique hydrogen production ability of green algae and cyanobacteria is due to the presence of certain enzymes that are absent in other plant species in which only CO_2 reduction takes place. These enzymes, named hydrogenase and nitrogenase, catalyze molecular hydrogen production. Green algae possess the hydrogenase enzyme, whereas cyanobacteria species can have both nitrogenase and hydrogenase enzymes. Two types of hydrogenase enzymes are active in the hydrogen production process: the reversible or classical hydrogenase, and membrane bound uptake hydrogenase [141]. The reversible hydrogenase oxidizes ferredoxin, or other low redox electron carriers, in a readily reversible reaction. The normal function of membrane bound uptake hydrogenase is to derive reductant from hydrogen; this enzyme does not produce hydrogen in measurable amounts. The normal function of nitrogenase in cyanobacteria is to convert nitrogen to ammonia [142,143], but it can also drive a hydrogen production reaction in the absence of nitrogen.

Hydrogenase was named by Stephenson and Stickland in 1931, discovered in their experiments using anaerobic colon bacterium Escherichia Coli that evolved hydrogen [144-146]. Hydrogenase enzymes are metalloproteins that contain sulfur and nickel and/or iron [147]. To date over 80 hydrogenase enzymes have been identified. These enzymes reversibly catalyze hydrogen production/uptake reactions:

$$2H^+ + e^- \leftrightarrow H_2 \qquad (2.4.4)$$

However, these enzymes favor either a forward (hydrogen evolution) or a backward (hydrogen uptake) reaction depending upon their redox partners provided by the host organism [148]. The reactions take place at a catalytically active region buried in the protein. These core regions access or release hydrogen through continuous hydrophobic channels connecting the surface and the active core region of the enzyme structural unit.

Depending upon the metal composition at the active sites, the hydrogenase enzymes are of three types: [Fe] hydrogenase, [NiFe] hydrogenase, and metal free or [FeS] cluster-free hydrogenase [149-152]. There is another type called [NiFeSe] hydrogenase, which is usually included in the [NiFe] hydrogenase family. Among the categories, [Fe] hydrogenase has the highest activity for hydrogen synthesis from protons and electrons [151]. This is the reversible hydrogenase present in green algae that generates hydrogen via biophotolysis [153]. The catalytically active core unit, called 'H cluster', consists of a six-iron cluster with a two-iron (binuclear iron) subcluster bound to carbon monoxide (CO) and cyanide (CN⁻) ligands [154-157]. The binuclear iron subcluster is unique to [Fe] hydrogenase and promotes oxygen dependent hydrogen uptake; hence oxygen inhibits the hydrogen evolution activity of the [Fe] hydrogenase. [NiFe] hydrogenase is the most abundant type of hydrogenase. These are found in conjunction with nitrogenases in cyanobacteria as well as other nitrogen-fixing prokaryotes [158]. The active unit, the H-cluster, in [NiFe] hydrogenase has a heterodinuclear active site with cysteine thiolates bridging a nickel ion with an iron center [159-161]. This enzyme generally promotes the hydrogen uptake reaction. Cyanobacteria possess both uptake and bi-directional hydrogenase [162]. The metal-free or [FeS] cluster free hydrogenase was discovered in the 1990s within methanogenic bacteria [163-165]. This type of hydrogenase does not contain iron or nickel at the active site but contains iron in the inactive region [165]. This enzyme catalyzes the reversible reduction of N^5, N^{10}-methenyltetrahydromethanopterin with H_2 to N^5, N^{10}-methylene-tetrahydromethanopterin [148]; the turn over number for hydrogenase enzyme approaches 10^6 per second.

Nitrogenase enzyme is a two-component protein system, consisting of dinitrogenase and a nitrogenase reductase, that uses the nitrogen and energy of ATP to produce ammonia and hydrogen [143,166]. The nitrogen fixation reaction is accompanied by the production of hydrogen as a side reaction the rate of which is about one-fourth that of the nitrogen fixation. In the nitrogen free ambient, the nitrogenase enzyme functions like a hydrogenase and metabolizes the photosynthetically generated reductants to molecular H_2 [167]. The production of hydrogen by nitrogenase is unidirectional as a byproduct in contrast with that by hydrogenase. The overall reaction can be given as

$$N_2 + 8H^+ + 8e^- + 16ATP \rightarrow 2NH_3 + H_2 + 16ADP + 16P_i \qquad (2.4.5)$$

where P_i represents inorganic phosphate.

Depending upon the metal content at the active site, there are three categories of nitrogenase enzymes which are named as MoFe, VFe and Fe nitrogenase [168,169]. The most common among these is MoFe nitrogenase. The two components in the MoFe nitrogenase protein system are the dinitrogenase (MoFe protein or protein I) and the dinitrogenase reductase (Fe protein or protein II) [162]. The protein II component in all the three categories are similar, whereas the molybdenum in protein I is replaced by vanadium and iron to form VFe and Fe nitrogenases respectively [169]. Although nitrogenases are sensitive to oxygen cyanobacteria have the inherent ability to protect it from oxygen evolution spatially or temporally. Nitrogenase catalyzed by hydrogen production is highly energy intensive due to the consumption of 16ATP. Hence these are less efficient compared to the hydrogenase-based reaction; the turn over number is less than 10 per second.

In general there are two routes, discussed below, in which hydrogen production can be accomplished using the noted enzymes in combination with sunlight and water. One route is called direct biophotolysis, in which hydrogen production can be considered a single-step water splitting reaction. The other route is called indirect biophotolysis, in which intermediate steps are involved in the process but the overall reaction gives hydrogen and oxygen from water [170,171].

Direct Biophotolysis

Direct biophotolysis is the biological process in which water is split into hydrogen and oxygen without involvement of intermediates; water is the electron donor in the process, and the energy that drives the process is obtained from visible radiation. The electron transfer route is the same as that in the case of NADPH formation (the Z-scheme discussed earlier) during photosynthesis with the exception that ferredoxin transfers electrons to the hydrogenase enzyme which in turn reduces protons to form molecular hydrogen. The hydrogenase catalyzed hydrogen production process in green algae is schematically represented in **Fig. 2.9**.

Under anaerobic conditions with a low partial pressure of hydrogen and under low intensity illumination, hydrogen evolution takes place and the overall reaction can be represented by (2.4.1) [172]; the electron transfer route is as follows [141]:

$$H_2O \rightarrow PSII \rightarrow PSI \rightarrow Ferredoxin \rightarrow Hydrogenase \rightarrow H_2$$
$$\downarrow$$
$$O_2$$

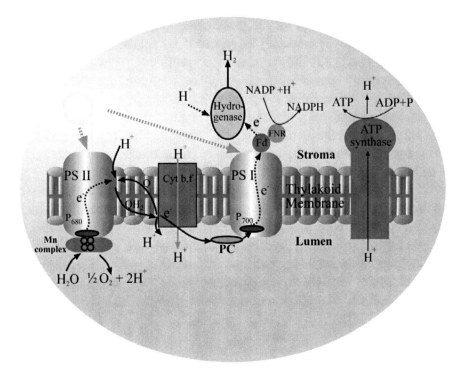

Fig. 2.9: Hydrogenase mediated hydrogen production process in green algae. See the text for abbreviations.

The energetics of electron transfer with the help of photoexcitated chlorophyll in PSII and PSI is shown in **Fig. 2.10**. Electrons from water can be considered traversing through PS I and PS II to Ferredoxin and then to hydrogenase, while protons are released to the lumen (**Fig. 2.9**). The energy from the absorbed light enables electron transport from water, at a potential of ~ +0.82 V, to ferredoxin at a potential of -0.44 V [142]. As indicated in **Fig. 2.9** two photons are responsible for transport of a single electron across the thylakoid membrane. Hence four electrons and eight photons are needed for the formation of a single hydrogen molecule. The need for such a number of electrons and photons is one of the factors that limit the maximum possible efficiency of hydrogen production through this process.

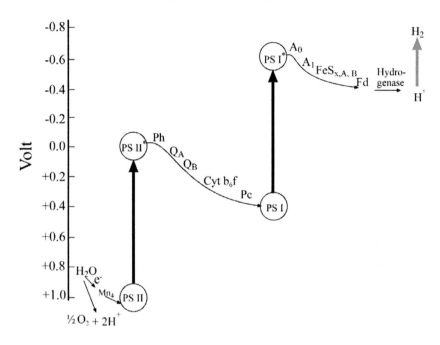

Fig. 2.10: Energetics of electron transfer in direct biophotolysis.

A necessary condition for hydrogen production is that the hydrogenase enzyme should be activated by exposure of the biophotolysis algal species to a dark, anaerobic condition for a period ranging from a few minutes to several hours depending upon the algal strain. In this condition the algal species activates hydrogenase to produce a small amount of hydrogen then used as a reductant for carbon dioxide. When exposed to light, again in anaerobic conditions, the hydrogen production resumes but at a much higher rate through the photosynthetic process. The process of hydrogen production creates a sustained flow of electrons across the thylakoid membrane that promotes the synthesis of ATP using protons [173].

Green alga evolve hydrogen through direct biophotolysis as they contain reversible hydrogenase [Fe hydrogenase]. Apart from *Scenedesmus obliquust* that was used by Gaffron and Rubin [137], green algae species including *Chlamydomonas reinhardtii* and *Chlamydomonas moewusii* have also proven useful [122,141,174]. Among these a maximum efficiency of hydrogen production has been observed in *Chlamydomonas reinhardii*. Several mutants of these species have been studied for extending the process to a commercial scale. While most studies have been performed using fresh water green algae, there are a few studies investigating the use of marine green algae [140,175,176].

The reversible hydrogenase responsible for the hydrogen evolution in these species is known for its oxygen sensitivity. The presence of oxygen (say, >0.1%) impairs the electron transport through ferredoxin and hydrogenase in turn inhibiting hydrogen evolution [177]. During their initial work, Gaffron and Rubin found that hydrogen evolution stops some time after exposing the species to light [137]; in the presence of light normal photosynthesis starts with the release of oxygen, inhibiting hydrogen evolution. Hence, sustained hydrogen generation by hydrogenase requires removal of the evolved oxygen.

Various methods have been sought for the sustained, continuous production of hydrogen via direct biophotolysis. Flushing the bioreactor continuously with an inert gas to remove the oxygen has proven to be one method; Greenbaum continuously purged a bioreactor with argon to remove the evolved oxygen thereby sustaining hydrogen production for several hours [178]. Although this method has been used in laboratory settings by several groups, scale-up of the bio-reactor for practical applications is difficult due to the needed inert gas flow rates, and the low partial pressure of the hydrogen produced [177]. Oxygen scavengers or absorbers have also been employed, but have yet to be found economically feasible; for example, glucose and glucose oxidase have both been found useful for facilitating sustained hydrogen production [179]. An approach of perhaps greater utility is the use of mutants exhibiting high oxygen tolerance [180]. However none of these techniques have yet found use in commercial scale hydrogen production.

Indirect Biophotolysis

As the name implies indirect biophotolysis involves more than one step, with the net effect being water splitting. Indirect biophotolysis can be effectively performed using either hydrogenase or nitrogenase, hence both green algae and cyanobacteria can be utilized. In this section, our discussion is limited to processes catalyzed by nitrogenase in cyanobacteria.

As a way of intrinsically protecting its nitrogen fixing enzyme, nitrogenase, from oxygen different cyanobacteria strains either spatially or temporally separate the oxygen evolving photosynthetic process from the nitrogen fixing process [179]. Carbohydrate preparation via photosynthesis takes place in one step, and nitrogenase-catalyzed nitrogen fixation occurs by consuming this carbohydrate in the second step. If provided with reductant and ATP in anaerobic and nitrogen deficient conditions nitrogenase switches its function to hydrogen production. Carbohydrate synthesis, assisted by carbon dioxide intake, act as an intermediate step between water oxidation and hydrogen evolution. Under illumination in anaerobic conditions this accumulated carbohydrate is used by nitrogenase to produce hydrogen. The overall effect is that water is consumed and oxygen and hydrogen are released whereas carbon dioxide is recycled. The reactions can be represented as [173]

$$CO_2 + H_2O \rightarrow CH_2O + O_2 \qquad (2.4.6)$$
$$CH_2O + H_2O \rightarrow CO_2 + 2H_2 \qquad (2.4.7)$$

With the net reaction

$$2H_2O \rightarrow 2H_2 + O_2 \qquad (2.4.8)$$

At 25°C and 1 bar the enthalpy change (ΔH) involved in reactions (2.3.5) and (2.3.6) is, respectively, 211.5 kJ and 361 kJ, hence that of reaction (2.3.7) is 572.5 kJ [173]; four ATP are consumed per nitrogen molecule fixed or hydrogen molecule evolved. Even in the presence of nitrogen, 25% of the ATP used is for hydrogen evolution. The whole process requires twenty photons and hence indirect biophotolysis using nitrogenase is not as energetically efficient as direct biophotolysis. A typical nitrogenase based hydrogen production process via indirect biophotolysis process is illustrated in **Fig. 2.11**.

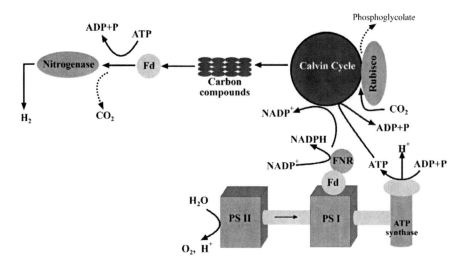

Fig. 2.11: Nitrogenase mediated indirect biophotolysis process in cyanobacteria.

Heterocystous cyanobacteria, which are filamentous blue-green algae, spatially separate the oxygen and nitrogen fixing (or hydrogen evolving) processes [181,141]. Heterocystous cyanobacteria possess two types of cells, vegetative cells and heterocysts. Photosynthesis and oxygen evolution take place in vegetative cells, which contain PS I and PS II; nitrogen fixation or hydrogen evolution takes place in the heterocysts. Heterocyst cells have thick walls with an envelope composed of a glycolipid layer and a polysaccharide layer [182]. This wall acts as an oxygen diffusion barrier and helps to maintain an anaerobic environment in the heterocysts. Further, these cells exhibit high respiration rates absorbing any residual oxygen [179]. The carbohydrate produced in the vegetative cells via photosynthesis is transferred to the heterocysts where it is used for nitrogen fixing or hydrogen evolution. Thus oxygen and hydrogen evolution can simultaneously take place in such species. A scheme of hydrogen evolution by heterocystous cyanobacteria is shown in **Fig. 2.11** [141].

$H_2O \rightarrow$ photosystems (vegetative cells) $\rightarrow [CH_2O]_2 \rightarrow$ Ferredoxin $\rightarrow$ Nitrogenase
$\downarrow$ $\uparrow$ $\downarrow$ NAPDH $\downarrow$
O_2 CO_2 recycle CO_2 H_2

Some of the heterocystous cyanobacteria strains used for hydrogen evolution include *Anabaena Azollae, Anabaena variabilis, Anabaena cylindrica, Nostoc muscorum, Nostoc spongiaeforme*, and *Westiellopsis prolifica* [174]. In the early 1970s Beneman and co-workers demonstrated that hydrogen and oxygen could be produced simultaneously for several hours using *Anabaena cylindrica* in a nitrogen deficient atmosphere [183,184] by flowing inert gas. In this process simultaneous oxygen and hydrogen evolution takes place, hence a gas separation step is required.

Some strains like unicellular and non-heterocystous filamentous diazotrophic cyanobacteria separate the oxygen and hydrogen evolution steps temporally [185]. Periods of O_2 evolution and CO_2 fixation alternate with periods of N_2 fixation or H_2 production [179], with the latter process taking place in an anaerobic environment. In many cases oxygen evolution occurs under illumination and nitrogen fixation (or hydrogen evolution) occurs in the dark [185], while carbohydrate accumulation occurs during the first phase and its consumption occurs in the second. The hydrogen evolution scheme by non-heterocystous cyanobacteria is given below [141]:

$H_2O \rightarrow$ photosystems (vegetative cells) $\rightarrow [CH_2O]_2 \rightarrow$ Ferredoxin $\rightarrow$ Nitrogenase
$\downarrow$ $\uparrow$ $\downarrow$ NAPDH   or   hydrogenase
O_2 CO_2 recycle CO_2 $\uparrow$ $\downarrow$
                                                         PS-I $\rightarrow$ ATP H_2

Some of the unicellular strains used for hydrogen evolution include *Plectonema boryanum, Oscillotoria Miami BG7, Oscillotoria limnetica, Synechococcus sp., Aphanothece halophytico, Mastidocladus laminosus*, and *Phormidium valderianum*. A fundamental advantage in using these strains is that there is no need of separating hydrogen and oxygen as these are evolved under different conditions at different times.

Cyanobacteria, in general, carry the three enzymes related to hydrogen production; nitrogenase, reversible hydrogenase, and uptake hydrogenase [186]. The presence and amount of these enzymes vary from strain to strain and according to the growth conditions. Uptake hydrogenase ionizes hydrogen produced by nitrogenase during nitrogen fixation so as to use the protons for the next cycle. The combined effect of the enzymes determines the overall hydrogen evolution rate by the cyanobacteria. For a high rate of hydrogen evolution the contribution from uptake hydrogenase should be reduced. As the turn over number of nitrogenase is about one-thousandth of that of hydrogenase, for the purposes of hydrogen production the strains with higher activity of reversible hydrogenase should be chosen [177]. Efforts have been made to reduce the activity of uptake hydrogenase, and enhance that of nitrogenase and reversible hydrogenase by manipulation of the metabolic scheme [174]. Uptake hydrogenase deficient strains were found to be useful for providing a higher hydrogen evolution rate [187].

Selective mutation and molecular cloning techniques have been used for developing strains useful for hydrogen production rather than nitrogen fixation [179]. For example, a mutant strain of Anabaena (AMC 414) in which the large subunit of the uptake hydrogenase (hupL) was inactivated by a deletion event, produced H_2 at a rate that was more than twice that of the parent strain, Anabaena PCC 7120 [187]. Asada and Kawamura reported on aerobic hydrogen production using an uptake hydrogenase deficient *Anabaena sp.* strain N7363 [188].

An advantage of using cyanobacteria is that it can be grown with minimal nutritional requirements, while fish can consume the waste bodies keeping environmental impact to a minimum [179]. Although the nitrogenase of the bacteria is not immediately useful for hydrogen production due to its low turnover number, the thermodynamic driving force (obtained from ATP) allows the enzyme to generate hydrogen against elevated pressures, up to about 50 atmospheres [173]. This is in contrast with direct photoelectrolysis where hydrogen can be produced only near normal atmospheric pressures.

Two-stage Indirect Biophotolysis

The process of growing organisms and utilizing them for hydrogen production in a single bioreactor, either by direct biophotolysis or nitrogenase based indirect biophotolysis, has several limitations; these processes require huge bioroeactors for specie growth, the simultaneous production of oxygen and hydrogen necessitates a separation step, and oxygen evolution may within a short span of time inhibit hydrogen production. Historically, it was realized early-on that such problems could hinder any possibility of large-scale production of hydrogen. Hence hydrogenase based indirect biophotolysis processes involving two or more stages, using either green algae or cyanobacteria, were developed [170]. In general, a two-stage process consists of growing microalgae, either green algae or cyanobacteria, in open ponds with their transfer to an enclosed photobioreactor where hydrogen evolution takes place [179,189]. In a two-stage indirect biophotolysis process conceptualized by Benemann [179], the algal growth takes place at first in open ponds where carbohydrates are prepared by carbon dioxide reduction and stored in the cells. The culture is then transferred to a dark anaerobic fermentation vessel in which activation of hydrogenase takes place with evolution of a small amount of hydrogen; the biomass is then illuminated, under anaerobic conditions, to produce hydrogen [179]. The three processes can be represented as

Carbohydrate storage:

$$O_2 \qquad\qquad\qquad\qquad CO_2$$
$$\uparrow \qquad\qquad\qquad\qquad\quad \downarrow$$
$$H_2O \rightarrow PSII \rightarrow PSI \rightarrow Ferredoxin \rightarrow RUBISCO \rightarrow [CHO]$$

Dark fermentation and formation of low amounts of hydrogen:

$$CO_2$$
$$\uparrow$$
$$[CHO] \rightarrow Ferredoxin \rightarrow Hydrogenase \rightarrow H_2$$

Light driven hydrogen evolution:

$$CO_2$$
$$\uparrow$$
$$[CHO] \rightarrow PSI \rightarrow Ferredoxin \rightarrow hydrogenase \rightarrow H_2$$

The overall effect is the decomposition of water. An advantage of this method is that only a low light energy input, about one quantum/H_2, is required for hydrogen evolution [179]. During illumination, a low activity of PSII is required to avoid simultaneous oxygen evolution, either by use of low light intensity, dense cultures, or inhibitors.

Melis and co-workers [177,190-192] observed that sulfur deprived cultures can inhibit simultaneous photosynthetic oxygen production and hence sustain hydrogen evolution. Sulfur deprived C. reinhardtii cultures kept in a sealed environment partially inactivate the function of photosystem II [193], hence the photosynthetic oxygen evolution rate drops below the rate of respiratory oxygen consumption leading to an intercellular anaerobiosis. In such conditions the algal cells maintain a low level of respiratory activity and start generating hydrogen from stored carbohydrates; continuous hydrogen production under illumination up to about 70 hours has been observed [173,177,193]. Kosourov and co-workers sustained hydrogen production for more than 140 hours by the re-addition of sulfur [194], as needed, obtaining a maximum yield of 6.6 mmol of hydrogen per hour per liter of culture; the C. reinhardtii culture-containing cell was illuminated (300 $\mu E/m^2/s$; μE = microEinsteins) on both sides by fluorescent lamps. Guan and co-workers grew marine green algae P. subcordiformis in sea water, then transferred the culture for sulfur deprivation where it was subjected to a dark anaerobic incubation period of 32 hours for hydrogenase activation before illuminating (160 $\mu E/m^2/s$) them for hydrogen production [195]. The hydrogen evolution was sustained for about ten hours with a peak evolution rate of 0.00162 mmol/per liter of culture/hour obtained after 2.5 hours of illumination.

Major Challenges

Both direct and indirect biophotolysis routes suffer from a number of challenges that currently prohibit them from becoming commercially viable processes [180]; these include, but are not limited to, solar energy conversion efficiency and the difficulty of bioreactor design. In particular the low solar energy to hydrogen conversion efficiency is a matter of major concern in the field of biophotolysis. The efficiency is calculated as [180,172]:

$$\varepsilon = \text{(Hydrogen production rate x Hydrogen energy content) / Light energy input} \qquad (2.4.9)$$

Microalgae utilize radiation for hydrogen production only in the wavelength region 400-700 nm; this is known as the photosynthetically active radiation (PAR). Only about 43% of the solar radiation falls in the PAR wavelength region. In some cases efficiency has been calculated using only the energy falling in the PAR region, with efficiency values reaching about 22% [196]. However, when commercial applications are considered in comparison to other means of producing hydrogen it is more appropriate to consider total solar energy input in the denominator of equation (2.4.9); as such calculations show a maximum solar conversion efficiency of 11% is possible with biophotolysis. However considering the amount of light reflected at different surfaces, and the low number of photons converted into hydrogen, upper values realistically practical are about 6-9% [197,198].

A major reason for this low efficiency is the light saturation effect in microalgae. The conversion efficiency of algae reaches a maximum value at low light intensities [180,199,200], typically at 10 to 20% of the maximum intensity, with the remaining portion of the energy wasted as heat or fluorescence [199]. This is due to the presence of a large number of antenna molecules per reaction center that collects photons in excess of that can be utilized by the photosystems for electron transfer across the thylakoid membrane [189]. A reduction in the number of antenna pigments per reaction center by genetic or metabolic means has been suggested as a solution to this problem; algal mutants with a reduced number of antennae pigments have been found to have saturations occurring at higher light intensities.

It is of course important to harvest as much light as possible to reduce needed bioreactor areas. Lower efficiencies mean greater land requirements for the collection of enough solar energy for meaningful amounts of hydrogen production. The annual average solar irradiance reaching earth is approximately 210 W/m^2. For an optimistic conversion efficiency of 10%, conversion of solar energy to hydrogen by biophotolysis equates to an energy equivalent of 5.5 US gallons of gasoline (121.8 MJ/gallon) per square meter of insolation, per year. In 2002 the United States used 135.4 billion gallons of gasoline; to satisfy this energy demand using 10% efficient biophotolysis (the above figures), would necessitate a square of land approximately 100 miles (157 km) by 100 miles (157 km) [173,187,194,210]. This amount of land, while large, appears a modest commitment in comparison to various consequences of continued fossil fuel use. Optimization of light intensity, pH, temperature and nutrient content while maintaining low partial pressures of H_2 and CO_2 are necessary to improve the hydrogen production rate. Additionally, the purity of the hydrogen produced is critical for its commercial use, particularly in fuel cells. Depending upon the technique used, the produced hydrogen may contain water vapor, oxygen, carbon dioxide, and various organic species. We note that 50%w/w potassium hydroxide has been used as a carbon dioxide absorbent, and an alkaline pyragallol solution for removing oxygen; the evolved gases would need to be passed through a dryer or condensation unit to remove moisture.

There are different types of reactor geometries used for hydrogen production through biological processes, such as flat panel and vertical column reactors, and it is necessary to properly design and maintain the photobioreactor appropriate for the technique used [139,180,202-204]. Safety is of paramount interest in designing the bioreactors, particularly if oxygen and hydrogen are simultaneously produced. There are as yet un-solved difficulties involving the transfer out of the reactor the hydrogen produced under water by the algae. In addition fouling of the bioreactor by unwanted species will quite possibly be a significant issue for continuously running reactors. Also reactor efficiency could be reduced by self-shadowing if proper attention is not given to the distribution of the biomass throughout the bioreactor [167].

2.5 Other Techniques for H_2 production via Water Splitting

In addition to the three major techniques discussed above other methods, that we now briefly consider, have been studied for hydrogen production via water splitting.

2.5.1 Hydrogen Production by Mechano-catalytic Water Splitting

In 1998 Domen and co-workers reported evolution of hydrogen and oxygen from distilled water when certain oxide powders were dispersed in it and then agitated using a magnetic stirrer [205]. Their experiments showed that these gases were generated according to the stoichiometric composition of water when a material like NiO, Co_3O_4, Cu_2O, Fe_2O_3, AWO_4 (A=Fe,Co,Ni,Cu) or $CuMO_2$ (M=Al, Fe, Ga) was used and that indeed the process was water splitting [206-210]. Rubbing of the particles by the stirring rod against the bottom wall of the reaction vessel apparently had a key role in the process.   It is believed that the electric charges created due to frictional forces together with catalytic activity of the oxide powder is responsible for the water splitting. Hence the phenomenon was named mechano-catalytic water splitting. In the process the conversion of energy takes place from mechanical to electrical and finally to chemical energy.

A typical experimental set up consists of a flat-bottomed pyrex glass cell that has provisions for evacuation and circulation of the evolved gases. In their experiments, the group generally used 200 ml of distilled water and 0.1 g of the desired oxide powder kept suspended by stirring [206]. The cell is initially evacuated prior to the reaction; the evolution rate of hydrogen and oxygen reduces when the over-head pressure increases due to the evolved gases, hence the cell needs to be periodically evacuated. The rate of evolved gases increases with stirring speed, saturating above 1500 rpm. Various means of stirring have been considered, with the shape, size and material of the stirring mechanism as well as reaction vessel material having an influence on the reaction and hence hydrogen and oxygen evolution rates. Pyrex glass was found superior to vessels made of quartz, sapphire, polytetrafluorethylene (PTFE) and alumina for driving the reaction, while a Teflon coated stirring rod was found highly appropriate for the reactions.

A hydrogen evolution rate of about 35 µmol/hr over the first few hours of reaction using NiO powder was reported. This decreased with time reaching a value of 20 µmol/h after 55 hours. The energy efficiency was calculated using the equation [206]

$$\varepsilon= (E_c/E_i)\times100$$

where E_c is the maximum useful work output that can be obtained from evolved hydrogen and E_i the mechanical energy input. An efficiency of up to 4.3% was obtained depending upon the shape and size of the stirring rod.

The stoichiometric evolution of hydrogen and oxygen was observed only on using the previously mentioned oxides. From the mass of the powder measured before and after the reaction, it was concluded that these materials are not undergoing any changes but acting only as a catalyst, hence the hydrogen and oxygen production was attributed to water splitting. Interestingly, oxides such as CuO, FeO, Fe_2O_3 and CoO evolved very small amounts of hydrogen without detectable evolution of oxygen, and photocatalytic materials like TiO_2, ZnO and WO_3 did not produce any gases. The results from a large number of oxides are given [206,209].

The group that discovered this technique tried to explain the phenomena based on triboelectricity [206]. According to this hypothesis, the friction between the particles and bottom of the vessel leads to charge separation, with the powder taking a negative charge and the reaction vessel becoming positively charged, with the triboelectric generated charges driving the redox reactions resulting in water splitting. However, their findings and conclusions were challenged by various researchers [211,212]; alternate explanations put forth included some kind of auto-redox reactions between water and oxides, electrical charging of the powders due to friction followed by local discharge as well as thermal decomposition due to friction induced localized increase in temperature (up to about 5000 K). To date there has been no consensus on the actual mechanism [213].

Tokio Ohta proposed a theory that explained several of the experimental findings [214], based on the observation that the oxides that showed mechano-catalytic water splitting are p-type semiconductors having an excess amount of oxygen, and that the surface of glass is susceptible to form micro cracks due to frictional rubbing [215]. According to the theory, friction between the stirring rod (R) and bottom of the glass cell (G) in the presence of the particles creates a number of micro-crevices on the surface of G [215-217]. The crevices have a high density of dangling bonds that accept electrons from R. Thus a frictional electrical capacitor is formed between the rod and glass. The coating (Teflon) on the magnetic stirrer, reaction vessel and distilled water are non-conductors, making formation of the frictional electrical capacitor possible. When the oxide particles trapped in the micro-crevices are contacted by R they are subject to high electric fields. The excess oxygen in the interstitials of the p-type oxide semiconductor receives electrons from neighboring metal ions and become O$^-$ (or O^{2-}), with a positive hole getting associated with this metal ion. The insulator coating on the stirring rod R gets positively charged due to the loss of electrons by friction.

The electrons and holes thus created are responsible for water splitting; R acts as the anode and G acts as the cathode. When R comes into contact with the particle the O$^-$ at the particle surface transfers electrons to the positively charged surface of R. These electrons are eventually transferred to G due to rubbing action. The positive holes associated with metal ions proceed via a hopping mechanism through successive metal ions in the particle toward the negatively charged G. The holes reach the surface of the particle where it encounters the potential barrier due to work-function difference. If the barrier height is not sufficiently low, tunneling of the holes could take place. This happens when the potential barrier is steep and the barrier height at the Fermi level is in the order of the wavelength associated with the carrier. These holes oxidize water and oxygen is evolved. The reaction is

$$2h^+ + H_2O \rightarrow 2H^+ + \tfrac{1}{2}O_2$$

The protons move towards the bottom of the glass cell G where they combine with electrons to generate hydrogen gas. Another possibility is the transfer of electrons from the particle surfaces to water, so that hydrogen is evolved according to the reaction [218]

$$2e^- + 2H_2O \rightarrow 2OH^- + H_2$$

OH$^-$ combines with the positive charge at the stirring rod R to evolve oxygen. The reaction is

$$2OH^- + R^{n+} \rightarrow H_2O + \tfrac{1}{2}O_2 + R^{+(n-2)}$$

At the same time holes created in the process hop through the particles finally combining with negative charge at the glass surface (G).

Ohta simplified the nomenclature and named the process as mechanolysis [214]. The theory puts restrictions on the particle size, which should match with the crevices and hence it should be between 1 to 10 μm. Although mechano-catalytic water splitting is simple and does not involve any toxic materials, a major limitation is the low yield of hydrogen. Most of the work in this technique was done between 1998 and 2000; no new results based on this technique have been seen since. It does seem an intriguing way to capture the energy of ocean waves, with reaction chambers left to tumble in the ocean surf.

2.5.2 Hydrogen Production by Water Plasmolysis

Electric discharge plasma technology is well developed and plays a crucial role in many industrial processes. In electrical discharges within gaseous media, the electrical energy is mainly transformed into electron kinetic energy, with the highly energetic electrons ionizing gas molecules/atoms by inelastic collisions sustaining the discharge. Within plasma various processes take place including ionization, dissociation and excitation of molecules and/or atoms. Hence discharge plasma processes are highly useful in carrying out thermodynamically non-spontaneous reactions ($\Delta G > 0$) such as water splitting, as well as hydrogen production by splitting methane and other hydrocarbons.

Plasmolysis involves the process of water splitting using electrical discharges. The plasma in general is classified as *cold plasma*, which includes glow discharge plasma or non-equilibrium plasma, or *hot plasma* created by electrical arcs. In both processes the electron temperatures in the plasma can reach several tens of thousands of degrees. In general, the process involves spraying water into or passing water vapor through a plasma created in a reactor and collecting the reaction products including hydrogen at the output. In water splitting using glow discharge plasma, a major portion of the energy is expended for vibrational excitation and dissociative attachment. The magnitude and nature of interaction between electrons and molecules depends upon the electron energy and electron energy distribution function. Depending upon the plasma conditions a number of species such as H$_2$O* (excited molecule), H, OH, H$^-$, H$^+$, H$_2$O$_2$, H$_2$ etc. can be found in the plasma. In order to increase the yield of hydrogen production certain reactions have to be favored and others inhibited. For example, OH radicals can combine with H$_2$ to form H and H$_2$O and hence it reduces the hydrogen yield. Catalysts are normally used to suppress unfavorable reactions and to selectively activate certain bonds [219-221]. Carbon dioxide is one such catalyst, reducing the OH free radical concentration and hence back reactions [222]. Such catalysts mitigate the need of having a high degree of ionization of water molecules for obtaining higher yields [223].

Chen and co-workers reported on water splitting conducted in a tubular reactor at atmospheric pressure using plasma and catalyst integrated technologies (PACT) [224,225]. The tubular PACT reactor consisted of a quartz tube fitted with an outer electrode, and an inner electrode passing through the center of the tube. The inner electrode was coated with a catalytic material like Ni, Pd, Rh or Au and the outer electrode was made of aluminum foil. Plasma was created in the quartz tube using a low frequency (8.1 kHz) electric field applied between the electrodes. Argon bubbled through water was used as the feed gas, containing 2.3 mol% water. Maximum hydrogen production was achieved using gold catalyst. At a voltage of about 2.5 kV, 14.2% of the water in the feed was converted into hydrogen (0.32 mol% H$_2$).

The energy conversion efficiency was calculated as

$$\varepsilon = (R_H \, \Delta G/P) \times 100$$

where R$_H$ is the rate of hydrogen production and P is the input power. Chen and co-workers have reported efficiencies up to 2.09%. The possible reactions in the plasma can be given as

$$H_2O + Ar^* \rightarrow H_2O^* + Ar$$
$$H_2O^* \rightarrow HO\cdot + H\cdot$$
$$HO\cdot + Ar^* \rightarrow H\cdot + O\cdot + Ar$$
$$2H\cdot \rightarrow H_2$$
$$2O\cdot \rightarrow O_2$$

The oxygen radicals get adsorbed on the catalyst surfaces and either combine to form oxygen or react with metal surfaces. This favors the decomposition reaction of H_2O and enhances the hydrogen yield. To date the hydrogen yield and efficiencies are not high enough for practical application.

2.5.3 Hydrogen Production by Water Magnetolysis

Magnetolysis is electrolysis but with the needed voltage created inside the electrolyzer by magnetic induction. The idea was suggested by Bockris and Gutmann in 1985 to eliminate the technical problems in supplying the low voltage, high current needed for electrolyzers [223]. An electrolyzer unit with a parallel configuration of individual cells operates at voltages between 1.5 V to 2.0 V dc requiring currents of several hundred amperes. Satisfying these electrical requirements requires transformers and rectifiers giving rise to higher capital expenditures, as well as losses in electrical energy. Hence, the idea was to generate the needed low voltage and high current directly inside the electrolyte via magnetic induction using a homopolar generator.

In a configuration used by Ghoroghchian and Bockris [226], a stainless steel disc of 30.48 cm diameter and 0.31 cm thickness was connected to a bearing isolated shaft; the disk was mounted vertically in an electrolytic cell containing a 35% potassium hydroxide electrolyte. The whole assembly was placed within a magnetic field, generated by an electromagnet, and the disk rotated using an electric motor. Hence a potential difference was created across the disk with the center acting as the anode and the rim as the cathode. Under a magnetic field of 0.86 T and at 2100 rpm, a potential difference of 2 V was created that resulted in hydrogen generation via electrolysis. The disk in such a configuration suffers viscous drag. To overcome this difficulty the magnetic field strength should be as large as possible, or the electrolyte should also be rotated so as to avoid any relative movement between the disk and the electrolyte. It was estimated that for producing hydrogen at minimum power consumption a magnetic field greater than 11 T is required [223].

The advances made over the past several decades to improve the performance of electronic circuitry, e.g. rectifiers, have made electrolyzers increasingly commercially viable. Furthermore in advanced electrolyzers a series cell configuration is used (bipolar filter press, SPE, *etc.*) and hence there is no need to work in low voltage high current mode with its inherent I^2R electrical losses. Consequently, the dormant magnetolysis field stands a good chance of remaining dormant.

2.5.4 Hydrogen Production by Water Radiolysis

Radiolysis of water involves the use of radioactive materials and/or highly energetic particles for water decomposition. The energy for dissociation can be obtained from γ rays or high-energy neutrons or charged particles like α and ß. Radiolysis of water has been known for over a century and is of considerable importance in nuclear reactor design where water is used as a coolant. Particles and radiation from the reactor core can decompose coolant water into hydrogen peroxide and oxygen, in addition to hydrogen, giving rise to corrosion problems. The technique received much attention when it was first used as a means to produce hydrogen in the 1950s [227] as means to effectively utilize nuclear waste emissions.

When water is irradiated by emissions from radioactive materials, a number of reactions leading to the production of a variety of species take place depending upon the nature and energy of the radiation. As the high-energy particles/radiations traverse the water they lose energy and eject electrons from the atomic shells. These high-energy electrons create low energy secondary electrons and help to initiate further reactions. Some of the prominent species formed as a result are H_2, H_2O_2, OH, H, HO_2, O and O_2 [228-230].

The effectiveness of interaction between a given type of radiation and water in yielding a particular species is represented by parameter G, the radiochemical or radiological yield. It is defined as the number of species created per unit of deposited energy (usually 100 eV is considered as the unit). It depends upon the energy transfer to the medium, called linear energy transfer (LET), from a given type of radiation [231]. The G value is about 0.45 for hydrogen when γ radiation is used. That is, 0.45 molecules of H_2 are produced per 100 eV energy absorbed by liquid water when a γ ray is passed through it. The yield is higher when energetic particles are used, with a value close to 1 when pure water is irradiated at room temperature [232]. The radiolysis of pure steam performed using Cm-244 alpha-particles gave a yield of 8 molecules/100 eV at 300°C whereas the yield was only 2 molecules/100 eV at 250°C [233]. With fast neutrons Sunaryo and co-workers reported a G value of 1.5 at 250°C [230]. The hydrogen yield is limited by back reactions between H_2 and OH.

Although the G value for hydrogen when pure water is irradiated by γ radiation is low, studies indicate that irradiation of water in the presence of solid materials improves the yield [234]. Irradiation of water adsorbed on solid oxides, mainly use of porous silica and zeolites have been studied, appreciably increases the yield [233,235-239]. Nakashima and Masaki obtained yields close to 1.5 molecules/100eV when water adsorbed zeolites (Type Y) were irradiated using ^{60}Co γ-rays at room temperature [238]. LaVerne and Tandon reported on irradiating water adsorbed on micron-sized particles of CeO_2 and ZrO_2 with γ radiation (dose rate 202 Gy/min) or 5 MeV α particles [239]. The yield was found greater for ZrO_2, with the yield dramatically increasing when the number of adsorbed layers of water was limited to one or two; they obtained 150 molecules/100eV when ZrO_2 having two layers of water molecules was irradiated with γ rays. However, this yield was calculated using the energy adsorbed by the water layers alone and not using the total energy absorbed by both oxide and water. Cecal and co-workers [233], from their study using oxides such as ZrO_2, TiO_2, BeO or SiO_2 dispersed in water, found that yield was highest for ZrO_2. Caer and co-workers also studied the effect of pore size on yield by irradiating nanoporous SiO_2 with 1-MeV electrons [240]; an increase in yield with reduction in pore size was observed. In general, electrons and holes formed inside the oxide that then migrate to the surface are believed responsible for the water splitting [235].

Radiolysis of water is, at least for today, inherently limited due to the use of radioactive materials by which the product stream could be contaminated by radioactive species. Consequently while of scientific interest this technique is not considered to have the potential to compete with other major water splitting techniques.

References

1. Funk JE (2001) Thermochemical hydrogen production: past and present. Int J Hydrogen Energy 26:185–190
2. Sato S (1979) Thermochemical hydrogen production. In: Ohta T (ed) Solar hydrogen energy systems. Pergamon Press, New York
3. Ihara S (1979) Direct thermal decomposition of water. In: Ohta T (ed) Solar   hydrogen energy systems. Pergamon press, New York
4. Casper MS (1978) Hydrogen manufacture by electrolysis, thermal decomposition and unusual techniques. Noyes Data Corporation, New Jersey, USA
5. Kreuter W, Hofmann H (1998) Electrolysis: the important energy transformer in a world of sustainable energy. Int J Hydrogen Energy 23:661–666
6. Ewan BCR, Allen RWK (2005) A figure of merit assessment of the routes to hydrogen. Int J Hydrogen Energy 30:809–819
7. Divisek J (1990) Water electrolysis in a low and medium temperature regime. In: Wendt H (ed) Electrochemical hydrogen technologies - Electrochemical production and combustion of hydrogen. Elsevier, New York, pp 137–212
8. Donitz W, Erdle E, Streicher R (1990) High temperature electrochemical technology for hydrogen production and power generation. In: Wendt H (ed.) Electrochemical hydrogen technologies - Electrochemical production and combustion of hydrogen. Elsevier, New York, pp 213–259
9. Barbir F (2005) PEM electrolysis for production of hydrogen from renewable energy sources. Sol Energy 78:661-669
10. JANAF (1971) Thermochemical tables QD511.D614
11. Rossmeisl J, Logadottir A, Norskov JK (2005) Electrolysis of water on (oxidized) metal surface. Chem Phys 319:178–184

12.   Takahashi T (1979) Water electrolysis. In:Solar hydrogen energy systems. Ohta T (ed.) Pergamon Press, New York

13.   Burstein GT (2005) A hundred years of Tafel's equation: 1905-2005. Corrosion Sci 47:2858–2870

14.   Esaki H, Nambu T, Morinaga M, Udaka M, Kawasaki K (1996) Development of low hydrogen overpotential electrodes utilizing metal ultra-fine particles. Int J Hydrogen Energy 21:877–881

15.   Nagai N, Takeuchi M, Kimura T, Oka T (2003) Existence of optimum space between electrodes on hydrogen production by water electrolysis. Int J Hydrogen Energy 28:35–41

16.   de Jonge RM, Barendrecht E, Janssen LJJ, van Stralen SJD (1982) Gas bubble behavior and electrolyte resistance during water electrolysis. Int J Hydrogen Energy 7:883–894

17.   Sillen CWMP, Barendrecht E, Janssen LJJ, van Stralen SJD (1982) Gas bubble behavior during electrolysis. Int J Hydrogen Energy 7:577–587

18.   Roy A, Watson S, Infield D (2006) Comparison of electrical energy efficiency of atmospheric and high-pressure electrolyzers, Int J Hydrogen Energy 31:1964–1979

19.   Dickson EM, Ryan, JW, Smulyan MH (1977) The hydrogen energy economy: a realistic appraisal of prospects and impacts. Praeger, New York, USA.

20.   Ulleberg O (2003) Modeling of advanced alkaline electrolyzers: a system simulation approach. Int J Hydrogen Energy 28:21–33

21.   Wendt H, Imarisio G (1988) Nine years of research and development on advanced water electrolysis. A review of research program of the commission of the European communities. J Applied Electrochem 18:1–14

22.   Dutta S (1990) Technology assessment of advanced electrolytic hydrogen production. Int J Hydrogen Energy 15:379–386

23.   LeRoy RL (1983) Industrial water electrolysis: Present and future, Int J Hydrogen Energy. 8:401–417

24. LeRoy RL, Bowen CT, LeRoy DJ (1980) The thermodynamics of aqueous water electrolysis. J Electrochem Soc 127:1954–1962

25. Onda K, Kyakuno T, Hattori K, Ito K (2004) Prediction of production power for high-pressure hydrogen by high-pressure water electrolysis. J Power Sources 132:64–70

26. Chen L, Lasia A (1991) Study of the kinetics of hydrogen evolution reaction on nickel-zinc Alloy electrodes. J Electrochem Soc 138:3321–3328

27. Rosalbino F, Maccio D, Angelini E, Saccone A, Delfino S (2005) Electrocatalytic properties of Fe-R (R=rare earth metal) crystalline alloys as hydrogen electrodes in alkaline water electrolysis. J Alloys Compd 403:275–282

28. Bockris JOM (1956) Kinetics of activation controlled consecutive electrochemical reactions: anodic evolution of oxygen. J Chem Phys 24:817–827

29. Chapman EA (1965) Production of hydrogen by electrolysis. Chem Process Eng 46:387–393

30. Dutta S (1990) Technology assessment of advanced electrolytic hydrogen production. Int J Hydrogen Energy 15:379–386

31. Yazici B, Tatli G, Galip H, Erbil M (1995) Investigation of suitable cathodes for the production of hydrogen gas by electrolysis. Int J Hydrogen Energy 20:957-965

32. Suffredini HB, Cerne JL, Crnkovic FC, Machado SAS, Avaca LA (2000) Recent developments in electrode materials for water electrolysis. Int J Hydrogen Energy 25:415–423

33. Nagarale RK, Gohil GS, Shahi VK (2006) Recent developments on ion-exchange membranes and electro-membrane processes. Adv Colloid Interface Sci 119:97–130

34. Singh RN, Pandey JP, Anitha KL (1993) Preparation of electrodeposited thin films of Nickel-Iron alloys on mild steel for alkaline water electrolysis. Part I: Studies on oxygen evolution. Int J Hydrogen Energy 18:467–473

35. Kaninski MPM, Stojic DLJ, Saponjic DP, Potkonjak NI, Miljanic SS (2006) Comparison of different electrode

materials- Energy requirements in the electrolytic hydrogen evolution process. J Power Sources 157:758-764

36. Hu W, Cao X, Wang F, Zhang Y (1997) Short Communication: a novel cathode for alkaline water electrolysis. Int J Hydrogen Energy 22:621–623

37. Stojic DL, Maksic AD, Kaninski MPM, Cekic BD, Mijanic SS (2005) Improved energy efficiency of the electrolytic evolution of hydrogen - Comparison of conventional and advanced electrode materials. J Power Sources 145:278–281

38. Raney M (1925), U.S. patent 1563787; (1927) 1628191; (1933) 1915473. From E. Endoh E et al. (1987) New Raney nickel electrode. Int. J. Hydrogen Energy 12:473–47939

39. Singh SP, Singh RN, Poillearat G, Chartier P (1995) Physiochemical and electrochemical characterization of active films of LaNiO$_3$ for use as anode in alkaline water electrolysis. Int J Hydrogen Energy 20:203–210

40. Rosa VM, Santos MBF, da Silva EP (1995) New materials for water electrolysis diaphragms. Int J Hydrogen Energy 20:697–700

41. http://americanhistory.si.edu/fuelcells/pem/pemmain.htm

42. http://www.chemsoc.org/chembytes/ezine/2000/kingston_jun00. htm

43. Han SD, Park KB, Rana R, Singh KC (2002) Developments of water electrolysis technology by solid polymer electrolyte. Ind J Chem 41A:245–253

44. Hijikata T (2002) Research and development of international clean energy network using hydrogen energy (WE-NET). Int J Hydrogen Energy 27:115–129

45. Paddison SJ (2003) Proton conduction mechanism at low degrees of hydration in sulfonic acid-based polymer electrolyte membranes. Ann Rev Mater Res 33:289–319

46. Rasten E, Hagen G, Tunold R (2003) Electrocatalysts in water electrolysis with solid polymer electrolyte. Electrochimica acta 48:3945–3952

47. Linkous CA, Anderson HR, Kopitzke RW, Nelson GL (1998) Development of new proton exchange membrane

electrolytes for water electrolysis at higher temperatures. Int J Hydrogen Energy 23:525–529

48.   Linkous CA (1993) Development of solid polymer electrolytes for water electrolysis at intermediate temperatures. Int J Hydrogen Energy 18:641–646

49.   Grigoriev SA, Porembsky VI, Fateev VN (2006) Pure hydrogen production by PEM electrolysis for hydrogen energy. Int J Hydrogen Energy 31:171–175

50.   Millet P, Andolfatto F, Durand R (1996) Design and performance of a solid polymer electrolyte water electrolyzer. Int J Hydrogen Energy 21:87–93

51.   Herring JS, Brien JEO, Stoots CM, Hawkes GL, Hartvigsen JJ, Shagnam M (2007) Progress in high temperature electrolysis for hydrogen production using planar SOFC technology. Int J Hydrogen Energy 32:440–450

52.   Hino R, Haga K, Aita H, Sekita K (2004) R&D on hydrogen production by high temperature electrolysis of steam. Nucl Eng Des 233:363–375

53.   Dutta S, Morehouse JH, Khan JA (1977) Numerical analysis of laminar flow and heat transfer in a high temperature electrolyzer. Int J Hydrogen Energy 22:883–895

54.   Donitz W, Erdle E (1985) High temperature electrolysis of water vapor-status of development and perspectives for application. Int J Hydrogen Energy 10:291–295

55.   Yildiz B, Kazimi MS (2006) Efficiency of hydrogen production systems using alternative nuclear energy technologies. Int J Hydrogen Energy 31:77-92

56.   Kharton VV, Marques FMB, Atkinson A (2004) Transport properties of solid oxide electrolyte ceramics: a brief review. Solid State Ionics 174:135–149

57.   Hong HS, Chae US, Choo ST, Lee KS (2006) Microstructure and electrical conductivity of Ni/YSZ and NiO/YSZ composites for high temperature electrolysis prepared by mechanical alloying. J Power Sources 149:84-89

58.   Utgikar V, Thiesen T (2006) Life cycle assessment of high temperature electrolysis for hydrogen production via nuclear energy. Int J Hydrogen Energy 31:939–944

59. Liepa MA, Borhan A (1986) High-temperature steam electrolysis: Technical and economic evaluation of alternative process designs. Int J Hydrogen Energy 11:435

60. Funk JE, Reinstrom RM (1966) Energy requirements in the production of hydrogen from water. Ind Eng Chem Process Des Dev 5:336-342

61. Funk JE (1976) Thermochemical production of hydrogen via multistage water splitting processes, Int J Hydrogen Energy 1:33–43

62. Engels H, Funk JE, Hesselmann K, Knoche KF (1987) Thermochemical Hydrogen-Production. Int J Hydrogen Energy 12:291–295

63. Rosen MA (1996) Thermodynamic comparison of hydrogen production processes. Int J Hydrogen Energy 21:349–365

64. Scott DS (2003) Exergy. Int J Hydrogen Energy 28:369–375

65. Struck BD, Schutz GH, Van Velzen D, (1990) Cathodic hydrogen evolution in thermochemical-electrochemical hybrid cycles. In: electrochemical hydrogen technologies-Electrochemical production and combustion of hydrogen Wendt H (ed), Elsevier, New York, pp 213-259

66. Schultz K, Herring S, Lewis M, Summers WA (2005) The hydrogen reaction. Nucl Eng Int 50:10–15

67. Abanades S, Charvin P, Flamant G, Neveu P (2006) Screening of water-splitting thermochemical cycles potentially attractive for hydrogen production by concentrated solar energy. Energy 31:2805–2822

68. Perkins C, Weimer AW (2004) Likely near-term solar-thermal water splitting technologies. Int J Hydrogen Energy 29:1587–1599

69. Yildiz B Kazimi MS (2006) Efficiency of hydrogen production systems using alternative nuclear energy technologies. Int J Hydrogen Energy 31:77–92

70. Kodama T (2003) High temperature solar chemistry for converting solar heat to chemical fuels. Prog Energy & Combust. Sci 29 567–597

71. Steinfeld A (2005) Solar thermochemical production of hydrogen – a review. Sol Energy 78:603–615

72. Kalogirou SA (2004) Solar thermal collectors and applications. Prog Energy Combust Sci 30:231–295
73. Sakurai M, Bilgen E, Tsutsumi A, Yoshida K (1996) Solar UT-3 thermochemical cycle for hydrogen production. Sol Energy 57:51-58
74. Baykara SZ (2004) Hydrogen production by direct solar thermal decomposition of water, possibilities for improvement of process efficiency. Int J Hydrogen Energy 29:1451–1458
75. Baykara SZ (2004) Experimental solar water thermolysis. Int J Hydrogen Energy 29:1459–1469
76. Lede J, Lapicque F, Villermaux J (1983) Production of hydrogen by direct thermal decomposition of water. Int J Hydrogen Energy 8:675–679
77. Kogan A, Spiegler E, Wolfshtein M (2000) Direct solar thermal splitting of water and on-site separation of the products. III. Improvement of reactor efficiency by steam entrainment. Int J Hydrogen Energy 25:739–745
78. Fletcher EA (2001) Solar thermal processing: A Review. J Solar Energy Eng 123:63–74
79. Kogan A (1998) Direct thermal splitting of water and on-site separation of the products—II. Experimental feasibility study. Int J Hydrogen Energy 23:89–98
80. Nakamura T (1977) Hydrogen production from water utilizing solar heat at high temperatures. Sol Energy 19:467–475
81. Ross RT (1966) Thermodynamic limitations on the conversion of radiant energy into work. J Chem Phys 45:1–7
82. Kogan A (1998) Direct thermal splitting of water and on-site separation of the products—II. Experimental feasibility study. Int J Hydrogen Energy 23:89–98
83. Sturzenegger M, Ganz J, Nuesch P, Schelling T (1999) Solar hydrogen from a manganese oxide based thermochemical cycle. J PhyS. IV:JP 9:3–331
84. Kaneko H, Gokon N, Hasegawa N, Tamaura Y (2005) Solar thermochemical process for hydrogen production using ferrites. Energy 30:2171–2178

85.   Kodama Y, Kondoh Y, Yamamoto R, Andou H, Satou N (2005) Thermochemical hydrogen production by a redox system of ZrO_2-supported Co(II)-ferrite. Sol Energy 78:623–631

86.   Alvani C, Ennas G, La Barbera A, Marongiu G, Padella F, Varsano F (2005) Synthesis and characterization of nanocrystalline $MnFe_2O_4$: advances in thermochemical water splitting. Int J Hydrogen Energy 30:1407–1411

87.   Kodoma T, Kondoh Y, Kiyama A, Shimizu K (2003) Hydrogen production by solar thermochemical water-splitting/methane-reforming process. International Solar Energy Conference pp 121–128

88.   Lede J, Ricart EE, Ferrer M (2001) Solar thermal splitting of zinc oxide: A review of some of the rate controlling factors. J Solar Energy Eng 123:91–97

89.   Agrafiotis C, Roeb M, Konstandopoulos AG, Nalbandian L, Zaspalis VT, Sattler C, Stobbe P, Steele AM (2005) Solar water splitting for hydrogen production with monolithic reactors. Sol Energy 79:409–421

90.   Wegner K, Ly HC, Weiss RJ, Pratsinis SE, Steinfeld A (2006) In situ formation and hydrolysis of Zn nanoparticles for H_2 production by the 2-step ZnO/Zn water-splitting thermochemical cycle. Int J Hydrogen Energy 31:55–61

91.   Steinfeld A (2002) Solar hydrogen production via a two-step water-splitting thermochemical cycle based on Zn/ZnO redox reactions. Int J Hydrogen Energy 27:611–619

92.   Haueter P, Moeller S, Palumbo R, Steinfeld A (1999) The production of zinc by thermal dissociation of zinc oxide-solar chemical reactor design. Sol Energy 67:161–167

93.   Berman A, Epstein M (2000) The kinetics of hydrogen production in the oxidation of liquid zinc with water vapor. Int J Hydrogen Energy 25:957–967

94.   Perret R, Chen Y, Besenbruch G, Diver R, Weimer A, Lewandowski A, Miller E (2005) High-temperature thermochemical: solar hydrogen generation research. DOE Hydrogen Program Progress Report

95.   Yalcin S (1989) A review of nuclear hydrogen production. Int J Hydrogen Energy 14:551–561

96.   Huang CP, Raissi AT (2005) Analysis of sulfur-iodine thermochemical cycle for solar hydrogen production. Part I: decomposition of sulfuric acid. Sol Energy 78:632–646

97.   Beghi GE (1981) Review of thermochemical hydrogen production. Int J Hydrogen Energy 6:555–566

98.   Norman JH, Mysels KJ, Sharp R, Williamson D (1982) Studies of the sulfur-iodine thermochemical water-splitting cycle. Int J Hydrogen Energy 7:545–556

99.   Keefe DO, Allen C, Besenbruch G, Brown L, Norman J, Sharp R (1982) Preliminary results from bench-scale testing of a sulfur-iodine thermochemical water splitting cycle. Int J Hydrogen Energy 7:381–392

100.   Kasahara S, Hwang GJ, Nakajima H, Choi HS, Onuki K, Nomura M (2003) Effects of chemical engineering parameters of the IS process on total thermal efficiency to produce hydrogen from water. J Chem Eng Jpn 36:887–899

101.   Onuki K, Inagaki Y, Hino R, Tachibana Y (2005) Research and development on nuclear hydrogen production using HTGR at JAERI. Prog Nucl Energy 47:496–503

102.   Kubo S, Kasahara S, Okuda H, Terada A, Tanaka N, Inaba Y, Ohashi H, Inagaki Y, Onuki K, Hino R (2004) A pilot test plan of the thermochemical water-splitting iodine-sulfur process. Nucl Eng Des 233:355–362

103.   Kubo S, Nakajima, Kasahara HS, Higashi S, Masaki T, Abe H, Onuki (2004) A demonstration study on a closed-cycle hydrogen production by the thermochemical water-splitting iodine-sulfur process. Nucl Eng Des 233:347–354

104.   Huang CP, Raissi AT (2005) Analysis of sulfur-iodine thermochemical cycle for solar hydrogen production. Part I: decomposition of sulfuric acid. Sol Energy 78:632–646

105.   Gorensek MB, Summers WA, Buckner MR (2004) Model-based evaluation of thermochemical nuclear hydrogen processes, Trans Am Nucl Soc 91:107–108

106.   Sakurai M, Miyake N, Tsutsumi A, Yoshida K (1996) Analysis of a reaction mechanism in the UT-3

thermochemical hydrogen production cycle. Int J Hydrogen Energy 21:871–875

107.   Sakurai M, Bilgen E, Tsutsumi A, Yoshida K (1996) Adiabatic UT-3 thermochemical process for hydrogen production. Int J Hydrogen Energy 21:865–870

108.   Teo ED, Brandon NP, Vos E, Kramer GJ (2005) A critical pathway energy efficiency analysis of the thermochemical UT-3 cycle, Int J Hydrogen Energy 30:559–564

109.   Lemort F, Lafon C, Dedryvere R, Gonbeau D (2006) Physicochemical and thermodynamic investigation of the UT-3 hydrogen production cycle: a new technological assessment. Int J Hydrogen Energy 31:906–918

110.   Tadokoro Y, Kajiyama T, Yamaguchi T, Sakai N, Kameyama H, Yoshida K (1997) Technical evaluation of UT-3 thermochemical hydrogen production process for an industrial scale plant. Int J Hydrogen Energy 22:49–56.

111.   Lewis MA, Serban M, Basco JK (2004) A progress report on the chemistry of the low temperature Cu-Cl thermochemical cycle. Trans Am Nucl Soc 91:113–114

112.   Dokiya M, Kotera Y (1976) Hybrid cycle with electrolysis using Cu-Cl system. Int J Hydrogen Energy 1:117-121

113.   Deneuve F, Roncato JP (1981) Thermochemical or hybrid cycles of hydrogen production- technolo-echonomical comparison with water electrolysis. Int J Hydrogen Energy 6:9-23

114.   Beghi GE (1985) Development of thermochemical and hybrid processes for hydrogen production. Int J Hydrogen Energy 10:431-438

115.   Deneuve F, Roncato JP (1981) Thermochemical or hybrid cycles of hydrogen production- technico-economical comparison with water electrolysis. Int J Hydrogen Energy 6:9-23

116.   Summers WA, Buckner MR (2005) Hybrid sulfur thermochemical process development, DOE Hydrogen Program Progress Report

117.   Carty R, Cox K, Funk J, Soliman M, Conger W (1977) Process sensitivity studies of the Westinghouse sulfur cycle

for hydrogen generation, International journal of hydrogen energy 2:17-22

118.   Bilgen E (1988) Solar hydrogen production by hybrid thermochemical processes, Solar Energy 41:199-206

119.   Volkov AG, Volkova-Gugeshashvili MI, Brown-McGauley CL, Osei AJ (2007) Nanodevices in nature: electrochemical aspects. Electrochim Acta 52:2905–2912

120.   Volkov AG (1989) Oxygen evolution in the course of photosynthesis: molecular mechanisms. Bioelectrochem Bioenergetics 21:3–24

121.   Giardi MT, Pace E (2005) Photosynthetic proteins for technological applications. Trends Biotechnol 23:257–263

122.   Weaver PF, Lien S, Seibert M (1980) Photobiological production of hydrogen. Sol Energy 24:3–45

123.   Prince RC (1996) Photynthesis: the Z-scheme revised. TIBS:121–122

124.   Kramer DM, Avenson TJ, Edwards GE (2004) Dynamic flexibility in the light reactions of photosynthesis governed by both electron and proton transfer reactions. Trends Plant Sci 9:349–357

125.   Bukhov NG (2004) Dynamic light regulation of photosynthesis (A review). Russ J Plant physiol 51:742–753

126.   Howell JM, Vieth WR (1982) Biophotolytic membranes: simplified kinetic model of photosynthetic electron transport, J Mol Catal 16:245–298

127.   Renger G (2001) Photosynthetic water oxidation to molecular oxygen: apparatus and mechanism, Biochim Biophys Acta 1503:210–228

128.   Kraub N (2003) Mechanisms for photosystems I and II. Curr Op Chem Biol 7:540–550

129.   Fromme P, Jordan P, Kraub N (2001) β Structure of photosystem I. Biochim Biophys Acta 1507:5-31

130.   Fairclough WV, Forsyth A, Evans MCW, Rigby SEJ, Purton S, Heathcote P (2003) Bidirectional electron transfer in photosystem I: electron transfer on the PsaA side is not essential for phototrophic growth in *Chlamydomonas*. Biochim Biophys Acta 1606:43–55

131. Setif P (2001) Ferredoxin and flavodoxin reduction by photosystem I. Biochim Biophys Acta 1507:161–179

132. Onda Y, Hase T (2004) FAD assembly and thylakoid membrane binding of ferredoxin: $NADP^+$ oxidoreductase in chloroplasts. FEBS Lett 564:116–120

133. Reeves SG, Hall DO (1978) Photophosphorylation in chlorplasts. Biochim Biophys Acta 463:275–297

134. Joliot P, Joliot A (2006) Cyclic electron flow in C3 plants. Biochim Biophys Acta 1757:362–368

135. Griffin KL, Seemann JR (1996) Plants, CO_2 and photosynthesis in the 21st century. Chem Biol 3:245–254

136. Jackson DD, Ellms JW (1896) On odors and tastes of surface waters with special reference to *Anabaena*, a microscopial organism found in certain water supplies of Massachusetts, Report of the Massachusetts State Board Health 410–420

137. Gaffron H, Rubin J (1942) Fermentative and photochemical production of hydrogen in algae. J Gen Physiol 219–240

138. Kruse O, Rupprecht J, Mussgnug JH, Dismukes GC, Hankamer B (2005) Photosynthesis: a blueprint for solar energy capture and biohydrogen production technologies, Photochem Photobiol Sci 4:957–969

139. Rupprecht J, Hankamer B, Mussgnug JH, Ananyev G, Dismukes C, Kruse O (2006) Perspectives and advances of biological H_2 production in microorganisms. Appl Microbiol Biotechnol 72:442–449

140. Miura Y (1995) Hydrogen-Production by Biophotolysis Based on Microalgal Photosynthesis. Process Biochem 30:1–7

141. Das D, Veziroğlu TN (2001) Hydrogen production by biological processes: a survey of literature. Int J Hydrogen Energy 26:13–28

142. Rao KK, Hall DO (1984) Photosynthetic production of fuels and chemicals in immobilized systems. Trends Biotechnol. 2:124–129

143. Howard JB, Rees DC (1996) Structural basis of biological nitrogen fixation. Chem Rev 96:2955–2982

144. Adams MWW (1990) The structure and mechanism of iron-hydrogenases. Biochim Biophys Acta 1020:115–145
145. Adams MWW, Mortenson LE, Chen JS (1981) Hydrogenase, Biochim Biophys Acta 594:105-176
146. Krasna AI (1979) Hydrogenase: properties and applications, Enzyme Microb. Technol. 1:165-172
147. Darensbourg MY, Lyon EJ, Smee JJ (2000) The bio-organometallic chemistry of active site iron in hydrogenases. Coord Chem Rev 206-207:533–561
148. Vignais PM, Billoud B, Meyer J (2001) Classification and phylogeny of hydrogenases. FEMS Microbiol Rev 25:455–501
149. Mertens R, Liese A (2004) Biotechnological applications of hydrogenases, Curr Op Biotechnol 15:343–348
150. Das D, Dutta T, Nath K, Kotay SM, Das AK, Veziroglu TN (2006) Role of Fe-hydrogenase in biological hydrogen production. Curr Sci 90:1627–1637
151. Happe T, Hemeschemeier A, Winkler M, Kaminski A (2002) Hydrogenases in green algae: do they save the algae's life and solve our energy problems? Trends Plant Sci 7:246–250
152. Frey M (2002) Hydrogenases: Hydrogen-activating enzymes. Chem Biochem 3:153–160
153. Happe T, Kaminski A (2002) Differential regulation of the Fe-hydrogenase during anaerobic adaptation in the green algae Clamydomonas reinhardtii. Eur J Biochem 269:1022–1032
154. Adams MWW, Stiefel EI (1998) Biological hydrogen production: Not so elementary, Science 282:1842–1843
155. Thauer RK, Klein AR, Hartmann GC (1996) Reactions with molecular hydrogen in microorganisms: evidence for a purely organic hydrogenation catalyst. Chem Rev 96:3031–3042
156. Ma Y, Balbuena PB (2007) Density functional theory approach for improving the catalytic activity of a biomimetic model based on the Fe-only hydrogenase active site. J Electroanal Chem (in press)

157. Nicolet Y, Lemon BJ, Fontecilla-Camps JC, Peters JW (2000) A novel FeS cluster in Fe-only hydrogenases. TIBS 25:138–143

158. Appel J, Schulz R (1998) Hydrogen metabolism in organisms with oxygenic photosynthesis: hydrogenases as important regulatory devices for a proper redox poising. J Photochem Photobiol 47:1–11

159. Bauwman E, Reedijk J (2005) Structural and functional models related to the nickel hydrogenases. Coord Chem Rev 249:1555–1581

160. Volbeda A, Charon MH, Piras C, Hatchikian EC, Frey M, Fontecilla-Camps JC (1995) Crystal structure of the nickel-iron hydrogenase from Desulfovibrio gigas. Nature 373:580–587

161. Albracht SPJ (1994) Nickel hydrogenases: in search of the active site. Biochim Biophys Acta 1188:167–204

162. Tamagnini P, Axelsson R, Lindberg P, Oxelfelt F, Wunschiers R, Lindblad P (2002) Hydrogenases and hydrogen metabolism of cyanobacteria. Microbiol Mol Biol Rev 66 (2002) 1–20

163. Korbas M, Vogt S, Meyer-Klaucke W, Bill E, Lyon EJ, Thauer RK, Shima S (2006) The iron-sulfur cluster-free hydrogenase (Hmd) is a metalloenzyme with a novel iron binding motif. J Biol Chem 281:30804–30813

164. Lyon EJ, Shima S, Burman G, Chowdhuri S, Batschauer A, Steinback K, Thauer RK (2004) UV-A/blue-light inactivation of the 'metal-free' hydrogenase (Hmd) from methanogenic archaea. Eur J Biochem 271:195–204

165. Berkessel A, Thauer RK (1995) On the mechanism of catalysis by a metal-free hydrogenase from methanogenic archaea: enzymatic transformation of H_2 without a metal and its analogy to the chemistry of alkanes in superacidic solution. Angew Chem Int Ed 34:2247

166. Rees DC, Howard JB (2000) Nitrogenase: standing at the crossroads. Curr Op Chem Biol 4:559–566

167. Asada Y Miyake J (1999) Photobiological hydrogen production. J Biosci Bioeng 88:1–6

168. Peters JW, Szilagyi RK (2006) Exploring new frontiers of nitrogenase structure and mechanism, Curr Op Chem Biol 10:101–108

169. Eady RR (2003) Current status of structure function relationships of vanadium nitrogenase, Coord Chem Rev 237:23–30

170. Benemann JR (2000) Hydrogen production by microalgae. J Appl Phycol 12 (2000) 291–300

171. Ni M, Leung DYC, Leung MKH, Sumathy K (2006) An overview of hydrogen production from biomass. Fuel Process Technol 87:461–472

172. Akkerman I, Janssen M, Rocha J, Wijffels RH (2002) Photobiological hydrogen production: photochemical efficiency and bioreactor design. Int J Hydrogen Energy 27:1195–1208

173. Prince RC, Kheshgi HD (2005) The photobiological production of hydrogen: Potential efficiency and effectiveness as a renewable fuel. Crit Rev Microbiol 31:19–31

174. Vijayaraghavan K, Soom MAM (2006) Trends in bio-hydrogen generation: A review, Environmental Sciences 3: 255-271.

175. Miura Y, Akano T, Fukatsu K, Miyasaka H, Mizoguchi T, Yagi K, Maeda I, Ikuta Y, Matsumoto H (1995) Hydrogen-Production by Photosynthetic Microorganisms. Energy Conv Manag 36:903–906

176. Miura Y, Akano T, Fukatsu K, Miyasaka H, Mizoguchi T, Yagi K, Maeda I, Ikuta Y Matsumoto H. (1997) Stably sustained hydrogen production by biophotolysis in natural day/night cycle. Energy Conv Manag 38:S533–S537

177. Hallenbeck PC, Benemann JR (2002) Biological hydrogen production; fundamental and limiting processes. Int J Hydrogen Energy 27:1185–1193

178. Greenbaum E (1980) Simultaneous photoproduction of hydrogen and oxygen by photosynthesis. Biotechnol Bioeng Symp 10:1–13

179.   Benemann JR (1997) Feasibility analysis of photobiological hydrogen production. Int J Hydrogen Energy 22:979–987

180.   Zaborsky OR (1998) Biohydrogen, Plenum Press, New York

181.   Benemann JR, Miyamoto K, Hallenbeck PC (1980) Bioengineering aspects of biophysics. Enzyme Microb. Technol. 2:103–111

182.   Adams DG (2000) Heterocyst formation in cyanobacteria. Curr Op Microbiol 3:618–624

183.   Weissman JC, Benemann JR (1977) Hydrogen production by nitrogen-starved cultures of Anabaena Cylindrica. Appl Environ Microbiol 33:123–131

184.   Benemann JR, Weare NM (1974) Hydrogen evolution by nitrogen-fixing Anabaena cylindrical cultures. Science 184:174–175

185.   Hansel A, Linblad P (1998) Towards optimization of cyanobacteria as biotechnologically relevant producers of molecular hydrogen, a clean and renewable energy source. Appl Microbiol Biotechnol 50:153–160

186.   Smith GD, Ewart GD, Tucker W (1992) Hydrogen-Production by Cyanobacteria. Int J Hydrogen Energy 17:695–698

187.   Levin DB, Pitt L, Love M (2004) Biohydrogen production: prospects and limitations to practical application. Int J Hydrogen Energy 29:173–185

188.   Asada Y, Kawamura S (1986) Aerobic hydrogen accumulation by a nitrogen-fixing cyanobacterium. Anabaena sp. Appl Envrion Microbiol 51:1063–1066

189.   Benemann JR (1994) Photobiological Hydrogen Production. Intersociety Energy Conversion Engineering Conference Proceedings, pp. 1636–1640

190.   Melis A, Zhang L, Forestier M, Ghirardi ML, Seibert M (2000) Sustained photobiological hydrogen gas production upon reversible inactivation of oxygen evolution in the green alga Chlamydomonas reinhardtii. Plant Physiol 122:127–135

191.   Ghirardi ML, Zhang L, Lee JW, Flynn T, Seibert M, Greenbaum E, Melis A (2000) Microalgae: a green souce of renewable H_2. TIBTECH 18:506–511

192. Melis A (2002) Green alga hydrogen production: progress, challenges and prospects Int J Hydrogen Energy 27:1217–1228

193. Antal TK, Krendeleva TE, Laurinavichene TV, Makarova VV, Ghirardi ML, Rubin AB, Tsygankov AA, Seibert M (2003) The dependence of algal H_2 production on photosystem II and O_2 consumption activities in sulfur-deprived *Chlamydomonas reinhardtii* cells. Biochim Biophys Acta 1607:153–160

194. Kosourov S, Tsygankov A, Seibert M, Ghirardi ML (2002) Sustained hydrogen photoproduction by *Chlamydomonas reinhardtii*: effects of culture parameters. Biotechnol Bioeng 78:731–740

195. Guan YF, Deng MC, Yu XJ, Zhang W (2004) Two-stage photo-biological production of hydrogen by marine green alga Platymonas subcordiformis. Biochem Eng J,19:69–73

196. Greenbaum E (1998) Energetic efficiency of hydrogen photoevolution by algal water splitting. Biophys J 54:365–368

197. Bergene T (1996)The efficiency and physical principles of photolysis of water by microalgae Int J Hydrogen Energy 21:89–194

198. Hall DO (1978) Solar energy conversion through biology-could it be practical energy source? Fuel 57:322–333

199. Herron HA, Mauzerall D (1972) The development of photosynthesis in a greening mutant of Chlorella and an analysis of the light saturation curve. Plant physiology 50:141–148

200. Masukawa H, Mochimaru M, Sakurai H (2002) Hydrogenases and photobiological hydrogen production utilizing nitrogenase system in cyanobacteria. Int J Hydrogen Energy 27:1471-1474

201. Markov SA, Thomas AD, Bazin MJ, Hall DO (1997) Photoproduction of hydrogen by cyanobacteria under partial vacuum in batch culture or in a photobioreactor. Int J Hydrogen Energy 22:521–524

202.   Modigell M, Holle N (1998) Reactor development for a biosolar hydrogen production process. Renewable Energy 14:421–426

203.   Hoekema S, Bijmans M, Janssen M, Tramper J, Wijffels RH (2002) A pneumatically agitated flat-panel photobioreactor with gas re-circulation: anaerobic photoheterotrophic cultivation of a purple non-sulfur bacterium. Int J Hydrogen Energy 27:1331–1338

204.   Miyake J, Miyake M, Asada Y (1999) Biotechnological hydrogen production: research for efficient light energy conversion. J Biotechnol 70:89–101

205.   Ikeda S, Takata T, Kondo T, Hitoki G, Hara M, Kondo JN, Domen K, Hosono H, Kawazoe H, Tanaka A (1998) Mechano-catalytic overall water splitting, Chem Commun 2185–2186

206.   Ikeda S, Takata T, Komoda M, Hara M, Kondo JN, Domen K, Tanaka A, Hosono H, Kawazoe H (1999) Mechano-catalysis-a novel method for overall water splitting. Phys Chem Chem Phys 1:4485–4491

207.   Takata T, Ikeda S, Tanaka A, Hara M, Kondo JN, Domen K (2000) Mechano-catalytic overall water splitting on some oxides (II). Appl Catal A: Gen 200:255–262

208.   Domen K, Ikeda S, Takata T, Tanaka A, Hara M, Kondo JN (2000) Mechano-catalytic overall water-splitting into hydrogen and oxygen on some metal oxides. Appl Energy 67:159–179

209.   Hitoki G, Takata T, Ikeda S, Hara M, Kondo JN, Kakihana M, Domen K (2001) Mechano-catalytic overall water splitting on some mixed oxides. Catal Today 63:175–181

210.   Hara M, Hasei H, Yashima M, Ikeda S, Takata T, Kondo JN, Domen K (2000) Mechano-catalytic overall water splitting (II) nafion-deposited Cu_2O. Appl Catal A: Gen 190:35–42

211.   Ross DS (2004) Comment on "A study of Mechano-Catalysis for overall water splitting". J Phys Chem B 108:19076–19077

212.   (1998) Mechano-catalytic water splitting claimed, Chemical & Engineering News 76:36

213. Hara M, Domen K (2004) Reply to "Comment on A study of mechano-catalysts for overall water splitting. J Phys Chem B 108:19078

214. Ohta T (2000) Preliminary theory of mechano-catalytic water-splitting. Int J Hydrogen Energy 25:287–293

215. Ohta T (2000) On the theory of mechano-catalytic water-splitting system. Int J Hydrogen Energy 25:911–917

216. Ohta T (2000) Mechano-catalytic water-splitting. Appl Energy 67:181–193

217. Ohta T (2000) Efficiency of mechano-catalytic water-splitting system. Int J Hydrogen Energy 25:1151–1156

218. Ohta T (2001) A note on the gas-evolution of mechano-catalytic water splitting system. Int J Hydrogen Energy 26:401

219. Suib SL, Brock SL, Marquez M, Luo J, Matsumoto H, Hayashi Y (1998) Efficient catalytic plasma activation of CO_2, NO and H_2O. J Phys Chem B 102:9661–9666

220. Luo J, Suib SL, Hayashi Y, Matsumoto H (2000) Water splitting in low-temperature AC plasmas at atmospheric pressure. Res Chem Intermed 26:849–874

221. Luo J, Suib SL, Hayashi Y, Matsumoto H (1999) Emission spectroscopic studies of plasma-induced NO decompostion and water spltting. J Phys Chem A 103 (1999) 6151–6161

222. Givotov VK, Fridman AA, Krotov MF, Krasheninnikov EG, Patrushev BI, Rusanov VD, Sholin GV (1981) Plasmochemical methods of hydrogen production, Int J Hydrogen Energy 6:441–449

223. Bockris JOM, Dandapani B, Cocke D, Ghoroghchian J (1985) On the splitting of water. Int J Hydrogen Energy 10:179–201

224. Chen X, Suib SL, Hayashi Y, Matsumoto H (2001) H_2O splitting in tubular PACT (Plasma and catalyst integrated technologies) reactors. J Catal 201:198–205

224. Kabashima H, Einaga H, Futamura S (2003) Hydrogen evolution from water, methane and methanol with nonthermal plasma. IEEE Transactions on Industry Applications 39:340–345

225. Chen X, Marquez M, Rozak J, Marun C, Luo J, Suib SL, Hayashi Y, Matsumoto H (1998) H_2O splitting in tubular plasma reactors. J Catal 178:372–377

226. Ghoroghchian J, Bockris JOM (1985) Use of a homopolar generator in hydrogen production from water. Int J Hydrogen Energy 10:101–112

227. Harteck P, Dondes S (1956) Producing chemicals with reactor radiations. Nucleonics 14:22–25

228. Daniels M, Wigg E (1966) Oxygen as a primary species in radiolysis of water. Science 153:1533–1534

229. Wojcik DS, Buxton GV (2005) On the possible role of the reaction $H^{\bullet}+H_2O \rightarrow H_2+ {}^{\bullet}OH$ in the radiolysis of water at high temperatures. Rad Phys Chem 74:210–219

230. Sunaryo GR, Katsumura Y, Ishigure K (1995) Radiolysis of water at elevated temperatures-III. Simulation of radiolytic products at 25 and 250°C under the irradiation with γ rays and fast neutrons. Rad Phys Chem 45:703–714

231. Gervais B, Beuve M, Olivera GH, Galassi ME (2006) Numerical simulation of multiple ionization and high LET effects in liquid water radiolysis, Radiation Physics and Chemistry 75:493-513

232. Katsumura Y, Sunaryo G, Hiroishi D, Ishiqure K (1998) Fast neutron radiolysis of water at elevated temperatures relevant to water chemistry. Prog Nucl Energy 32:113–121

233. Cecal A, Goanta M, Palamaru M, Stoicescu T, Popa K, Paraschivescu A, Anita V (2001) Use of some oxides in radiolytical decomposition of water. Rad Phys Chem 62:333–336

234. Sawasaki T, Tanabe T, Yoshida T, Ishida R (2003) Application of gamma radiolysis of water for H_2 production. J Radioanal Nucl Chem 255:271–274

235. Petrik NG, Alexandrov AB, Vall AI (2001) Interfacial energy transfer during gamma radiolysis of water on the surface of ZrO_2 and some other oxides. J Phys Chem B 105:5935–5944

236. Laverne JA (2005) H_2 formation from the radiolysis of liquid water with zirconia, J Phys Chem B 109:5395–5397

237. Yamamoto TA, Seino S, Katsura M, Okitsu K, Oshima R, Nagata Y (1999) Hydrogen gas evolution from alumina nanoparticles dispersed in water irradiated with gamma ray. Nanostructured Mater 12:1045

238. Nakashima M, Masaki NM (1996) Radiolytic hydrogen gas formation from water adsorbed on type Y zeolytes. Rad Phys Chem 46:241–245

239. LaVerne JA, Tandon L (2002) H_2 produced in the radiolysis of water on CeO_2 and ZrO_2. J Phys Chem B 106:380–386

240. Caer SL, Rotureau P, Brunet F, Charpentier T, Blain G, Renault JP, Mialocq JC (2005) Radiolysis of confined water: Hydrogen production at a high dose rate. Chem Phys Chem 6:2585-2596

Chapter 3

PHOTOELECTROLYSIS

3.1 General Description of Photoelectrolysis

Electrolysis is a process of detaching or dissociating bonded elements and compounds by passing through them an electric current. Water electrolysis decomposes H_2O into hydrogen and oxygen gas. Care must be taken in choosing the correct electrolytes, nominally substances that contain free ions and hence behave as an electrically conductive medium. Electrolytes dissolve and dissociate into cations (positive ions, +) and anions (negative ions, −) that carry the current. As we have seen in Chapter 2, such processes can occur in an electrolysis cell, or electrolyzer, which consists of two electrodes, cathode and anode, where reduction and oxidation reactions simultaneously take place forming H_2 (at the cathode) and O_2 (at the anode). The fundamental problem in hydrogen production by water electrolysis is that today the electricity used to drive the process is primarily generated by the burning of fossil fuels.

Photoelectrolysis describes electrolysis by the direct use of light; that is to say, the conversion of light into electrical current and then the transformation of a chemical entity (H_2O, H_2S, etc.) into useful chemical energy (such as H_2) using that current. A photoelectro-chemical cell is used to carry out the various photoelectrolytic reactions, being comprised of a semiconductor device that absorbs solar energy and generates the necessary voltage to split water molecules. Photoelectrolysis integrates solar energy collection and water electrolysis into a single photoelectrode, and is considered the most efficient renewable method of hydrogen production. Our interest in hydrogen stems from it being an energy source that, like fossil fuels, are energy dense and can be readily transported and stored, but unlike fossil fuels is not of finite supply and its combustion does not result in pollution nor the release of climate altering gases.

3.1.1 Photoelectrolysis and the Hydrogen Economy

It appears the term "Hydrogen Economy" became part of the common vernacular in 1974 during the first international conference on hydrogen energy in Miami, Florida signifying the concept of a renewable, non-polluting energy infrastructure based on hydrogen [1]. The underlying premise of a hydrogen economy is the ability to renewably, cleanly, and efficiently produce hydrogen. Photoelectrolysis is a single step process in which sunlight is absorbed by a semiconductor, with the resulting photo-generated electron-hole pair splitting water into hydrogen and oxygen. At present non-renewable hydrogen production methods, such as steam reforming of methane (SMR), are less expensive than photoelectrolysis. Of course SMR results in CO_2 emissions, and is ultimately limited by the finite reserves of fossil fuels. In contrast, water photoelectrolysis does not result in CO_2 emissions, sunlight and water can be considered inexhaustible resources, and as solar-to-hydrogen efficiencies increase it can be expected that the intrinsic costs will continue to decrease. For example the relatively recent advances in nanotechnology have given the scientific community an opportunity to design and synthesize specific semiconductor nanostructures with previously unseen properties. As illustrated by **Fig. 3.1**, photoelectrolysis offers the world a permanent energy solution, one that is both sustainable and pollution free.

3.1.2 Background and Perspectives: Artificial Photosynthesis

The atmosphere of earth is largely composed of nitrogen gas (78%), with other major gases including oxygen (21%), argon (0.93%), and carbon dioxide (0.04%). Plants and some bacteria release oxygen through a process called *Photosynthesis* [2,3]. Photosynthesis, see reaction 3.1.1, is an efficient method of transforming solar energy into chemical energy in the form of starch or sugar.

$$6CO_2 + 12H_2O + \text{solar light} \rightarrow C_6H_{12}O_6 + 6O_2 + 6H_2O \qquad (3.1.1)$$

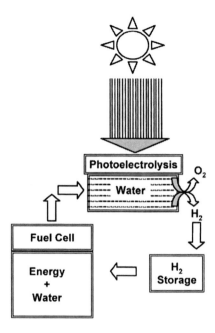

Fig. 3.1: Solar hydrogen production via photoelectrolysis, with enormous potential for providing a renewable and clean energy carrier.

Photosynthesis is the way a plant makes fuel, that is to say useful energy, for itself. There are two types of photosynthetic reaction centers, Photosystem I and Photosystem II, that signify the arrangement of *chlorophyll a* and other pigments packed in a thylakoid membrane. Photosystem I absorbs light at 700 nm and is commonly referred to as P_{700}, while Photosystem II absorbs light at 680 nm and is commonly referred to as P_{680}. Upon activation by solar light an electron is removed from P_{680}, making it sufficiently electronegative to withdraw electrons from water. This electron travels via a cascade of electron carriers to Photosystem I, where $NADP^+$ (oxidized nicotinamide adenine dinucleotide) is reduced to NADPH (reduced nicotinamide adenine dinucleotide). This process generates a redox potential, or energy rich state, across the thylakoid membrane. This potential helps drive the hydrogen ion through the protein channels leading to generation of ATP (adenosine triphosphate) from ADP (adenosine diphosphate). The outcome of this electron transport is that water is split into oxygen gas and hydrogen ions. Although the primary photoredox reactions in the chloroplast proceed with high quantum efficiency it saturates at modest light intensity, leading to an overall peak illumination photochemical conversion yield of about $\approx$ 6%. Taking into account seasonal variations, i.e. winter, and its affect on photosynthesis efficiency the result is that biological systems, in comparison to solar cells, are a rather low efficiency method of converting solar energy [4-6].

The idea of constructing an artificial device capable of converting solar energy by mimicking the natural photosynthesis conversion of sunlight into a useful energy is a major driving force in artificial photosynthesis research. Ideally, we seek efficient light induced reactions to split water into molecular oxygen and hydrogen, a process often referred to as artificial photosynthesis [7]. One wishes to be able to mimic the electrochemical energy conversion achievable in photosynthesis, illustrated in **Fig. 3.2a** [7-10]. When light falls on Chlorophyll P_{680} an electron moves from water to the acceptor plastquinone (Q) via pheophytin (Q_A) and secondary quinonone (Q_B) making, in combination with CO_2, sugar (carbohydrate). Electron transport is aided by the presence of four Manganese (Mn) metal atoms in Photosystem II. The photo-ejected electron from P_{680} is replenished by taking one from the Mn cluster through the redox active tyrosine linkage (or mediator), which in turn extracts an electron from water. P_{680} generates an oxidizing potential of + 1.2 V, sufficient to overcome the energy barrier required to oxidize water to molecular oxygen (0.87 V, pH = 6).

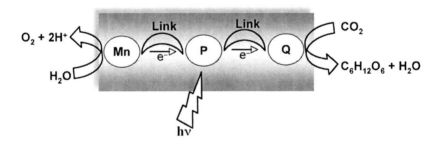

Fig. 3.2a: Electron transport in (natural) photosynthesis. P = chlorophyll that acts as a light sensitizer, from which a photogenerated electron travels to Q = Plastquinone that in combination with CO_2 forms a carbohydrate. The photo-ejected electron from P_{680} is replenished by taking one from the Mn cluster through the redox active tyrosine linkage (or mediator), which in turn extracts an electron from water.

Photosynthesis is a form of photoelectrolysis [11], with the molecular arrangement in the thylakoid membranes creating a photovoltage, resulting in water oxidation on one end of the membrane and reduction on the other. Attempts to successfully achieve artificial photosynthesis via a robust, durable and photoactive supermolecule capable of efficient electron transport, and thus water splitting, have yet to be realized, see **Fig. 3.2b**. The efficiency of artificial systems are currently limited by the poor light absorption achievable in a thin layer of photoactive molecules, while a thin layer is needed to prevent energy losses in electron transport through a membrane. This is in contrast to the thylakoid membranes, in which the P_{680} and P_{700} systems are embedded, that fold upon each other in a disk-like stack resulting in a structure that efficiently harvests sunlight while preventing back reactions.

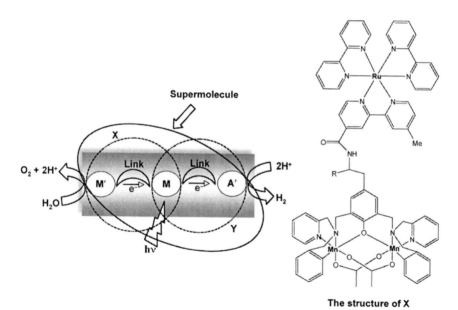

The structure of X

Fig. 3.2b: Electron transport in an artificial photosynthesis scheme. M = light sensitizer, M' = a water oxidation site, and A' = a reduction site.

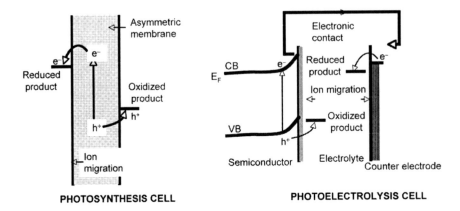

Fig. 3.3: A comparison between photosynthesis and photoelectrolysis in terms of electron transfer reactions.

Figure 3.3 illustrates the relationship between photosynthesis and photoelectrolysis in terms of the redox energy for water splitting. In a photosynthesis cell, both electronic and ionic currents pass through the membrane in parallel. Semiconductor electrodes form the basis of water splitting by photoelectrolysis, with a counter electrode collecting one of the charge carriers generated by an illuminated semiconductor.

3.2 Photoelectrochemical Cells

A cell that can convert light energy into a more useful energy product through light-induced electrochemical processes is commonly known as a *Photoelectrochemical Cell* (PEC). In a photoelectrochemical cell current and voltage are simultaneously produced upon absorption of solar light by one or more of the electrodes, with at least one of the electrodes a semiconductor. The output product is either electrical or chemical energy. Some PECs have been used to produce harmless chemicals from hazardous wastes [12-16].

A typical PEC, as depicted in **Fig. 3.4a** for water splitting, consists of three electrodes immersed in an electrolyte solution; namely the working electrode (WE) or anode, counter electrode (CE) or cathode, and reference electrode (RE). The working electrode, usually a semiconductor, is also called the photoelectrode or photoanode due to the light induced chemical reactions initiated at its surface. The counter electrode or cathode used for this cell is typically a corrosion resistant metal, commonly platinum, to prevent dissolution products from the counter electrode contaminating the solution.

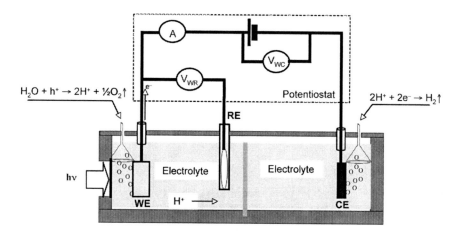

Fig. 3.4a: Schematic representation of a conventional three-electrode photoelectrochemical cell showing: WE = working electrode, RE = reference electrode, CE = counter electrode. If the working electrode is an n-type semiconductor and the counter electrode is a metal, then oxygen evolution occurs at the WE and hydrogen evolution occurs at the CE.

In a common photoelectrochemical cell design, an n-type semiconductor photoanode is used that upon illumination evolves oxygen, while hydrogen is evolved at a metal cathode. Photoelectrochemical cells of other configurations can also be prepared. A p-type semiconductor cathode, or photocathode, reduces the H^+ ion into H_2 upon solar light irradiation, while oxygen is evolved at a metal anode. In another photoelectrochemical cell design both electrodes are comprised of photoactive semiconductors. In this case, the n-type electrode will act as a photo-anode for water oxidation and release of the H^+ ion; the p-type electrode will act as the photo-cathode, where H^+ ions are reduced to H_2. A standard calomel electrode filled with a saturated KCl solution is commonly used as a reference electrode for these cells when an external bias is needed. During use KCl solution from the electrode flows extremely slowly into the cell through a fine channel in the base of the electrode, providing electrical contact to the cell for characterization without reference electrode contamination. The level of the KCl solution in the electrode needs to remain higher than that of the electrolyte solution in the cell.

3.2.1 Water Splitting

As shown in **Fig. 3.4b**, when a semiconductor electrode is illuminated with photons having an energy hv equal to or larger than the semiconductor bandgap the result is formation of electronic charge carriers, electrons in the conduction band and holes in the valence band, see equation (3.2.1).

$$2h\nu + \text{Semiconductor} \rightarrow 2h^+ + 2e^- \tag{3.2.1}$$
$$2h^+ + H_2O\ (l) \rightarrow \tfrac{1}{2}\ O_2(g) + 2H^+ \tag{3.2.2}$$
$$2H^+ + 2e^- \rightarrow H_2(g) \tag{3.2.3}$$

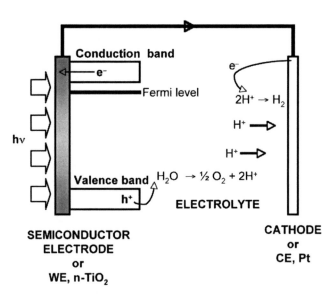

Fig. 3.4b: Illustration of the operating principle of a photoelectrochemical cell producing hydrogen and oxygen during water photoelectrolysis.

At the interface between the photoelectrode and electrolyte, the photogenerated holes h^+ react with water to form oxygen and hydrogen ions H^+. Gaseous oxygen is evolved at the photo-electrode, and the resulting hydrogen ions travel through the aqueous electrolyte (referred to as the internal circuit) to the cathode, equation (3.2.2). At the same time the photogenerated electrons, transferred through the external circuit to the cathode, react with hydrogen ions at the cathode - electrolyte interface reducing the hydrogen ions to gaseous hydrogen, equation (3.2.3).

Under standard conditions water can reversibly electrolyze at a potential of 1.23 V, a value derived from the following relationship;

$$\Delta G^0 = -nF \cdot \Delta E^0 \qquad (3.2.4)$$

where ΔG^0 and ΔE^0 are standard Gibbs free energy change and standard electric potential of the reaction. In any redox reaction, the energy released in a reaction due to movement of charged particles gives rise to a potential difference. The maximum potential difference is called the electromotive force (emf or ΔE). The overall reaction of a photoelectrochemical cell, expressed in the following form

$$H_2O + h\nu \rightarrow H_2 + 1/2O_2 \qquad (3.2.5)$$

can take place when the energy of the photon absorbed by the working electrode is equal to or larger than the water splitting threshold energy of 1.23 eV. Equation (3.2.5) is an endothermic process and involves a change in the Gibbs free energy, equation (3.2.4), which is the negative value of maximum electric work corresponding to 237.14 kJ/mol or 2.46 eV for equation (3.2.5). Since this is a two-electron redox process electrochemical decomposition of water is possible when the cell emf is equal to or greater than 1.23 V.

3.3 Types of Photoelectrochemical Devices

There are three general types of photoelectrochemical devices using semiconductor electrodes for the conversion of water into hydrogen [17-54].

3.3.1   Photoelectrolysis Cell [26-34]

In this type of cell both electrodes are immersed in the same constant pH solution.   An illustrative cell is [27,28]: n-SrTiO$_3$ photoanode|9.5-10 M NaOH electrolyte|Pt cathode. The underlying principle of this cell is production of an internal electric field at the semiconductor-electrolyte interface sufficient to efficiently separate the photogenerated electron-hole pairs. Subsequently holes and electrons are readily available for water oxidation and reduction, respectively, at the anode and cathode. The anode and cathode are commonly physically separated [31-34], but can be combined into a monolithic structure called a photochemical diode [35].

3.3.2   Photo-assisted Electrolysis Cell [36-48]

These cells operate under illumination in combination with a bias, which serves to either drive electrolytic reactions for which the photon energy is insufficient or to increase the rate of chemical energy conversion by reducing electron-hole recombination in the semiconductor bulk. Most commonly an electrical bias is provided to drive the reactions [36-41].

Chemically biased photo-assisted photoelectrolysis cell [42-45]

A chemical bias is achieved by using two different electrolytes placed in two half-cells, with the electrolytes being chosen to reduce the voltage required to cause the chemical splitting. An n-TiO$_2$ photoanode|4M KOH||4M HCl|Pt-cathode is one example of a chemically biased photoelectrochemical cell [44].

Dye sensitized photoelectrolysis cell [46-48]

This cell involves the absorption of light by dye molecules spread on the surface of the semiconductor, which upon light absorption will inject electrons into the conduction band of the n-type semiconductor from their excited state. The photo-oxidized dye can be used to oxidize water and the complementary redox process can take place at the counter electrode [46,47]. Tandem cells such as these are discussed in Chapter 8.

3.3.3   Photovoltaic Electrolysis Cell [48-54]

This cell employs a solid state photovoltaic to generate electricity that is then passed to a commercial-type water electrolyzer (see Chapter 2). An alternative system involves the semiconductor photovoltaic cell configured as a monolithic structure and immersed directly in the aqueous solution, see Chapter 8; this cell involves a solid-state p-n or schottky junction to produce the required internal electric field for efficient charge separation and the production of a photovoltage sufficient to decompose water [49-51].

3.4 Photoelectrolysis Principles

Photoelectrolysis is generally carried out in cells having similar configuration as electrolysis cells (discussed in Chapter 2) with at least one of the two electrodes comprised of a semiconductor material. Upon exposure to sunlight the semiconductor electrode, called photoelectrode, immersed in an aqueous electrolyte solution generates, in an ideal case, enough electrical energy to drive the oxygen and hydrogen evolution reactions respectively at the interfaces of anode and cathode within the electrolyte. A necessary condition for such a spontaneous water splitting process upon illumination is that the semiconductor conduction band edge should lie at a position more negative (NHE as reference) relative to the reduction potential of water while the valence band edge more positive compared to the oxidation potential. However, in many material-electrolyte systems the conduction band edge is located close to or more positive relative to the reduction potential of water. Such a situation in the case of titania and an aqueous electrolyte of pH=1 is depicted in **Fig. 3.5**. In most cases the photovoltage developed between the electrodes is less than 1.23V, the minimum voltage required for water splitting. For example, a widely studied photoanode material rutile TiO_2 [36-40,42-44], 3.0 eV bandgap, generates photovoltages of only 0.7-0.9 V under solar light illumination. Hence water splitting can be effectively performed only with the assistance of an external electrical bias or internal chemical bias (by creating anode and cathode compartments with different hydrogen ion concentrations).

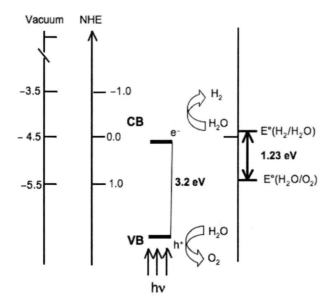

Fig. 3.5: Band position of anatase TiO$_2$, bandgap = 3.2 eV, in the presence of a pH = 1 aqueous electrolyte. The energy scale is indicated in electron volts (eV) using either normal hydrogen electrode (NHE) or vacuum level as reference showing the condition for water splitting.

For meaningful photoelectrochemical decomposition of water to occur three essential requirements must be met [11,17-20,55-65]. First, the conduction and valence band edges of the semiconductor materials must overlap, the energy levels of the hydrogen and oxygen reduction reactions, **see Fig. 3.5**. Second, the semiconductor system must be stable under photoelectrolysis conditions. Third, charge transfer from the surface of the semiconductor must be fast enough to prevent corrosion and also reduce energy losses due to overvoltage or overpotential.

3.4.1 Energy Levels in Semiconductors and Electrolytes

The electronic structure of the semiconductor electrodes is usually described in terms of energy bands that can effectively be considered a continuum of energy levels due to the small difference in energy between adjacent molecular orbitals [66,67]. The highest energy band comprised of occupied molecular orbitals is called the valence band, with its upper energy level denoted as E_{VB}. The lowest energy band comprised of unoccupied molecular orbitals, or empty energy states, is called the conduction band with its lower edge denoted as E_{CB}. The difference in energy between the upper edge of valence band and lower edge of the conduction band is the bandgap energy E_{BG}. Electrons can be thermally or photochemically excited from the valence band to the conduction band with the transfer of an electron, e⁻, leaving a positively charged vacancy in the valence band which is referred to as a hole, h⁺. Holes are considered mobile since holes are created by the migration of electrons. The Fermi energy level is defined as the energy level at which the probability of occupation by an electron is one-half. The Fermi energy level can be calculated by the following formula

$$E_F = \frac{1}{2}\left(E_{VB} + E_{CB}\right) + \frac{1}{2}\left(kT \ln \frac{N^*_{VB}}{N^*_{CB}}\right) \tag{3.4.1}$$

Where E_{CB} and E_{VB} are, respectively, the energy levels of the conduction and valence band edges, k (1.38×10^{-23} J/K) is the Boltzmann constant, and T (Kelvin scale, K) is the temperature. N^*_{VB} and N^*_{CB} are the effective density of (energy) states function in the valence and conduction bands

$$N^*_{VB} = 2\left[\frac{2\pi m^*_h kT}{h^2}\right]^{3/2} \tag{3.4.2}$$

$$N^*_{CB} = 2\left[\frac{2\pi m^*_e kT}{h^2}\right]^{3/2} \tag{3.4.3}$$

in which h is Plank's constant, and m^*_h and m^*_e are, respectively, the effective masses of holes and electrons that takes into account their ability to move through the atomic lattice [67].

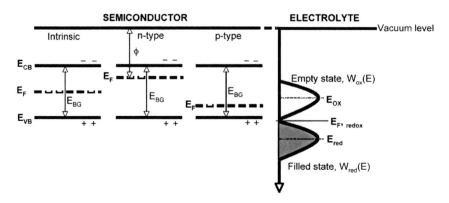

Fig. 3.6: A schematic representation of semiconductor energy band levels and energy distribution of the electrolyte redox system.

As shown in **Fig. 3.6**, for intrinsic (undoped) semiconductors the number of holes equals the number of electrons and the Fermi energy level E_F lies in the middle of the band gap. Impurity doped semiconductors in which the majority charge carriers are electrons and holes, respectively, are referred to as n-type and p-type semiconductors. For n-type semiconductors the Fermi level lies just below the conduction band, whereas for p-type semiconductors it lies just above the valence band. In an intrinsic semiconductor the equilibrium electron and hole concentrations, n_0 and p_0 respectively, in the conduction and valence bands are given by:

$$n_0 = N_{CB}^* \cdot e^{\left(-\frac{E_{CB}-E_F}{kT}\right)} \qquad (3.4.4)$$

$$p_0 = N_{VB}^* \cdot e^{\left(-\frac{E_F-E_{VB}}{kT}\right)} \qquad (3.4.5)$$

Multiplying equations (3.4.4) and (3.4.5) an equilibrium concentration can be expressed as [20]:

$$n_0 p_0 = N_{CB}^* N_{VB}^* \cdot e^{\left(-\frac{E_{CB}-E_{VB}}{kT}\right)} = n_i^2 \qquad (3.4.6)$$

n_i^2 is the intrinsic carrier concentration which exponentially decreases with increasing bandgap. Electron and hole concentrations can also be obtained as a function of donor and acceptor impurity concentrations. At equilibrium

$$n_0 = \frac{N_d}{2} + \sqrt{\left(\frac{N_d}{2}\right)^2 + n_i^2} \qquad (3.4.7)$$

$$p_0 = \frac{N_a}{2} + \sqrt{\left(\frac{N_a}{2}\right)^2 + n_i^2} \qquad (3.4.8)$$

From equations (3.4.4) and (3.4.5) one can determine the energy difference between the energy band edges and the Fermi level [67].

$$E_{CB} - E_{F,n} = kT \ln \frac{N_{CB}^*}{n} \qquad (3.4.9)$$

$$E_{F,p} - E_{VB} = kT \ln \frac{N_{VB}^*}{p} \qquad (3.4.10)$$

For an n-type semiconductor, if the donor impurity concentration is much greater than the intrinsic carrier concentration, $N_d >>> n_i$, then $n_0 \approx N_d$. Equation (3.4.9) can then be written as

$$E_F = E_{CB} - kT \ln \frac{N_{CB}^*}{N_d} \qquad (3.4.11)$$

From equation (3.4.11) we see that the energy gap between the conduction band edge and the Fermi energy level is a logarithmic function of donor concentration. As the donor concentration increases so does the electron concentration in the conduction band, with the Fermi level energy moving closer to the conduction band edge.

Similarly, one can derive an equation for a p-type semiconductor, where the distance between the Fermi energy level and the valence band is a logarithmic function of acceptor impurity concentration. As the acceptor impurity increases so too does the hole concentration in valence band, with the Fermi level moving closer to the valence band.

$$E_F = E_{VB} + kT \ln \frac{N_{VB}^*}{N_a} \qquad (3.4.12)$$

The Fermi energies are related to the electrochemical potential of electrons and holes, which are usually given with respect to a reference electrode, commonly the normal hydrogen electrode (NHE) or Standard Calomel electrode (SCE). Considering a simple redox couple, to correlate energy positions in relation to the electrolyte the electrode potential (V) must be converted into the free energy of electrons at the same electrostatic potential. The electrochemical potential of electrons in a redox system is equivalent to the Fermi level, $E_{F,redox}$ on an absolute scale [68]. Hence the electrochemical potential of a redox system, usually given with respect to NHE, is described by the following relationship:

$$E_{F,redox} = -eV_{redox} + const_{ref} \qquad (3.4.13)$$

Where V_{redox} is the redox potential *vs* NHE and $const_{ref}$ is the free energy of the electrons in the reference electrode with respect to vacuum level. Since the electrochemical scale is arbitrarily based on a reference electrode the connection between these two electrodes is given by a work function (ϕ) for the removal of an electron from the Fermi level of the reference electrode to the vacuum level. For NHE, the constant has a value between -4.5 to -4.7 eV. Thus equation (3.4.13) can be written as

$$E_{F,redox} = -4.5eV - eV_{redox} \qquad (3.4.14)$$

with respect to vacuum level [69,70].

Figure 3.6 shows the various relationships between the energy levels of solids and liquids. In electrolytes three energy levels exist, $E_{F,\ redox}$, E_{ox} and E_{red}. The energy levels of a redox couple in an electrolyte is controlled by the ionization energy of the reduced species E_{red}, and the electron affinity of the oxidized species E_{ox} in solution in their most probable state of solvation; due to varying interaction with the surrounding electrolyte, a considerable fluctuation in their energy levels occurs. The standard redox potential is an average of the ionization energy and the electron affinity. This energy level is attained with equal probability by fluctuations of the ionization energy of the reduced, and the electron affinity of the oxidized, species. As illustrated on the right side of **Fig. 3.6**, the Gaussian distribution of E_{ox} and E_{red} is a consequence of these energy level fluctuations of redox species in solution; for each single redox species a probability distribution of energy states can be described as follows;

$$W_{ox}(E) = e^{\left(-\frac{(E_{ox}-E)^2}{4kT\lambda}\right)} \tag{3.4.15}$$

$$W_{red}(E) = e^{\left(-\frac{(E_{red}-E)^2}{4kT\lambda}\right)} \tag{3.4.16}$$

The most probable energy levels E_{ox} and E_{red} are connected with the standard redox potential or by the standard redox Fermi level by the following symmetrical relation:

$$E_{red} = E_{F,redox} - \lambda \tag{3.4.17}$$

$$E_{ox} = E_{F,redox} + \lambda \tag{3.4.18}$$

where λ is the reorganization energy, defined as the energy needed to bring the solvation shell of one redox species from its most probable state into the most probable solvation structure of its redox counter part. The width of the distribution function in equations (3.4.15) and (3.4.16) is controlled by the reorganization energy λ.

3.4.2 The Semiconductor-Electrolyte Interface

Band bending at the interface

Helmholtz [71] first described the interfacial behavior of a metal and electrolyte as a capacitor, or so-called "electrical double layer," with the excess surface charge on the metallic electrode remaining separated from the ionic counter charge in the electrolyte by the thickness of the solvation shell. Gouy and Chapmen subsequently developed a diffuse ionic double layer model, in which the potential at the surface decreases exponentially due to chemisorbed counter ions from the electrolyte solution. Thus the movement of the ionic counter charges of an electrolyte, near the surface of a metal electrode, makes them lose part of their solvation shell [72,73].

Unlike metals, semiconductors do not possess high conductivity hence diffuse ionic double layer models can be used to describe the interfacial properties between a semiconductor electrode and a liquid [11,17-20,55-65,74-77]. Three different situations are depicted for n-type, **Fig. 3.7 (a-c)**, and p-type, **Fig. 3.7 (d-f)**, semiconductors. When a semiconductor photoelectrode is brought into contact with an electrolyte solution the excess charge does not lie on the surface, but rather extends into the electrode for approximately 1 μm; this region is called the space charge region (or space charge layer), and has an associated electric field. Charge transfer from the semiconductor to electrolyte leads to the formation of surface charge, which is then compensated by a charge of opposite sign induced in the electrolyte within a localized layer known as the Helmholtz layer. The Helmholtz layer is formed by oriented water molecule dipoles and adsorbed electrolyte ions at the electrode surface. The unusual charge distribution results in band bending at the semiconductor-electrolyte interface. For an electrolyte-immersed n-type semiconductor a depletion layer forms where the region is depleted of electrons, leaving a net positive charge balance behind that is compensated for at the interface by negative counter ions from the electrolyte, **Fig. 3.7(a)**. Similarly, an accumulation layer forms when the negative excess charge (electrons) of a semiconductor accumulates at the interface, which is compensated by the positive ions of an electrolyte, **Fig. 3.7(b)**. **Fig. 3.7(c)** shows an interface where no net excess charge on a semiconductor is observed, hence the bands are flat, reflecting the potential of zero charge for photoelectrode; this potential is called the flat band potential, V_{FB}.

For p-type semiconductors, an accumulation layer forms when excess positive charge (holes) accumulate at the interface, which is compensated by negative ions of an electrolyte, **Fig. 3.7(d)**. Similarly, a depletion layer forms when the region containing negative charge is depleted of holes, and thus positive counter ions from the electrolyte compensates charge at the interface, **Fig. 3.7(e)**. When no net excess charge is observed on an interface there is no band bending, reflecting the zero charge potential of the photoelectrode, **Fig. 3.7(f)**; this potential is called the flat band potential, V_{FB}.

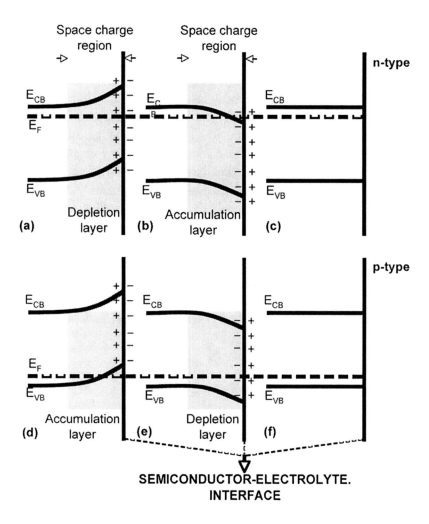

Fig. 3.7: Band bending at semiconductor-electrolyte interface.

Equilibrium between the two phases at a semiconductor-electrolyte interface, solid and liquid, can only be achieved if their electrochemical potential is the same, that is:

$$E_F = E_{F,\,redox} \tag{3.4.19}$$

The electrochemical potential of the solution and semiconductor, see **Fig. 3.6**, are determined by the standard redox potential of the electrolyte solution (or its equivalent the standard redox Fermi level, $E_{F,redox}$), and the semiconductor Fermi energy level. If these two levels do not lie at the same energy then movement of charge across the semiconductor - solution interface continues until the two phases equilibrate with a corresponding energy band bending, see **Fig. 3.8**.

For an n-type semiconductor electrode the Fermi level is typically higher than the redox potential of the electrolyte, hence electrons move from the electrode to electrolyte solution leaving positive charge behind in the space charge region reflected by upward band bending. Withdrawal of the semiconductor majority charge carriers from the space charge region ensures the formation of a depletion layer in this region, **Fig. 3.7(a)** and **Fig. 3.8(a)**. For a p-type semiconductor the Fermi level lies lower than the redox potential, therefore electrons move from solution to the semiconductor electrode to attain equilibrium. In this process, negative charge in the space charge region causes downward band bending, with the removal of holes from the space charge region ensuring formation of a depletion layer, **Fig. 3.7(e)** and **Fig. 3.8(b)**.

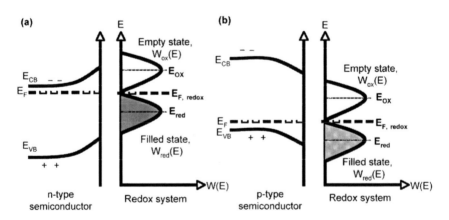

Fig. 3.8: Electron energy distribution at the contact between a semiconductor and a redox electrolyte for two different redox systems at equilibrium. **(a)** n-type semiconductor, and **(b)** p-type semiconductor.

Potential Distribution Across the Interface

Let us consider an n-type semiconductor in equilibrium with a redox couple, $E_F = E_{F,redox}$, with formation of the space charge layer leading to band bending as shown in **Fig. 3.9(a)**. A potential (V_E) is established between the working semiconductor electrode, under potentiostatic control, and the reference electrode. In the case of an ideally polarizable interface, i.e. when there is no exchange of charge with the electrolyte, the potential of the bulk semiconductor matches the potential of the redox couple at the reference electrode, **Fig. 3.9(b)**. This indicates that the difference in potential between the semiconductor working electrode and the reference electrode is dependent on the charge concentration in the space charge region.

At a semiconductor-electrolyte interface, if there is no specific interaction between the charge species and the surface an electrical double layer will form with a diffuse space-charge region on the semiconductor side and a plate-like counter ionic charge on the electrolyte side resulting in a potential difference ϕ across the interface. The total potential difference across the interface can be given by

$$V_E = \phi_H + \phi_{SC} + \phi_G + C \qquad (3.4.20)$$

where V_E is the electrode potential measured in relation to the reference electrode. ϕ_H is the interfacial difference potential between the solid and the liquid phases, commonly called the Helmhotz potential; ϕ_{SC} is the potential difference developed across the space charge layer; constant C depends on the nature of reference electrode. There also exists a diffuse double layer (Gouy region, ϕ_G) that can be neglected under sufficiently high redox electrolyte concentration.

A representative potential distribution across the interface is shown in **Fig. 3.9(c)**, taking the potential of the bulk solution as zero. The potential difference across the space charge region (ϕ_{SC}) occurs over a larger distance than that of the Helmholtz layer (ϕ_H). For an n-type semiconductor, ϕ_{SC} results from the excess positive charge of ionized donors in the bulk of the space charge region within the solid, and ϕ_H is due to the accumulation (~1 nm thick layer) of negative ions in the electrolyte solution from the solid surface, **Fig. 3.9(d)**.

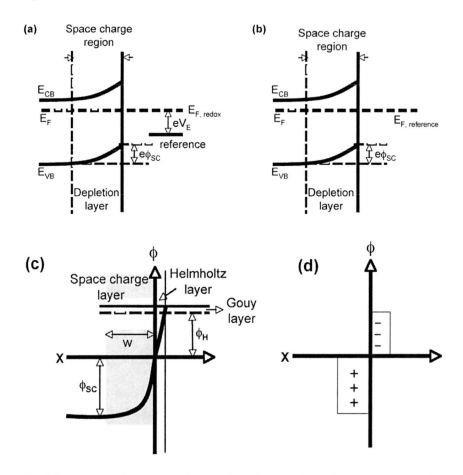

Fig. 3.9: Energy diagram of the semiconductor-electrolyte interface under equilibrium. **(a)** The Fermi level E_F is equal to redox potential energy, $E_{F, redox}$ **(b)** The Fermi level, E_F is equal to reference electrode energy, $E_{reference}$. **(c)** Potential distribution. **(d)** Charge across the interface.

For a p-type semiconductor, ϕ_{SC} results from the excess negative charge associated with the ionized acceptors in the space charge region of the solid, and ϕ_H is due to the accumulation of a positive ion layer (~1 nm thick, from the solid surface) in the electrolyte solution. A Helmholtz layer a few Angstroms thick is formed from oriented solvent molecule dipoles and electrolyte ions adsorbed at the semiconductor electrode surface, while the space charge layer is formed by the distribution of the semiconductor counter charges such as electrons, holes, ionized donors and ionized acceptor states over a finite distance below the semiconductor. The amount of chemisorbed ionic charge at the surface depends mainly on the electrolyte composition, and to a much lesser degree the excess charge on the semiconductor. Since the charge in both regions is equal but of opposite sign, comparison of the capacitance of the space charge region (C_{sc}) with the Helmholtz capacitance (C_H) shows that C_{sc} usually much smaller. Under these conditions ϕ_H remains essentially constant, with changes in the applied potential between the semiconductor electrode and reference electrode appearing in ϕ_{SC} so that:

$$V_E = \phi_{SC} + V_{FB} \tag{3.4.21}$$

V_{FB} is the flat band potential, and at this potential the surface concentration of charge is equal to that of the bulk. The effect of applied potential on the band edges of bulk n-type and p-type semiconductors are shown in **Fig. 3.7**. When the applied potential is more positive than the flat band potential ($V_E > V_{FB}$) a depletion layer forms in an n-type semiconductor, while for a p-type semiconductor the surface concentration of electrons decreases creating an accumulation layer of holes, and the bending of the bands at the surface to higher energies. Similarly, when the applied potential is more negative than the flat band potential ($V_E < V_{FB}$) an accumulation layer forms in the n-type semiconductor due to an excess of electrons in space charge layer, while for a p-type semiconductor the additional electrons produce a depletion layer, bending the energy bands at the surface downward to lower energies.

The electronic charge distribution in a semiconductor varies with applied electrode potential (V_E), which in turn determines the differential capacitance at the interface [11,78]. Relating charge density and electric field, the capacitance of a space charge (or depletion) region can be quantitatively derived. For an n-type semiconductor Poissons' equation can be written:

$$\frac{\partial^2 \phi}{\partial x^2} = \frac{eN_D}{\varepsilon\varepsilon_0} \qquad (3.4.22)$$

where e is electronic charge, N_D is the electron donor concentration, ε_0 is the permittivity of space, and ε is the relative dielectric constant of the medium. Assuming the electric field $\frac{d\phi}{dx}$ is zero at x_0 ($0 \leq x \leq w$) integrating equation (3.4.22) across the length of the space charge layer w we find:

$$\phi_{SC} = \frac{eN_D}{2\varepsilon\varepsilon_0} \cdot w^2 \qquad (3.4.23)$$

An electric field is generated in the depletion region due to ionized donors and a gradient in electron concentration. From Gauss's law we find

$$\left(\frac{d\phi}{dx}\right)_{\phi \to \phi_{SC}} = \frac{Q_{SC}}{\varepsilon\varepsilon_0 A} \qquad (3.4.24)$$

where A is the electrode area. The total charge moving through the depletion region is

$$Q_{SC} = e \cdot N_D \cdot A \cdot w \qquad (3.4.25)$$

The capacitance of the space charge region is given by

$$C_{SC} = \frac{dQ}{d\phi} = e \cdot N_D \cdot A \cdot \frac{dw}{d\phi} \qquad (3.4.26)$$

Using equations (3.4.23)-(3.4.26), for $\phi \to \phi_{SC}$, two equations can be derived.

$$C_{SC} = \frac{\varepsilon\varepsilon_0 A}{w} \qquad (3.4.27)$$

$$\frac{1}{C_{SC}^2} = \left(\frac{2}{e\varepsilon\varepsilon_0 N_D}\right)\left(\phi_{SC} - \frac{kT}{e}\right) \qquad (3.4.28)$$

Equation (3.4.28) is commonly known as Mott-Schottky equation.

The space charge layer capacitance is inversely proportional to the width of the depletion layer w. As the width of the depletion layer approaches zero the capacitance approaches infinity, hence $\frac{1}{C_{SC}^2}$ will be zero at the flat band potential. Using equation (3.4.21) to modify equation (3.4.28) the Mott-Schottky relation can be described as:

$$\frac{1}{C_{SC}^2} = \left(\frac{2}{e\varepsilon\varepsilon_0 N_D}\right)\left((V_E - V_{FB}) - \frac{kT}{e}\right) \qquad (3.4.29)$$

Donor density can be calculated from the slope of $\frac{1}{C_{SC}^2}$ vs applied potential, and the flat band potential determined by extrapolation to $\frac{1}{C_{SC}^2} = 0$. **Fig. 3.10** represents the Mott-Schottky plot for both n-type and p-type GaAs in an ambient temperature molten salt electrolyte made up of AlCl$_3$/n-butylpyridinium chloride [79]. The capacitance dependence on applied potential is an important means of characterizing a semiconductor-electrolyte interface [80-84]. The flat band potential is the difference between the Fermi level of the semiconductor and the Fermi level of the reference electrode. Using equations (3.4.11) and (3.4.12) one can then determine the conduction and valence band edges of the semiconductor with reference to vacuum energy level, and thus the bandgap of the semiconductor.

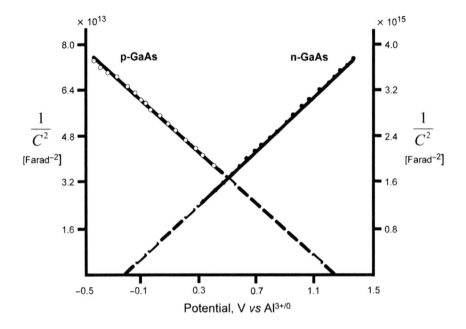

Fig. 3.10: Mott-Schottky plot for n-type and p-type semiconductor of GaAs in $AlCl_3$/n-butylpyridinium chloride molten-salt electrolyte [79].

The acidic or basic character of a semiconductor surface gives rise to interaction with H^+ or OH^- ions of an aqueous electrolyte:

$$M–O + H_2O \leftrightarrow HO–M^+ + OH^- \qquad \text{(A)}$$
$$M–O + H_2O \leftrightarrow M–(OH)_2 \leftrightarrow HO–M–O^- + H^+ \qquad \text{(B)}$$

where M stands for metal ions. Since the charge balance across the solid-liquid interface is potential dependent, the Helmholtz double layer varies with the change in the H^+ or OH^- concentrations, that is to say the *pH* of the electrolyte. This behavior can be expressed:

$$\phi_H = const. + 0.059pH \qquad (3.4.30)$$

In many cases the flat band potential, and hence the semiconductor band edge varies with the pH of the aqueous electrolyte solution, as illustrated in **Fig. 3.11**.

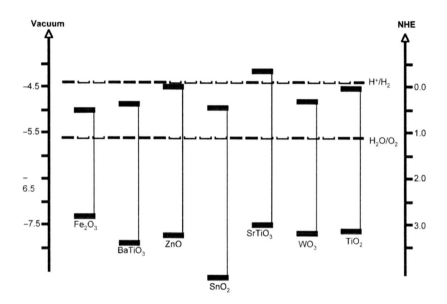

Fig. 3.11a: Band edge positions of several oxide semiconductors in contact with a pH 1 aqueous electrolyte.

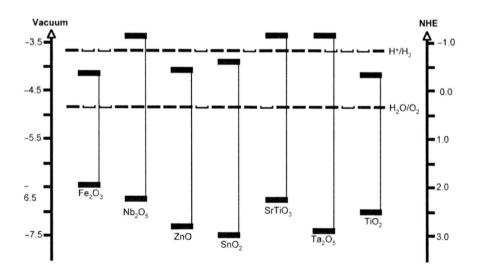

Fig. 3.11b: Band edge positions of several oxide semiconductors in contact with a pH 7 aqueous electrolyte.

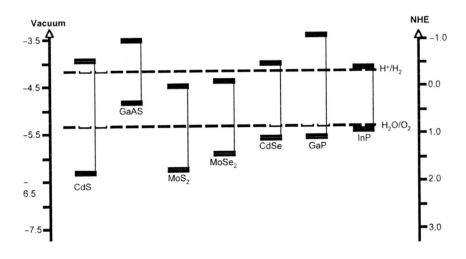

Fig. 3.11c: Band edge positions of several non-oxide semiconductors in contact with a pH 13 aqueous electrolyte.

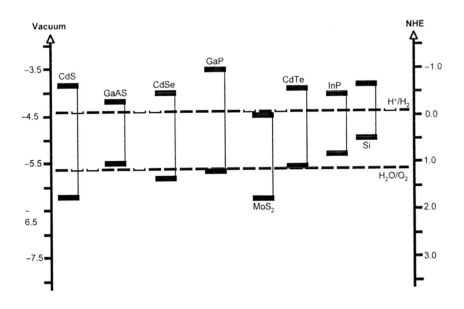

Fig. 3.11d: Band edge positions of several non-oxide semiconductors in contact with a pH 1 aqueous electrolyte

Charge Transfer Processes at the Interface

The Fermi level of a liquid immersed metal electrode largely defines its electrochemical potential; as long as one remains in equilibrium with the empty acceptor state or with the filled donor state of the solution redox species electron transfer occurs in an energy range close to the Fermi level [85,86]. In the case of a metal electrode the density of states at the Fermi energy level, and the corresponding charge carrier concentration, is very high therefore the rate constant at the metal electrode/solution interface can be easily controlled through the potential difference across the Helmholtz layer. Modulation of this potential difference readily enables characterization of the electron-transfer kinetics at the metal electrode/solution interface.

In contrast to metal electrodes, for a semiconductor-electrolyte interface most of the potential drop is located in the semiconductor making it difficult to study interfacial processes using potential perturbation techniques [11,20,55,58,60-65,75-78]. H. Gerischer [76] proposed a model in which electrons and holes are considered as individual interfacial reactants. Distinct and preferential electron transfer reactions involve either the conduction band or valence band as dependent on the nature of the redox reactants of the electrolyte, with specific properties dependent upon the energy state location.

The energy level of an electrolyte redox couple is governed by the ionization potential of the reduced species and the electron affinity of the oxidized species. As shown in equations (3.4.17)-(3.4.19) the Fermi level at a semiconductor surface adjusts itself to the position of the redox Fermi level by a factor λ due to the appropriate charging of the electric double layer, thus an equilibrium between the two can be achieved. Unless there is a considerable change in the Helmholtz layer potential (ϕ_H) due to variation of the electrical double layer charge, the semiconductor band edge energies do not change with respect to the redox couple energy level. Rather it is the electron and hole surface concentrations that vary with the double layer charge. For a given applied voltage, equilibrium between a semiconductor and an electrolyte is established by the adjustment of electron and hole concentrations.

Let us consider an n-type semiconductor in equilibrium with three different redox couples. In **Fig. 3.12(a)** the Fermi level of the redox couple is close to conduction band edge forming an accumulation layer at equilibrium; in this case electrons are available at the semiconductor surface showing a high rate of electron transfer between the redox system and the conduction band. As shown in **Fig. 3.12(b)** the semiconductor has formed a depletion layer. For this case electrons are not available at the surface for transitions to the oxidized species, while electrons in the reduced species cannot reach the conduction band energy; the result is no electron exchange. **Fig. 3.12(c)** shows an even higher barrier height for electrons that precludes electron exchange with the conduction band. Since the Fermi level of the redox couple overlaps with the valence band it is clear that reduction and oxidation of this system are valence band processes better described by the exchange of (valence band) holes across the interface. Formation of an inversion energy barrier layer between the bulk electrons and holes on or near the surface retards the hole exchange process at the electrode surface.

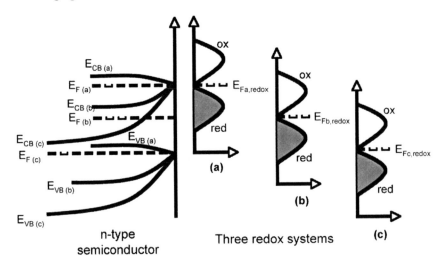

Fig. 3.12: Three different redox systems in equilibrium with an n-type semiconductor: **(a)** exchange of electrons with conduction band, **(b)** no or negligible electron exchange, and **(c)** exchange of electrons with valence band.

With reference to the Gerischer model [11,76,77], the charge transfer reaction of a semiconductor electrode in contact with a redox electrolyte is generally considered a bimodal reaction taking into account the concentration of redox ions in solution and the concentration of surface electrons/holes in the solid. The rate of electron transfer in the conduction band J_C is based on kinetics that considers both electrons and holes as reactants. The current due to conduction band transfer processes can be expressed as:

$$\frac{J_C}{J_{C,0}} = \left\{ \frac{[C_{red}]}{[C_{red,0}]} - \frac{n_s}{n_{s,0}} \cdot \frac{[C_{ox}]}{[C_{ox,0}]} \right\} \tag{3.4.31}$$

where C_{red}, C_{ox}, and n_s represent, respectively, concentrations of the reduced component, concentration of the oxidized component, and surface concentration of electrons. $C_{red,0}$, $C_{ox,0}$, and $n_{s,0}$ are the same parameters at equilibrium. $J_{C,0}$ is the exchange current density in the conduction band at equilibrium. Note that the exchange current density is a function of the electrode material(s) and the electroactive specie(s) in solution that at equilibrium are described in the general form:

$$J_{C,0} = e \cdot n_{s,0} \cdot K_{ET} \cdot [A] \tag{3.4.32}$$

where e denotes electronic charge, K_{ET} is the rate constant for electron transfer from the conduction band to the redox acceptor in solution, and A the redox acceptor concentration. The units of the rate constant are $cm^4 \cdot s^{-1}$ due to the second-order kinetics.

From equation (3.4.31), if the ratio $n_s/n_{s,0}$ is unity there will be no net current flow across the interface; this condition is depicted in **Fig. 3.13(a)** for an n-type semiconductor. Under this equilibrium state surface electrons can undergo isoenergetic electron transfers to the redox species due to a built-in potential, equal to the difference of potential between E_{CB} and E_{redox}. Equilibrium can be perturbed, with a resulting observable transient current flow, by varying the concentrations of the redox species.   The surface electron concentration n_s is related to the bulk concentration n_0 by the potential difference of the space charge layer as follows:

$$n_s = n_0 \cdot e^{\frac{e\phi_{SC}}{kT}} \tag{3.4.33}$$

The rate of forward electron transfer is equal to the rate of back electron transfer under this equilibrium condition, with no applied bias on the electrodes.

As shown in **Fig. 3.13(b)** and **3.13(c)** when ratio $n_s/n_{s,0}$ is less than or greater than 1 the system is in non-equilibrium resulting in a net current, with the electron transfer kinetics at the semiconductor-electrolyte interface largely determined by changes in the electron surface concentration and the application of a bias potential. Under reverse bias voltage, $V_{E1} > 0$ and $n_{s,0} > n_s$ as illustrated in **Fig. 3.13(b)**, anodic current will flow across the interface enabling oxidized species to convert to reduced species (reduction process). Similarly, under forward bias, $V_{E2} < 0$ and $n_s > n_{s,0}$ as illustrated in **Fig. 3.13(c)**, a net cathodic current will flow.

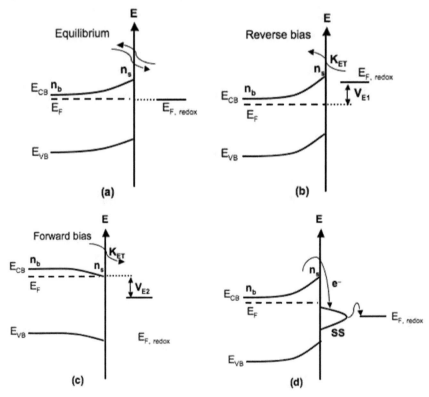

Fig. 3.13: Semiconductor-electrolyte interface **(a)** at equilibrium, **(b)** under reverse bias **(c)** under forward bias. Arrows denote direction of current flow [reduction reaction; ox + e⁻ → red]. **(d)** Electron transfer mediated through surface states.

For an interface described by a constant Helmholtz potential ϕ_H, there is no electron exchange between the semiconductor and redox electrolyte solution. The result is that $dV = d\phi_{SC}$, and for a non-equilibrium system one can obtain the current-voltage relation:

$$J_C = J_{C,0}\left[1 - e^{\frac{e(V_E - V_0)}{kT}}\right]$$

(3.4.34)

$(V_E - V_0)$ is the overpotential, the potential required to initiate reactions at the electrode surface, the difference between the equilibrium potential V_0 (no current flowing) and operating potential V_E (current flowing). The above kinetics indicate that the rate of electron transfer from the n-type semiconductor to the redox system depends on the surface electron concentration, while electron injection from the redox system into the conduction band is constant independent of applied potential [11,76,77]. If the Helmholtz layer potential ϕ_H varies across the interface the description of electron transfer becomes considerably more complicated requiring a charge transfer coefficient in equation (3.4.34).

For hole transfer and hole injection between a p-type semiconductor and a redox system the corresponding rate of electron transfer in the valence band can be expressed as

$$\frac{J_V}{J_{V,0}} = \left\{ \frac{p_s}{p_{s,0}} \cdot \frac{[C_{red}]}{[C_{red,0}]} - \frac{[C_{ox}]}{[C_{ox,0}]} \right\}$$

(3.4.35)

$J_{V,0}$ is the exchange current density in the valence band at equilibrium. $p_{s,0}$ denotes the equilibrium surface concentration of holes. p_s represents the dynamic surface concentration of holes related to bulk surface hole concentration p_0 by:

$$p_s = p_0 \cdot e^{\frac{e\phi_{SC}}{kT}}$$

(3.4.36)

Under non-equilibrium conditions, the current-voltage relation can be described as

$$J_V = J_{V,0}\left[e^{\frac{e(V_E - V_0)}{kT}} - 1\right] \tag{3.4.37}$$

Using equations (3.4.31), (3.4.33) (3.4.35) and (3.4.36), a relationship between the current densities in the valence and conduction bands is described:

$$\ln\left(\frac{J_{C,0}}{J_{V,0}}\right) = \ln\left(\frac{N_{CB}^*}{N_{VB}^*}\right) - \left(\frac{E_{BG} + 2\lambda}{2\lambda kT}\right)\cdot\left(\frac{E_{CB} + E_{VB}}{2} - E_{F,redox}\right) \tag{3.4.38}$$

N_{CB}^* and N_{VB}^* are the effective charge carrier densities at the conduction and valence band edges, λ the reorganization energy, and $E_{F, redox}$ the Fermi level of the redox system. By this model electron transfer occurs at the conduction band edge energy without loss. The rate constant is determined by the relative position of the energy states across the semiconductor-electrolyte interface, a topic extensively reviewed [11,55,58,64,65,87-96]. Various approaches to describe electron transfer reactions at semiconductor/liquid interfaces have been detailed, including the Lewis "electron-ball" model [89,90], Gerischer "half-sphere" model [91], tight binding approach [92], and adiabatic and non-adiabatic [91,93-95] transfer reactions.

Surface states can form due to abrupt distortion of the semiconductor crystal lattice. Charge transfer processes between surface states and the electrolyte have been analyzed in relation to water photoelectrolysis application [96]. Electron transfer mediated through surface states for an n-type semiconductor under dynamic equilibrium is shown in **Fig. 3.13(d)**.

Light Activity at the Interface

Light absorption by a semiconductor results in the creation of electron-hole pairs, by means of either direct or indirect momentum transitions depending on the crystal structure. Momentum is conserved in direct transitions, while a change in momentum is required for indirect transitions. The light absorption coefficient (α) of such transitions is described by [97,98]:

Direct transition: $\alpha = A' \sqrt{hv - E_{BG}}$ (3.4.39)

Indirect transition: $\alpha = A' \left(hv - E_{BG}\right)^2$ (3.4.40)

where A' is a constant of proportionality. With light absorption by a semiconductor surface the free energy of both the minority and majority charge carriers shift leading to a non-equilibrium condition, quantified by quasi-Fermi energy levels E_F^*, that results in photocurrent and photovoltage generation at a semiconductor-electrolyte interface [11,55,65,75-77]. **Figure 3.14(a)** shows an n-type semiconductor in contact with an electrolyte showing photogeneration in the depletion region. The photocurrent of an n-type semiconductor absorbing monochromatic light with an absorption coefficient (α) is given by:

$$I_{ph} = eI_0 \left[1 - \left(1 + \alpha L_p\right) \cdot e^{-\alpha w}\right]$$ (3.4.41)

Where

$$L_p = \left(D_p \tau_p\right)^{1/2} = \left(kT \cdot \mu_p \tau_p\right)^{1/2}$$ (3.4.42)

The space charge region is denoted by length w, while L_P is the hole (minority carrier) diffusion length. τ_p is the minority carrier (hole) lifetime, μ_p the (minority carrier) hole mobility, and D_p the minority carrier diffusion coefficient.

$$w = L_D \left[\frac{\left(V - V_{FB}\right)}{kT}\right]^{1/2}$$ (3.4.43)

$$L_D = \left(\frac{2\varepsilon\varepsilon_0 kT}{eN_D}\right)^{1/2}$$ (3.4.44)

L_D is the Debye length of the space charge layer; a depth or distance greater than this leads to electron-hole pair recombination. Thus the magnitude of the quantum yield depends on the light penetration length $1/\alpha$ and the Debye length L_D [87,99]. At the flat band potential a finite current flows due to diffusion of minority carriers despite non-inclusion of surface recombination in this model. Schottky barriers play an important role in preventing electron-hole recombination by removing the majority charge carriers from the interface. For current transfer across the semiconductor-electrolyte interface a positive bias is needed for an n-type semiconductor, and negative bias for a p-type semiconductor. At equilibrium the initial amount of band bending between the bulk and surface in the space charge layer is given by:

$$\phi_{SC,0} = E_{F,redox} - V_{FB} \qquad (3.4.45)$$

Using equation (3.4.13), equation (3.4.45) can be modified as

In the dark: $\qquad\qquad \phi_{SC,0} = V_{redox} - V_{FB} \qquad (3.4.46)$

Under illumination, I: $\qquad \phi_{SC,I} = V_{redox} - V_{FB} \qquad (3.4.47)$

Due to relatively slow rates of electron transfer between the semiconductor and the redox system, and/or trapping of minority carriers at surface states an additional driving force relative to the flat band potential is needed for the efficient photocurrent generation. A photovoltage arises with light illumination due to generation of electron-hole pairs that in turn decreases the band bending; ideal conditions for photocurrent generation from an n-type semiconductor, in the dark and under illumination, are shown in **Figs. 3.14(b)** and **3.14(c)** respectively. Correlating these two figures in the dark (equilibrium state) and under illumination (non-equilibrium) respectively with equations (3.4.46) and (3.4.47), the Gerischer model describes a threshold in illumination intensity as the driving force for water photoelectrolysis.

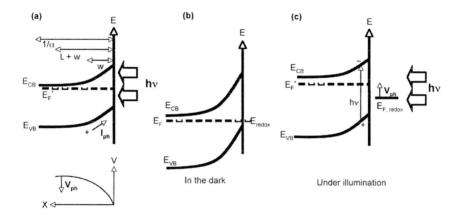

Fig. 3.14: (a) Photocurrent generation by electron-hole separation in an n-type semiconductor. Photovoltage arises through a decrease in the band bending across the depletion layer under light illumination. Energy bands of an n-type semiconductor-electrolyte interface: **(b)** in the dark, and **(c)** in the presence of light, with photogenerated free energy at open circuit represented by $eV_{ph} = E_{F,n}^* - E_{F,redox}$.

Under open circuit conditions the photovoltage of the interface can be expressed as:

$$V_{ph} = \frac{kT}{e} \cdot \ln \frac{I_{ph}}{I_0} = \frac{kT}{e} \cdot \frac{n_s^*}{n_{s,0}} \qquad (3.4.48)$$

where I_{ph} is the minority carrier current and I_0 is the photon flux. Equation (3.4.48) shows a logarithmic dependence of the photogenerated free energy on the illumination intensity under equilibrium conditions.

Salvador [100] introduced a non-equilibrium thermodynamic approach taking entropy into account, which is not present in the conventional Gerischer model, formulating a dependence between the charge transfer mechanism at a semiconductor-electrolyte interface under illumination and the physical properties thermodynamically defining the irreversible photoelectrochemical system properties. The force of the resulting photoelectrochemical reactions are described in terms of photocurrent intensity, photoelectochemical activity, and interfacial charge transfer mechanisms. The following relationships are given for a photoelectrochemical reaction based on hole transfer

$$h^+ + A^- \xrightarrow{K_P} A \qquad (3.4.49)$$

where K_P is the rate constant for forward hole transfer reaction. With I_{ph} and V_{ph} representing the photocurrent and open-circuit photovoltage we have:

$$I_{ph} = e \cdot K_P \cdot p_{s,0}[A^-]\left\{ e^{\frac{a_{ph}}{kT}} - 1 \right\} \qquad (3.4.50)$$

$$V_{ph} = \frac{1}{a_{ph}} \cdot \frac{kT}{e} \cdot \ln\left\{ \frac{n_{s,0} \cdot K_{ET}[A]}{p_{s,0} \cdot K_P[A^-]} \right\} \qquad (3.4.51)$$

$$\sigma = \frac{1}{w} \cdot \frac{I_{ph}}{e} \cdot \frac{a_{ph}}{T} \qquad (3.4.52)$$

where $\qquad a_{ph} = kT \cdot \ln\dfrac{p_s^*}{p_{s,0}} \qquad (3.4.53)$

a_{ph} is the thermodynamic driving force of the photoelectrochemical reaction; K_{ET} is the rate constant for forward electron transfer reaction; σ is the entropy production with respect to stationary photocurrent (I_{ph}) under illumination; $n_{s,0} \approx n_s^*$, $p_{s,0} \approx p_s^*$ and $a_{ph} \geq E_{CB} - E_{redox}$. p_s^* (or n_s^*) are the surface excess population of holes under illumination (or electrons under illumination for electron transfer reactions).

3.5 Photoelectrochemical Cell Band Model

A common photoelectrolysis cell structure is that of a semi-conductor photoanode and metal cathode, the band diagrams of which are illustrated in **Fig. 3.15** together with that of electrolyte redox couples. In **Fig. 3.15(a)** there is no contact between the semiconductor anode and metal cathode (no equilibrium effects communicated through the electrolyte). As seen in **Fig. 3.15(b)**, contact between the two electrodes (no illumination) results in charge transfer from the semiconductor anode having a lower work function to the metal cathode having a higher work function until the work functions of both electrodes equilibrate. The result of this charge transfer is band bending by energy E_B. The energy levels of **Fig. 3.15(b)** are not favorable for water decomposition since the H^+/H_2 energy level is located above the cathode Fermi level. Under illumination, **Fig. 3.15(c)**, the photoanode surface potential and the (H^+/H_2) water reduction potential are each lowered, but the (H^+/H_2) water reduction potential still remains above the cathode Fermi level. Anodic bias is thus needed to elevate the Fermi level of cathode above the water reduction potential, see **Fig. 3.15(d)**, making the water splitting process feasible. This applied bias provides overvoltage at the metal cathode necessary to sustain the current flow, and increases the semiconductor band bending to maintain the required electric field driven charge separation in the semiconductor.

As illustrated by the example of **Fig. 3.15** external bias is an exceedingly useful tool for enabling operation of the photoelectrolysis cell to produce H_2 and O_2. The bias can be provided either by an external voltage (power) source, or by immersing the anode in a basic solution and the cathode in acidic solution. As illustrated by **Fig. 3.16**, several oxide semiconductors have flat band potentials above the H^+/H_2 level therefore no external bias is needed to produce H_2 and O_2. Unfortunately these oxide semiconductors have relatively large bandgap energies that result in low optical absorption, and hence low visible spectrum photoconversion efficiencies. In the first report on water photoelectrolysis by Fujishima and Honda [26], using an n-type TiO_2 anode and Pt cathode, it was observed that the photovoltage generated in the cell was not sufficient to carry out the water photolysis. An additional bias voltage of 0.25-0.50 V was required to achieve simultaneous oxygen evolution at the TiO_2 electrode and hydrogen evolution at the Pt cathode.

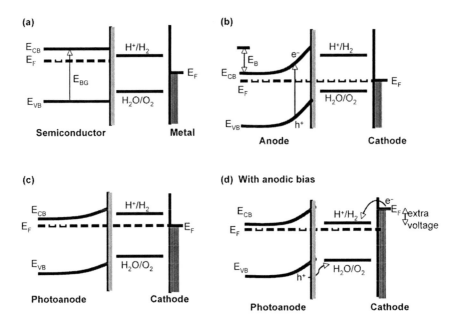

Fig. 3.15: Energy diagram of semiconductor-metal photoelectrolysis cell. **(a)** No contact and no chemical potential equilibrium; **(b)** galvanic contact in dark; **(c)** effect of light illumination; **(d)** effect of light illumination with bias. **(e)** Light illumination without bias, however in this case the semiconductor band edges straddle the redox potential for water photoelectrolysis.

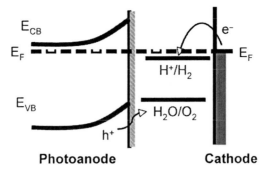

Fig. 3.16: Energy diagram of semiconductor-metal photoelectrolysis cell with light illumination without bias, however in this case the semiconductor band edges straddle the redox potential for water photoelectrolysis.

Another photoelectrolysis cell structure of considerable interest is that with both electrodes comprised of semiconductors, one n-type and the other p-type [31-35]. If the electron affinity of the n-type electrode is greater than that of the p-type electrode, when both electrodes are simultaneously illuminated the available electron-hole potential for driving chemical reactions is enhanced. Thus two photons, one at each electrode, are absorbed for the generation of one (minority carrier) electron-hole pair, a hole in the n-type semiconductor and an electron in the p-type semiconductor. Although the majority carriers recombine at the ohmic contacts a greater amount of potential energy is available to drive the chemical reactions. The energy diagram of such a system is shown in **Fig. 3.17**. The advantage in two semiconductor electrodes is that one can use a semiconductor of relatively smaller band gaps, able to capture a larger amount of solar spectrum energy, since the Fermi level of the majority carrier in the illuminated semiconductor does not have to be of energy suitable for driving the counter electrode reaction. In contrast to a single semiconductor electrode cell, a p-n photoelectrolysis cell does not require external bias when the flatband potential is below the H^+/H_2 level. Several p-n photoelectrolysis cells are reported using various semiconductor combinations such as TiO_2/p-GaP, n-TiO_2/p-CdTe, n-$SrTiO_3$/p-CdTe, n-$SrTiO_3$/p-GaP and n-Fe_2O_3/p-Fe_2O_3 [31-34].

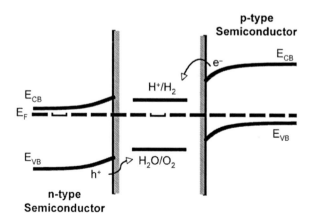

Fig. 3.17: Energy diagram for p-n photoelectrolysis cell.

The desire to eliminate the need for bias in a photoelectrolysis cell leads to another configuration where the anode and cathode are combined into a monolithic structure called a 'photochemical diode' [35]. A photochemical diode generally consists of a sandwich-like structure of an n-type/p-type bilayer connected through ohmic contacts. **Figure 3.18** shows the energy diagram for photochemical diodes consisting of a p-type/n-type bilayer electrode. This system resembles power generation in photosynthesis where two photoredox reactions are series coupled to drive water oxidation and CO_2 reduction. Both systems, photosynthesis and the photochemical diodes, require absorption of two photons to produce one useful electron-hole pair [11]. The n- and p-type semiconductors are analogous to photosystem II and photosystem I, and the majority carrier recombination at the ohmic contacts analogous to the recombination of photogenerated electrons from excited pigment II with the photogenerated holes in photosystem I. Unfortunately to date such devices generally face serious photocorrosion stability problems since both semiconductors must be stable in the same aqueous electrolyte solution, in the dark and under illumination.

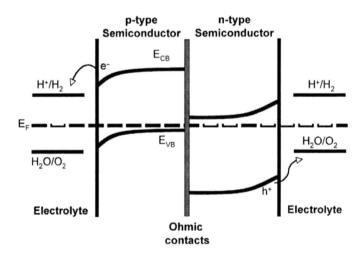

Fig. 3.18: Energy diagram for p-n photochemical diode for water photoelectrolysis.

3.6 Efficiency of Water Splitting in a Photoelectrochemical Cell

The usefulness of a water photoelectrolysis cell is primarily determined by the efficiency with which it converts light energy into chemical energy, which is stored in the form of hydrogen. In the case of water splitting, the redox properties of the electrolyte are fixed and hence the light harvesting and utilization properties of the light absorber, photoelectrode(s) or dispersed particles, decide cell efficiency. Since electric charges play an intermediate role in these cells, a knowledge about efficiencies related to production of electric charges by photons, charge injection into the external circuit and production of chemical energy by electric charges are useful for independently evaluating the corresponding functions of the absorber.

The light-to-electrical energy conversion efficiency measurements have their basis in photovoltaics and there are standard ways to determine these efficiencies. However, a sense of confusion prevails in the calculation of the light-to-chemical energy conversion, i.e. the overall photo conversion efficiency, due to the different definitions and cell configurations used by various research groups. Despite some isolated efforts seen in the literature to give a clear definition and methodology this is particularly significant in the case where an electrical bias is used to assist the water splitting. Different definitions given to efficiency and the correct methodology are discussed in this section.

The overall photoconversion efficiency is defined as the ratio of the maximum energy output that can be obtained from the final products, hydrogen and oxygen, to the energy supplied in the form of light to produce them. In terms of power, it can be defined as the ratio of the power density that can be obtained from hydrogen to the power density of the incident light. Since any practical photoelectrolysis system should operate using solar energy, before going to the various formula and methodologies used to calculate the efficiency, it is worth looking at limitations on the solar energy to hydrogen energy conversion efficiency in an ideal case.

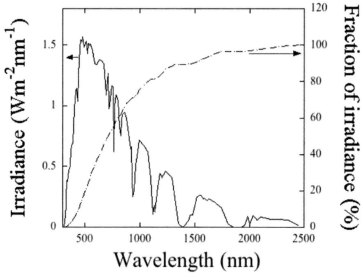

Fig. 3.19a: Solar spectral irradiance (global AM1.5) and the fraction of the irradiance above λ_{min} where λ_{min} is the minimum wavelength at which the spectral irradiance has a measurable value [101].

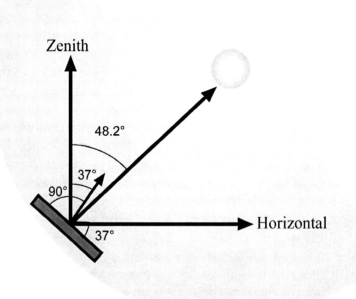

Fig. 3.19b: Configuration for AM1.5 solar illumination.

The solar energy available for conversion depends upon the relative position of the sun to the absorbing plane, and atmospheric conditions. **Figure 3.19a** shows a representative solar spectral irradiance under the global AM 1.5 (air mass 1.5) condition [101]. AM 1.5 represents the average atmospheric conditions in the United States. AM 1.5 corresponds to a situation when the absorber plane is inclined 37° towards the equator and sun is at a zenith angle of 48.19° (AM=1/COS48.19=1.5)as shown in **Fig. 3.19b**. 37° is the average latitude for the 48 contiguous states of the U.S.A. In such a configuration, the light travels through the atmosphere 1.5 times the distance it travels when the sun is at 0° zenith angle (AM 1.0). The global irradiance includes both direct and diffused components of light from the sun. The total irradiance can be calculated as:

$$P_t = \int_0^\infty P(\lambda) \, d\lambda \qquad (3.6.1)$$

The total irradiance is about 964.1 W/m^2. However, considering the effect of variations in the atmospheric conditions (such as cloudiness, dust particles and relative humidity), the spectrum is normalized to 1000 W/m^2. This is generally regarded as 1 sun. Photoelectrochemical water splitting is represented by the reaction:

$$H_2O_{(liquid)} + 2h\nu \rightarrow \tfrac{1}{2} O_{2(gas)} + H_{2(gas)} \qquad (3.6.2)$$

As discussed in Chapter 2, a minimum energy of 1.229eV per electron should be supplied to split water. A minimum of two quanta (photons) are needed to generate one molecule of hydrogen. 1.229eV corresponds to 1010 nm and hence about 77% of the solar energy is available for water splitting (see the fraction of irradiance above λ_{min} given in **Fig. 3.19a**).

The basic parameter deciding the light harvesting ability of the photoelectrode is the bandgap E_g of the material. There are inherent losses associated with any solar energy conversion processes involving materials [102-105]. These losses include: {1} Only the energy $E \geq E_g$ (or the photons with wavelength $\lambda \leq \lambda_g$ where λ_g is the wavelength corresponding to the bandgap) is absorbed and the rest is lost [102,105]. {2} Of the absorbed energy, the excess energy which is the difference in the energy of the absorbed photon and the band gap energy (E - E_g) is lost as heat during the relaxation of the absorber to the level of E_g. {3} The energy of the excited state is thermodynamically an internal energy that has the entropy term involved in it and only a fraction (up to about 75%) of this energy can be converted into work (electrical energy) or stored as chemical energy. {4} Although commonly negligible, losses due to spontaneous emissions such as fluorescence also contributes to the efficiency limitations.

The limiting efficiency ε_{limit} of a solar energy conversion process is [102,103]:

$$\varepsilon_{\lim it} = \frac{F_g \, \Delta\mu_x \, \phi}{P_t} \qquad (3.6.3)$$

where F_g is the absorbed photon flux given by

$$F_g = \int_{\lambda_{min}}^{\lambda_g} \frac{P(\lambda)}{(hc / \lambda)} \, d\lambda \qquad (3.6.4)$$

$\Delta\mu_x$ is the chemical potential of the excited state relative to the ground state and ϕ is the internal quantum efficiency which is the fraction of the excited states utilized for the generation of a useful product. The chemical potential is related Gibbs energy by $G = \sum_i^n \mu_i \, N_i$ where μ_i is the chemical potential of the ith state and N_i is the number of ith states, which is the maximum energy available to do work or to be stored as chemical energy. The ideal limiting efficiency in the case of single bandgap devices (devices involving single photosystem) is shown in **Fig. 3.20** [102,103]. A maximum of 33% is possible at about 900 nm for single bandgap devices. This value is higher for dual photosystems. The maximum value of solar irradiance efficiency ε_g corresponds to semiconductors of $1.0 \le E_g \le 1.4$ ev [106,107]. Hanna and Nozik [108] have recently calculated the limiting efficiency of a two band gap tandem device without carrier multiplication (i.e., when more than one electron-hole pair are generated by a single photon) as 40% and with carrier multiplication as 46%.

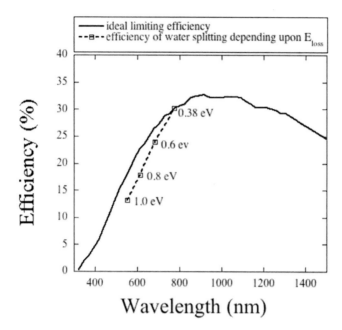

Fig. 3.20: The ideal limiting solar conversion efficiency for single bandgap devices. The dotted line shows efficiency of photoelectrolysis cells at different values of E_{loss} in relation (3.6.8) [102].

In the case of photoelectrolysis cells involving semiconductors, the three major processes leading to water splitting are the absorption of photons of energy $E \geq E_g$, conversion of absorbed photons into electric charges (or excited states) and utilization of electrical charges (conversion of excited states) for water splitting. The overall solar energy conversion efficiency ε_o can be written as the product of the efficiencies of the cell in performing these processes:

$$\varepsilon_o = \varepsilon_g \, \phi \, \varepsilon_c \qquad (3.6.5)$$

ε_g is solar irradiance efficiency, ϕ is the quantum efficiency and ε_c the chemical efficiency. ε_g is defined as the fraction of the incident solar irradiance with photon energy $E \geq E_g$, given by:

$$\varepsilon_g = \frac{F_g E_g}{P_t} = E_g \frac{\int_{\lambda_{min}}^{\lambda_g} \frac{P(\lambda)}{(hc/\lambda)} d\lambda}{\int_0^\infty P(\lambda)\, d\lambda} \quad OR \quad \varepsilon_g = E_g \frac{\int_{E_g}^\infty N(E)\, dE}{\int_0^\infty E\, N(E)\, dE} \qquad (3.6.6)$$

N(E) is the distribution of photons with respect to their energy. ϕ is given by

$$\phi = \frac{N_E}{N_T} \qquad (3.6.7)$$

N_E is the number of photons utilized for electron-hole pair generation and N_T is the total number of absorbed photons. $\phi = 1$ in the ideal case where all the photons of energy $E \geq E_g$ are utilized for carrier generation.

The chemical efficiency, ε_c, is the fraction of excited state energy converted to stored chemical energy and given by:

$$\varepsilon_c = \frac{E_g - E_{loss}}{E_g} \qquad (3.6.8)$$

E_{loss} is the actual energy loss per molecule involved in the overall light energy to chemical energy conversion process. E_{loss} always has a value greater than zero due to the entropy change (the term $T\Delta S$) involved in the process. In the ideal case ($\phi=1$) it has a value $E_g - (\Delta G^0/n)$ where n is the number of photons required to drive reaction (1) and ΔG^0 is the standard Gibbs energy of the reaction. Considering (3.6.6)-(3.6.8), the ideal ε_o ($\phi=1$) can be written as [109]

$$\varepsilon_o = \frac{F_g \Delta G (1 - \phi_{loss})}{P_t} \quad \text{if } E_g \geq \Delta G^0 + E_{loss} \qquad (3.6.9)$$

where ϕ_{loss} is the radiative quantum yield which is the ratio of re-radiated photons to absorbed photons. ε_o is 0 for $E_g < \Delta G^0 + E_{loss}$ in the ideal case. The ϕ_{loss} corresponding to the maximum value of efficiency is given by $\phi_{loss} \approx [\ln (F_g/F_{bb})]^{-1}$ where F_{bb} is the blackbody photon flux at wavelengths below the bandgap wavelength.

Although the minimum bandgap needed for water splitting is 1.23 eV, the entropy consideration in E_{loss} necessitates materials with significantly higher bandgaps. The maximum photoconversion efficiencies corresponding to different values of E_{loss} for single bandgap devices are shown in **Fig. 3.20** [102,103]. For a device involving single photosystem (single bandgap) where minimum two photons are required for the reaction (3.6.1), the ideal limit of ε_o is 30.7% corresponding to a wavelength of 775 nm (1.6eV) and E_{loss} 0.38 eV (**Fig. 3.20**). That is, a minimum bandgap of 1.6 eV is required for the photelectrode used for water splitting using solar radiation. In the case of dual photosystems involving two bandgaps and absorption of four photons, the ideal limit is 41%. In practical systems E_{loss} takes values greater than 0.8 eV due to other loss factors. These losses include those due to transport of electrons within the electrode during charge separation, transport of electrons from photo-electrode to counter electrode (for n-type materials this is from photoanode to cathode), transport of electrons/holes to the photoelectrode/electrolyte interface, Joule heating due to the electron flow through external circuit and the cathodic and anodic overpotentials. The overvoltages associated with electrodes are functions of mechanisms of the electrode reactions, current density, structures of the electrodes, surface properties of the electrodes, temperature, composition of electrolyte and similar factors [106]. Considering all these losses, E_{loss} has a value [110]:

$$E_{loss} \geq 0.5 + e\ (\eta_a + \eta_c + IR)\ eV$$

Where η_a and η_c are, respectively, the anodic and cathodic overpotentials. Considering all these losses an optimum bandgap of 2.0 to 2.25 eV is required for the materials used as photoelectrodes for water photoelectrolysis. In practical cases, a reasonable value of overall solar efficiency is 10% for single bandgap devices involving two photons and 16% for dual photosystem devices involving 4 photons [102,103,110,111].

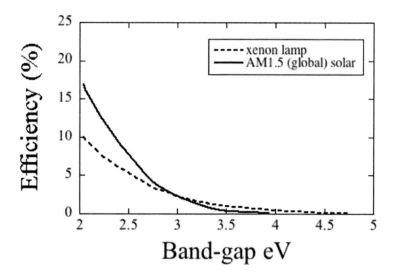

Fig. 3.21: Maximum efficiency possible depending upon semiconductor bandgap, under xenon arc lamp and AM1.5 solar illuminations.

In research laboratories, different types of light sources are used instead of solar radiation. In most cases the simulated spectrums have considerable deviation from the solar spectrum. Based on equation (3.6.9) Murphy et al [109] analyzed the maximum possible efficiencies for different materials according to their band gap in the case of solar global AM 1.5 illumination and xenon arc lamp, see **Fig. 3.21**. For example, anatase titania with a bandgap of 3.2 eV has a maximum possible efficiency of 1.3% under AM 1.5 illumination, and 1.7% using Xe lamp without any filter. For rutile titania these values are 2.2% and 2.3% respectively.

In practical cases the absorption edge is not sharp and absorption coefficients of semiconductors decrease as the wavelength approaches the bandgap wavelength. The absorbed photon flux can then be written as [109]

$$F_x = \int_0^{\lambda_g} \alpha(\lambda)P(\lambda)d\lambda \text{ where } \alpha(\lambda)=1-e^{-k(\lambda)h} \qquad (3.6.10)$$

k is the absorption coefficient and h is the thickness of the semiconductor. Hence for lower thickness electrodes where path length is not enough for complete light absorption the efficiency decreases. Reflections from the sample and electrolyte containers, as well as absorption in electrolyte also reduce the efficiency of the cell [109].

The difference between photoelectrochemical solar cells and photoelectrochemical water splitting is that in the former case free energy (electrical energy) is produced but the net gain is zero whereas in photoelectrochemical water splitting there is a net gain in free energy (from hydrogen). In both cases, free energy appears as the photo voltage between the electrodes that drives the electric charges through the circuit or supplies carriers for hydrogen/oxygen generation during illumination; at zero current this is termed the open circuit voltage. When short-circuited, the photovoltage goes to zero and current becomes maximum. Under dark conditions, the Fermi levels of the semiconductor photoelectrode and counter electrode, redox potential of the electrolyte equalize when short circuited and the valence band and conduction band are bent up to E_f - E_{redox} where E_f is the Fermi level of the semiconductor and E_{redox} is the potential of the redox couples in the solution [112]. The barrier height represents the upper limit of the open circuit voltage that can be achieved under high irradiance. Upon irradiation, say in the case of n-type semiconductors, the conduction band population of electrons increases and the Fermi level shifts up and conduction band bending reduces. Now the difference between the electrochemical potential of electrons in the semiconductor (Fermi level) and the chemical potential of electrons in the solution (the redox potential) gives the open circuit voltage. V_{oc} cannot exceed $|E_f$ - $E_{vb}|$ for photoanodes where E_{vb} is the valence band energy. Therefore, the open circuit potential can never be as high as the bandgap potential [102,112]. As mentioned earlier in the discussion on the fundamental limitations on attainable efficiency, this is a consequence of the fact that the energy of the excited state is thermodynamically an internal energy and not Gibbs energy due to the entropy term ($\Delta G = U + PV - T\Delta S$). Up to about 75% of the internal energy can be converted into free energy [102]. Hence, higher band gap materials are useful for water splitting as these can supply higher V_{oc} even exceeding 1.229 eV and hence the possibility of water splitting without supplying any additional electrical energy.

However the higher band gap materials cannot effectively utilize solar energy. Therein of course lies the difficulty. In practical cases, V_{oc} values up to about 1 V are common.

Though equation (3.6.5) is more useful for analyzing the performance of photoelectrolysis cells, for practical purposes the photoconversion efficiency (solar conversion efficiency if sunlight is used) is calculated by modifying (3.6.5) in the form

$$\varepsilon_o = \frac{\Delta G^0 \, R_{H_2}}{P_t} \qquad\qquad (3.6.11)$$

R_{H2} is the rate of production (moles/s) of hydrogen in its standard state per unit area of the photoelectrode. The standard Gibbs energy $\Delta G^0 = 237.2$ kJ/mol at 25°C and 1 bar, and P_t is the power density (W/m^2) of illumination. The numerator and denominator have units of power and hence, as in the case of photoelectrochemical solar cells, the photoconversion efficiency is the ratio of power output to the power input.

Equation (3.6.11) is based on the assumption that the free energy ΔG^0 can be completely retrieved in an ideal fuel cell run by the products from the photoelectrolysis cell for which the relation is applied. Instead of the free energy ΔG^0, the enthalpy (heat) of water splitting ΔH^0 also has been used in some cases. Here, it is assumed that the heat of water splitting is completely retrieved by burning hydrogen. At 25°C and 1 bar, $\Delta H^0 = 285$ kJ/mol. However, due to the similarity in functioning of photoelectrolysis cells and photoelectrochemical solar cells where the free energy term V_{oc} is used to calculate the efficiency, ΔG^0 is commonly used in (3.6.11).

If I is the current density responsible for the generation of hydrogen at the rate of R_{H2} in (3.6.11), then under 100% Faradaic conversion (that is, all the carriers are utilized only for generating hydrogen/oxygen), $R_{H2} = I/nF$. The voltage corresponding to the Gibbs energy is $V_{rev} = \Delta G^0/n\,F = 1.229$ V as n, the number of moles of electrons used for generating one mole of hydrogen, is 2.

$$\varepsilon_o = \frac{1.229 \, I}{P_t} \qquad\qquad (3.6.12)$$

When ΔH^0 is used in (3.6.11), 1.229 V in (3.6.12) needs to be replaced by 1.482 V.

Spontaneous water-splitting upon illumination needs semiconductors with appropriate electron affinity and flat band conditions. The flat band positions shift with electrolyte pH. Hence, an external bias needs to be applied between the electrodes in most cases to effect water splitting. The external bias can be either electrical or chemical. This external bias contribution should be subtracted from (3.6.11) or (3.6.12) to get the overall photoconversion efficiency. In the case of an external electrical bias, the efficiency can be defined as:

$$Efficiency \ \varepsilon_0 = \frac{energy \ stored \ as \ hydrogen - Energy \ input \ from \ power \ supply}{Light \ energy \ input}$$

OR

$$\varepsilon_o = \frac{\Delta G^0 \ R_{H_2} - V_{bias} \ I}{P_t} \qquad (3.6.13a)$$

$$\varepsilon_o = \frac{(1.229 - V_{bias}) \ I}{P_t} \qquad (3.6.13b)$$

The basis of this definition is that a fuel cell run by the products from the photoelectrolysis cell supplies a part of its output to the photoelectrolysis cell as electrical bias. The combined system must have a significant positive energy output to be considered as useful.

Practical photoelectrolysis cells consist of two electrodes immersed in the electrolyte and the bias voltage is applied between the working and counter electrodes [113]; the overall chemical reaction in such a cell is made of two independent half-reactions. In laboratory water photoelectrolysis experiments, to understand the chemical changes at the photoelectrode a three-electrode geometry is used to measure photocurrent. This geometry involves a working electrode (photocathode or photoanode), a counter electrode which generally is platinum, and a reference electrode. The internationally accepted primary reference is the standard hydrogen electrode (SHE) or normal hydrogen electrode (NHE) which has all components at unit activity $[(Pt/H_2 \ (a=1)/H^+ \ (a=1, \ aqueous))]$. However, using such an electrode is impractical and hence other reference electrodes such as silver-silver chloride (Ag/AgCl/KCl) and saturated calomel electrodes (SCE) $(Hg/Hg_2Cl_2/KCl)$ are generally used. Their potentials may then be converted in terms of Normal Hydrogen Electrode potential (NHE). The potential of Ag/AgCl electrode is 0.197 V vs NHE and that of SCE is 0.242 V vs NHE [114].

Equation (3.6.13) gives a thermodynamical measure of the efficiency and is generally applied in a two-electrode configuration. Nevertheless, different approaches have been followed by various groups to find the efficiency, especially when a three- electrode geometry is used [115-117]. This makes a direct comparison of reported efficiency values meaningless. Some of these approaches are discussed below.

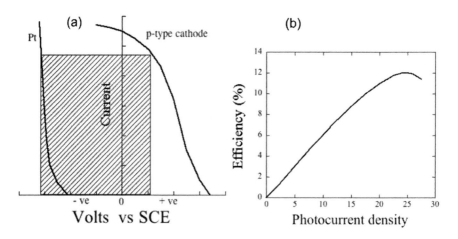

Fig. 3.22: (a) Current – voltage characteristics of a p-type photocathode and that when it is replaced by a platinum shown to illustrate the efficiency calculation using the power saving approach. The shaded area represents the maximum power saving as a result of photoelectrolysis. **(b)** The efficiency at various photocurrent densities obtained useing the base graph.

One approach to find efficiency uses the electrical power saved due to the use of light in a photoelectrolysis process compared to an electrolysis process using light-insensitive electrodes [118].

$$\varepsilon_o = \frac{P_{saved}}{P_t} = \frac{V_{save} I}{P_t} \tag{3.6.14}$$

The method involves recording the current from a semiconductor photoelectrode (anode or cathode) at various applied potentials and then repeating the experiment by replacing this semiconductor electrode with a metal electrode like platinum. For example **Fig. 3.22(a)** shows the current-voltage relationship of a hydrogen evolving p-type cathode in 1 M $HClO_4$, and the same relationship when the electrode is replaced by a platinum cathode [119,120]. The power saved P_{saved} for a particular current density I, is $V_s I$ where V_s is the difference between the corresponding voltages for semiconductor and metal cathodes. A representative graph showing the relation between photocurrent density and efficiency [equation (3.6.14)] is given as **Fig. 3.22(b)**.

The power saving is maximum when both electrodes operate at a current density (I_{max}) and a voltage corresponding to the maximum power conversion point in the photoelectrolysis process. The maximum power saved as a result of photoelectrolysis is shown as the shaded area in **Fig. 3.22(a)**.

$$\varepsilon_{o\,max} = \frac{\Delta V_{max} I_{max}}{P_t} \tag{3.6.15}$$

Here $\Delta V_{max} = V_{save(max)}$ represents the difference in voltages at the semiconductor electrode and metal electrode at the maximum power conversion point. For example, in their experiment using a p-type InP photocathode, Heller and Vadimsky [120] obtained a current 23.5 mA/cm^2 at maximum power point. A voltage of 0.11V vs SCE was applied in the case of InP electrode and -0.33V vs SCE in the case of platinum electrode, to obtain this current. Thus, the maximum saved voltage $\Delta V_{max}= 0.11-(-0.33)$ V$= 0.43$V. Therefore, $P_{saved}=0.43$ V x 23.5 mA/cm$^2 = 10.1$mW/cm^2. As they used a solar illumination of 84.7 mW/cm^2, the efficiency is 11.9%.

An issue in using this approach is that equations (3.6.14) and (3.6.15) involve overpotential losses. Hence highly catalytic metal electrodes with low overpotential are required for comparison. If a metal electrode with a low catalytic activity is used these equations yield exaggerated values for photoconversion efficiency.

An approach similar to this avoids the use of a comparative noble metal electrode and neglects overpotential losses at the electrodes. In this method, the potential applied at the hydrogen (or oxygen) electrode (in a three electrode configuration) is compared with the potential generated at an ideal fuel cell anode (or cathode). In the case of a n-type semiconductor photoanode:

$$\varepsilon_0 = \frac{\left(V_{ox}^0 - V_{app}\right) I}{P_t} \quad \text{(oxygen evolution using photoanodes)} \quad (3.6.15a)$$

or

$$\varepsilon_o = \frac{\left(V_{app} - V_{H_2}^0\right) I}{P_t} \quad \text{(hydrogen evolution using photocathodes)} \quad (3.6.15b)$$

V_{ox}^0 and $V_{H_2}^0$ represent the standard potentials of oxygen and hydrogen electrodes respectively. V_{ox}^0 takes a value of +0.401 V vs NHE in alkaline electrolytes (pH=14) and +1.229V vs NHE in acidic electrolytes (pH=0). The corresponding values of $V_{H_2}^0$ are -0.828 V vs NHE and 0 V vs NHE. For example, Ang and Sammells [116] reported on obtaining a photocurrent of 24 mA in a p-type InP cathode at –0.65V vs SCE (i.e. 0.408 V vs NHE) in KOH electrolyte. Thus the efficiency calculated using equation (3.6.15b) ($V_{H_2}^0 = -0.828$V vs NHE) is 10.1%.

Another form of this definition [equation (3.6.15)] has sparked much debate in the scientific community [121-124]. In this approach V_{app} (or V_{bias}) is taken as the absolute value of the difference between the potential at the working electrode measured with respect to a reference electrode (V_{meas}) and the open circuit potential (V_{oc}) measured with respect to the same reference electrode under identical conditions (in the same electrolyte solution and under the same illumination). In the case of a semiconductor photoanode where oxygen evolution takes place the efficiency is calculated as:

$$\varepsilon_0 = \frac{I_p \left(V_{rev}^0 - \left|V_{app}\right|\right)}{P_t}$$

$$\varepsilon_o = \frac{I \left(V_{rev}^0 - \left|V_{meas} - V_{aoc}\right|\right)}{P_t}$$

(3.6.16)

$V^0_{rev} = 1.229V$ is the standard state reversible potential for the water splitting reaction and V_{aoc} is the anode potential at open circuit conditions. Term $V_{meas}-V_{aoc}$ arises from the fact that V_{oc} represents the contribution of light towards the minimum voltage needed for water splitting potential (1.229V) and that the potential of the anode measured with respect to the reference electrode V_{meas} has contributions from the open circuit potential and the bias potential applied by the potentiostat (i.e. $V_{meas}= V_{app}+V_{aoc}$). The term $V_{meas}-V_{aoc}$ makes relation (3.6.16) independent of the electrolyte pH and the type of reference electrode used. Thus the use of V^0_{rev} in relation (3.6.16) instead of V^0_{ox} or V^0_{H2} as in the case of relation (3.6.15) is justified.

Although this approach has received wide attention and is being commonly used, there is skepticism that the efficiency values obtained using relation (3.6.16) gives exaggerated photoconversion efficiency values. In either the two electrode or three electrode geometry the voltage measured between the working and the counter electrodes gives the actual bias voltage $V_{app(wc)}$ applied (voltage in excess of the open circuit voltage). However in practice, where a potentiostat is used to apply an external bias to the photoelectrode, this actual voltage ($V_{app(wc)}$) may exceed the bias voltage measured as $V_{app}=V_{meas}-V_{aoc}$ with respect to the reference electrode [125]. Thus the use of the latter term in (3.6.16) can show a higher value for efficiency than when the term $V_{app(wc)}$ is directly used as in the case of relation (3.6.13).

Figure 3.23 is plotted to demonstrate the difference in the efficiency values calculated by different approaches represented by relations (3.6.13), (3.6.15a) and (3.6.16). A two electrode geometry was used for photocurrent measurements and efficiency calculation using relation (3.6.13) and a three electrode geometry was employed for relations (3.6.15a) and (3.6.16). A 6 μm long titania nanotube array film (polycrystalline) on titanium foil was used as the photoanode. Platinum served as the counter electrode and the 1M KOH solution as the electrolyte. In the three electrode configuration, an Ag/AgCl reference electrode was used to measure the photoanode potentials. As the titania bandgap is 3.0 to 3.2 eV (depending upon whether the crystalline phase is rutile or antase), a near UV light illumination (320-400nm) from a 50W metal-hydride lamp was used. For the two electrode geometry a Keithley source meter (model 2400) and for the three electrode geometry a potentiostat (CH Instruments, model CHI 600B) was used to apply the bias. The x-axis of the plot (**Fig. 3.23**) shows the voltage measured between the working (titania nanotube array) and counter electrodes for relation (3.6.13) and the potential at the photoanode anode measured with respect to Ag/AgCl reference electrode for relations (3.6.15a) and (3.6.16).

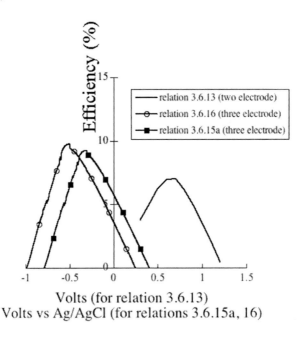

Volts (for relation 3.6.13)
Volts vs Ag/AgCl (for relations 3.6.15a, 16)

Fig. 3.23: Efficiency under near UV illumination of a photoelectrochemical cell comprised of a titania nanotube array photoanode and Pt counter electrode. For the calculation of efficiency using equation (3.6.13), a two electrode geometry was used while for the calculation using equations (3.6.15a) and (3.6.16), a three electrode geometry was used.

It can be seen from **Fig. 3.23** that the relation (3.6.16) yielded the highest efficiency of about 9.5% whereas the efficiency calculated using relation (3.6.13) employing a two-electrode geometry has a maximum value of about 7%. Although the relative values may vary with respect to the experimental setup and measurement conditions, this exercise demonstrates that a certain degree of exaggeration could occur in the efficiency values calculated using (3.6.16) in a three electrode geometry. Also, note that the efficiency shown is not the solar photoconversion efficiency. The value of solar photoconversion efficiency is much lower (discussed later).

Figure 3.23 shows that the efficiency calculated using (3.6.13) in a two-electrode geometry is maximum at a bias voltage of about +0.65 V. As discussed above, in the case of a three electrode geometry, the bias is calculated as $V_{app} = V_{meas} - V_{aoc}$. With a V_{aoc} of about -0.98 V vs Ag/AgCl and V_{meas} of about −0.5 V vs Ag/AgCl corresponding to the point of maximum efficiency, V_{app} takes a value of about +0.48 V. Thus, the actual bias applied between the working and the counter electrodes is higher (by 0.17 V) compared to the bias measured with respect to the reference electrode.

Replacement of V^0_{rev} in relation (3.6.16) by a term $V^0_{rev} + V_{overvoltage}$ was also suggested [126]. That is, overvoltage losses are added to the work output from hydrogen. This gives unrealistic values for efficiency as the efficiency increases with increased overvoltage losses at the working and counter electrodes. Furthermore, even by burning hydrogen such an amount of energy cannot be retrieved. However in some cases it is considered that the hydrogen is burnt to retrieve the energy, and the thermoneutral potential 1.48 V is used in equation (3.6.13) instead of 1.229 V [127].

A criticism often seen in the literature regarding use of relation (3.6.13) and its different forms is that it gives negative efficiency values. It should be noted that negative efficiency signifies that the bias voltage exceeded 1.229 V and the regime has changed from photoelectrolysis to direct electrolysis. That is, the minimum energy needed for water splitting is completely provided by the external power supply [122]. In a carefully done work, Murphy et al. [109] measured photocurrent using a three electrode geometry but the efficiency was calculated using (3.6.13) with the V_{bias} taken as the voltage between working and counter electrodes. It appears that any meaningful efficiency calculation should use (3.6.13) in three or two electrode geometry with V_{bias} measured between counter electrode and working electrode [117,128-130].

In addition to (3.6.13), (3.6.15) and (3.6.16) other approaches exist to calculate the efficiency. However, these are not considered to effectively represent the actual ability of the cell to convert light energy into chemical energy. One such definition for efficiency [131] is

$$\varepsilon_o = \frac{V_{rev}^0 \, I}{\left(P_t + I V_{bias}\right)} \tag{3.6.17}$$

However this relation does not yield a solar conversion efficiency but gives the throughput efficiency for the device [115]. Furthermore, it is insensitive to the contribution from light and only approaches zero at a bias voltage much higher than 1.229 V that can be regained in an ideal fuel cell. For the case where the photoelectrolysis cell supplies the input for the fuel cell and a part of the electrical energy output from the fuel cell is used to bias the phototoelectrolysis cell, relation (3.6.17) is modified as [131]:

$$\varepsilon_o = \frac{\left(V_{rev}^0 - V_{bias}\right) I}{\left(P_t + I V_{bias}\right)} \tag{3.6.18}$$

Recently, Raja et al. [132] proposed a method to calculate efficiency. In this method, the power output by the three-electrode photoelectrochemical cell is calculated by considering the voltage increase between the anode and cathode due to light illumination under external bias conditions.

$$\varepsilon_o = \frac{I_p \, \Delta E}{P_t} \tag{3.6.19}$$

I_p is the photocurrent density in mA cm^{-2}, ΔE the potential difference between working electrode and counter electrode under illumination minus the potential difference between the same electrodes without illumination (dark). That is, ΔE is the photovoltage with dark voltage subtracted from it. This equation is misleading and has no thermodynamics basis. ΔE does not necessarily represent the sample behavior but it depends upon the experimental conditions. Furthermore the hydrogen produced at current I_p can yield a power output higher than $I_p \Delta E$.

When a chemical bias is used instead of electrical bias, Ghosh and Maruska [133] defined efficiency in terms of heat of combustion of 1 mole of hydrogen (285.6 kJ) and heat of neutralization of 2 moles of H^+ (117.6kJ) [134] as:

$$\varepsilon_o = \frac{\dfrac{I}{nF}(285600 - 117600)}{P_t} \qquad (3.6.20)$$

where $n = 2$, $F = 96485$ Coulombs.   In the case of two-photoelectrode geometry involving an illuminated p-type anode and an illuminated n-type cathode, where light energy is converted to both chemical and electrical energy, Kainthla et al. [135] used the formula

$$\varepsilon_o = \frac{1.23 I + I V_{cell}}{2P_t} \qquad (3.6.21)$$

The chemical energy output is given by $1.23I$ while the electrical energy output is given by IV_{cell}. The factor of 2 is included to take into account the simultaneous illumination of both electrodes of equal area.

A very useful parameter for evaluating the performance of a photoelectrolysis cell is the incident photon to current conversion efficiency (IPCE).   This is a measure of the effectiveness in converting photons incident on the cell to photocurrent flowing between the working and counter electrodes. IPCE is also called the external quantum efficiency.

$$IPCE = \frac{I_p(\lambda)}{e F(\lambda)} \qquad (3.6.22a)$$

or $$IPCE = \left(\frac{hc}{e}\right)\left(\frac{I_p(\lambda)}{P(\lambda)\,\lambda}\right) \qquad (3.6.22b)$$

or $$IPCE = 1240 \frac{I_p(\lambda)}{P(\lambda)\lambda} \qquad (3.6.22c)$$

$I_p(\lambda)$ is the photocurrent density at wavelength λ. IPCE becomes 100% when all photons generate electron-hole pairs. However, in practical situations IPCE is always less than 100% due to the losses corresponding to the reflection of incident photons, their imperfect absorption by the semiconductor and recombination of charge carriers within the semiconductor, etc.

IPCE is calculated by measuring the current in a cell when a particular wavelength or a small group of wavelengths (band pass, usually up to 12 nm) with a known power density $P(\lambda)$ incident on it. It is usually measured at a bias voltage corresponding to the maximum power point (voltage corresponding to the peak efficiency in **Fig. 3.23**). The IPCE of a titania nanotube array biased at 0.5 V illuminated using wavelengths from a 300 W xenon arc lamp and monochromator (cornerstone 130) at a band pass of 4 nm is given in **Fig. 3.24**. The nanotube has a band gap of about 3.0 - 3.2 eV corresponding to antase-rutile mixed phase and hence the IPCE reduces to zero at the wavelength corresponding to the bandgap.

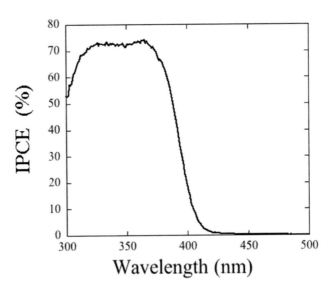

Fig. 3.24: Incident photon to current efficiency (IPCE) spectrum of a titania nanotube array photoelectrode.

IPCE enables the estimation of the total photocurrent as well as efficiency of a photoelectrolysis cell under any type of illumination, say for example global sunlight [136]. If $I_p(\lambda)$ is the current density corresponding to wavelength λ, the photocurrent spectrum corresponding to a particular energy distribution can be obtained by multiplying the IPCE with the photon flux density of that distribution.

$$I_p(\lambda) = IPCE\ F(\lambda)d\lambda = IPCE\ P(\lambda)\ \lambda\left(\frac{e}{hc}\right) \qquad (3.6.23)$$

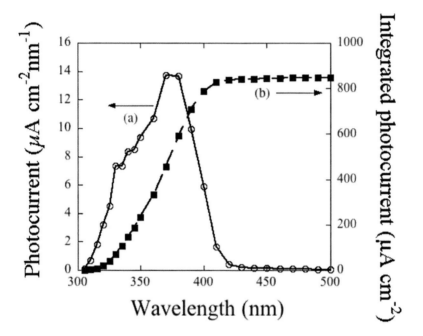

Fig. 3.25: **(a)** The solar photocurrent spectrum of a titania nanotube array obtained using data from **Fig. 3.19** and **Fig. 3.24**. **(b)** The total solar photocurrent obtained by integrated the photocurrent of **(a)**.

For example, the solar photocurrent spectrum of a titania nanotube array photoanode calculated using relation (3.6.23) is given in **Fig. 3.25**. The solar irradiance P (λ) for the calculation is taken from **Fig. 3.19**. The total photocurrent that can be obtained from this electrode when exposed to global sunlight is

$$I_p = \int_{\lambda_{min}}^{\infty} I_p(\lambda)d\lambda \qquad (3.6.24)$$

Curve (b) of **Fig. 3.25** is obtained by the integral of Curve (a). As can be seen from Curve (b) of **Fig. 3.25**, I_p= 847 $\mu A/cm^2$. Outdoor measurements yielded values agreeing well with this calculated photocurrent value (State college, Pennsylvania; latitude 40.79°N, longitude 77.86°W, on April 3, 2007 at 3:00 PM, clear sky, incident irradiance 950 W/m^2).

Solar photoconversion efficiency can be calculated using relation (3.6.24) in (3.6.13).

$$\varepsilon_0 = \left(\frac{e}{hc}\right)\left(V_{rev}^0 - V_{bias}\right)\left(\frac{\int_{\lambda_{min}}^{\infty} IPCE(\lambda)\, P(\lambda)\, \lambda\, d\lambda}{P_t}\right) \qquad (3.6.25)$$

This modified form of relation (3.6.13) is the most acceptable relation for calculating the photoconversion efficiency [109,123,137]. With V_{bias}= 0.51 V, the solar photoconversion efficiency of titania nanotube (6 μm length) array photoelectrodes was calculated as 0.6 %.

Another parameter of interest, used mainly in photoelectrochemical solar cells is the absorbed photon to current conversion (APCE) efficiency. This is also called the internal quantum efficiency. APCE is defined as the number of electrons (or holes) collected per absorbed photon. It is calculated after considering the losses in the incident photons like reflection, scattering, absorption, etc. APCE and IPCE are related by [138]:

$$IPCE(\lambda) = LHE(\lambda)\, \phi_{inj}\, \eta \qquad (3.6.26)$$

LHE is the light harvesting efficiency or absorptance, defined as LHE = 1-10^{-A} where A is the absorbance, ϕ_{inj} is the quantum yield of charge injection, and η is the efficiency of transporting injected electrons in to the external circuit. Equation (3.6.26) can be written as:

$$APCE = \frac{IPCE}{1-10^{-A}} \qquad (3.6.27)$$

IPCE and APCE can have values close to 100%. As discussed before, the maximum attainable photoconversion efficiency in a single bandgap photoelectrolysis cell is 30.7%. Although stable, the photoconversion efficiencies of most oxide semiconductors are low (<2% except the case of 8.35% reported for carbon modified titania [121]) due to their large band gap.

References

1. Veziroğlu TN (2000) Quarter century of hydrogen movement. 25:1143–1150
2. Govindjee (Ed.)(1975)   Bioenergetics of Photosynthesis. Academic Press, New York
3. Blankenship RE (2002) Molecular Mechanism of photosynthesis. Blackwell Science, Publishers, USA
4. Hall DO (1978) Solar energy conversion through biology—could it be a practical energy source? Fuel 57:322-333
5. Cuendet P, Grätzel M (1982) Artificial photosynthetic systems. Cellular and Molecular Life Sciences 38:223-228
6. Grätzel M (1982) Artificial photosysnthesis, energy-and light-driven electron transfer in organized molecular assemblies and colloidal semiconductors. Biochim Biophys Acta 683:221–244
7. Collings AF, Critchley C (Ed.) (2004) Artificial Photosynthesis-from basic biology to industrial application. Willey-VCH, Weinheim
8. Sun L, Hammarström, L, Akermark B, Styring S (2001) Towards artificial photosynthesis; ruthenium-manganese chemistry for energy production, Chem Soc Rev 30:36–39
9. Sun L, Akermark B, Hammarström, L, Styring S. (2003) Towards solar energy conversion into fuels; design and synthesis of ruthenium-manganese supramolecular complexes to mimic the function of photosystem II. In: utilization of Green house gases Liu CJ, Mallinson RG, Aresta M (Eds.) Amer Chem Soc Book Dept, Symposium Series No. 852, Washington USA, pp. 219–244
10. Lomoth R, Magnuson, A, Sjödin, Huang P, Styring S, Hammarström L (2006) Mimicking the electron donor side

of the photosystem II in artificial photosysnthesis. Photosysnthesis Res 87:25–40

11. Gerischer H (1979) Solar Photoelectrolysis with semiconductor electrodes. In: Solar energy conversion: Solid-state physics aspects, Seraphin BO (Ed). pp.115–172 Springer-Verlag New York

12. Vinodgopal K, Hotchandani S, Kamat PV (1993) Electrochemically assisted photocatalysis - TiO_2 particulate film electrodes for photocatalytic degradation of 4-chlorophenol. J Phys Chem 97:9040-9044

13. Hoffmann MR, Martin ST, Choi W, Bahneman DW (1995) Environmental applications of semiconductor photocatalysis. Chem Rev 95:69-96

14. Byrne JA, Eggins BR, Byers W, Brown NMD (1999) Photoelectrochemical cell for the combined photocatalytic oxidation of organic pollutants and the recovery of metals from waste waters. Appl Catal B: Environ 20:L85-89

15. Solarska R, Santato C, Jorand-Sartoretti C, M. Ulmann and J. Augustynski (2005) Photoelectrolytic oxidation of organic species at mesoporous tungsten trioxide film electrodes under visible light illumination. J Appl Electrochem 35:715–721

16. Quan X, Yang S, Ruan X, Zhao H (2005) Preparation of titania nanotube and their environmental applications as electrode. Envion Sci Technol 39:3770-3775

17. Nozik AJ (1980) Photoelectrochemical Cells. Phil Trans Royal Soc London Series 295:453-470

18. Heller A (1981) Conversion of sunlight into electric power and photoassisted electrolysis of water in photoelectronchemical cells. Acc Chem Res 14:154-162

19. Memming R (1988) Photoelectrochemical solar energy conversion. Top Curr Chem 143:79-112

20. Nozik AJ, Memming R (1996) Physical Chemistry of semiconductor-liquid interface. J Phys Chem 100:13061–13078

21. Hill R, Archer MD (1990) Photoelectrochemical cells- a review of progress in the past 10 years. J Photo Chem Photo Biol A: Chem 51:45–54

22. Bolton JR (1996) Solar photoproduction of hydrogen: review. Sol Energy 57:37–50

23. Tryk, DA, Fujishima A, Honda K. Recent topics in photoelectrochemistry: achievement and future prospect (2000) Electrochim Acta 45:2363-2376

24. Bak T, Nowotny, J, Rekas M, Sorrell CC. (2002) Photoelectrochemical hydrogen generation from water using solar energy. Materials-related aspects. Int J Hydrogen Chem 991–1022

25. Aroutiounian VM, Arakelyan VM, Shahnazaryan GE (2005) Metal oxides photoelectrode for hydrogen generation using solar water radiation driven water splitting. Sol Energy 78:581–592

26. Fujishima A, Honda K (1972) Electrochemical photolysis of water at semiconductor electrode. Nature 238:37-38

27. Wrighton MS, Ellis AB, Wolczanski PT, Morse DL, Abrahamson HB, Ginley DS (1976) Strontium titanate photoelectrodes. Efficient photoassisted electrolysis of water at zero applied potential. J Am Chem Soc 98:2774–2779

28. Mavroides JG, Kafalas JA, Kolesar DF (1976) Photoelectrolysis of water in cells with $SrTiO_3$ anodes. Apl Phys Lett 28:241–243

29. Bicelli LP, Razzini G (1985) Photoelectrochemical performance of anodic n-TiO_2 films submitted to hydrogen reduction. Int J Hydrogen Energy 10:645–649

30. Jaramillo TF, Baeck SH, Shwarsctein AK, Choi KS, Stucky GD, McFarland EW (2005) Automatated electrochemical synthesis and photoelectrochemical characterization of $Zn_{1-x}Co_xO$ thin film for solar hydrogen production. J Comb Chem 7:264–271

31. Nozik AJ (1976) p-n photoelectrolysis cell. Appl Phys Lett 29:150–153

32. Ohashi K, McCann J, Bockris JOM (1977) Stable photoelectrochemical cell for splitting of water. Nature 266:610-611

33. Lee J, Fujishima A, Honda K, Kumashiro Y (1985) Photoelectrochemicl behaviour of p-type boron phosphide photoelectrode in acidic solution. Bull Chem Soc Jpn 58:2634-2637

34. Kainthala RC, Zelenay B, Bockris JOM (1987) Significant efficiency increase in self-driven photoelectrochemical cell for water photoelectrolysis. J Electrochem Soc 134:841–845

35. Nozik AJ (1977) Photochemical diodes. Appl Phys Lett 30:567–569

36. Nozik AJ (1975)Photoelectrolysis of water using semiconducting TiO_2 crystals. Nature 257:383–386

37. Mavroides JG, Tchernev DI, Kafalas JA, Kolesar DF (1975) Photoelectrolysis of water in cells with TiO_2 anodes. Mater Res Bull 10:1023–1030

38. Ohnishi T, Nakato Y, Tsubumura H (1975) Quantum yield of photolysis of water on titanium oxide. Ber Bunsenges Phys Chem 79:523–525

39. Kung HH, Jarrett HS, Sleight AW, Ferretti A (1977) Semiconducting oxide anodes in photoassisted electrolysis of water. J Appl Phys 48:2463–2469

40. Giordano N, Antonucci V, Cavallaro S, Lembo R, Bart JCJ (1982) Photoassisted decomposition of water over modified rutile electrodes. Int J Hydrogen Energy 7:867–872

41. Khan SUM, Al-shahry M, Ingler Jr. WB (2002) Efficient photochemical water splitting by a chemically modified n-TiO_2. Science 297:2243–2245

42. Fujishima A, Kohayakawa K, Honda K (1975). Hydrogen production under sunlight with an electrochemical photocell. J electrochem Soc 122:1487–1489

43. Akikusa J, Khan SUM (1997) Photo response and AC impedance characterization of n-TiO_2 during hydrogen and oxygen evolution in an electrochemical cell. Int J Hydrogen Energy 22:875–882

44. Bak T, Nowotny J, Rekas M, Sorrell CC (2002) Photoelectrochemical properties of TiO_2-Pt system in aqueous solutions. Int J Hydrogen Energy 27:19–26

45. Heller A and Vadimsky RG (1981) Efficient solar to chemical conversion: 12% efficient photoassisted electrolysis in the [p-type InP(Ru)/HCl-KCl/Pt(Rh)] cell. Phys Rev Lett 46:1153–1156

46. El Zayat MY, Saed MO, El Dessouki MS (1998) Photoelectrochemical properties of dye-sensitized Zr-doped SrTiO$_3$ electrodes. Int J Hydrogen Energy 23:259-266

47. Grätzel M (2001) The photoelectrochemical Cells. Nature 414:338-344

48. Carpetis C (1982) A study of water electrolysis with photovoltaic solar energy conversion. Int J Hydrogen Energy 7:287–310

49. Murphy OJ, Bockris JOM (1984) Photovoltaic electrolysis: Hydrogen and electricity from water and light. Int J Hydrogen Energy 9:557-561

50. Fischer M (1986) Review of hydrogen production with photovoltaic electrolysis system. Int J Hydrogen Energy 11:495–501

51. Siegel A, Schott T (1988) Optimization of photovoltaic hydrogen production. Int J Hydrogen Energy 13:659–675

52. Khaselev O, Turner JA (1998) A monolithic photovoltaic-photoelectrochemical device for hydrogen production via water splitting, Science 280:425–427

53. Rocheleau RE, Miller EL, Misra A (1998) High efficiency photoelectrochemical hydrogen production using multijunction amorphous photoelectrode. Energy & Fuels 12:3-10

54. Licht S, Ghosh S, Tributsch, H, Fiecher (2002) High efficiency solar energy water splitting to generate hydrogen fuel: probing RuS$_2$ enhancement of multiple band electrolysis. Sol Energy Mater Sol Cells 70:471-480

55. Harry LS, Wilson RH (1978) Semiconductor for photoelectrolysis Annu Rev Mat Sci 8:99-134

56. Bard AJ (1979) Photoelectrochemistry and heterogeneous photocatalysis at semiconductor. J Photochem 10:59-75

57. Wrighton MS (1979) Photoelectrochemical conversion of optical energy to electricity and fuels. Acc Chem Res 12:303-310

58. Nozik AJ (1978) Photoelectrochemistry: Applications to solar energy conversion. Annu Rev Phys Chem 29:189-122

59. Gerischer H (1981) The principles of photoelectrochemical energy conversion. In: Cardon F, Gomes WP, Dekeyser W

(Eds) Photovoltaic and photoelectrochemical solar energy conversion, Plenum, New York, pp. 199-245

60.   Nozik AJ (1981) Photoelectrochemical devices for solar energy conversion. In: Cardon F, Gomes WP, Dekeyser W (Eds) Photovoltaic and photoelectrochemical solar energy conversion, Plenum, New York, pp. 263-312

61.   Heller A (1984) Hydrogen-evolving solar cells. Science 233:1141-1148

62.   Lewis NS (1990) Mechanistic studies of light-induced charge separation at semiconductor/ liquid interfaces. Acc Chem Res 23:176-183

63.   Pleskov YV (1990) Solar energy conversion: a photoelectrochemical approach, Springer-Verlag, Berlin

64.   Koval CA, Howard JN (1992) Electron transfer at semiconductor electrode-liquid electrolyte interfaces. Chem Rev 92:411-433

65.   Memming R (1994) Photoinduced charge transfer processes at semiconductor electrodes and particles. Top Curr Chem 169:105-181

66.   Sze SM (1981) Physics of semiconductor devices. John Wiley and Sons, New York

67.   Neamen DA (2002) Semiconductor Physics and devices: basic principles 3rd Ed, Mc-Graw Hill professional, New York

68.   Memming R (1983) Comprehensive treaties electrochemistry V. 7, Plenum press, New York

69.   Gerischer H (1975) Electrochemical photo and solar cell principles and some experiments. J Electroanal Chem:Interfacial Electrochem 58:263-274

70.   Gomer R, Tryson G (1977) An experimental determination of absolute half-cell emf's and single ion free energies of solvation. J Chem Phys 66:4413-4424

71.   von Helmholtz HLF (1879) Studies of electric boundary layers Ann Phys Chem 7:337-382

72.   Parsons R (1990) The electrical double layer: recent experimental and theoretical developments. Chem Rev 90:813-826

73.   Bockris JOM, Khan SUM (1993) Surface Electrochemistry. A molecular level approach. Plenum Press, New York

74.   Green M (1959) Electrochemistry of the semiconductor-electrolyte electrode. I The electrical double layer. J Chem Phys 31:200-203

75.   Memming R (2002) Semiconductor electrochemistry. Wiley-VCH, Weinheim

76.   Gerischer H (1970) Physical Chemistry: an advanced treatise. V.9A and V 4 Academic Press, New York

77.   Gerischer H (1990) The impact of semiconductors on the concept of electrochemistry. Electrochim Acta 35:1677–1690

78.   Memming R, Schwandt (1967) Potential and charge distribution at semiconductor electrolyte interface. Angew chem. Int Ed 6:851-861

79.   Thapar R, Rajeshwar K (1983) Mott-Schottky analyses on n- and p-GaAs/room temperature chloroaluminate molten-salt interfaces. Electrochim Acta 28:195-198

80.   Morrison SR (1980) Electrochemistry at semiconductor and oxidized metal electrodes. Plenum Press, New York

81.   Bard AJ, Faulkner LR (1980) Electrochemical methods: Fundamental and applications, John Wiley and Sons, New York

82.   Finklea HO (Ed.) (1988) Semiconductor electrodes; Studies in physical and theroretical chemistry. V. 55, Elsevier, NewYork

83.   Chazalviel J (1988) Experimental techniques for the study of the semiconductor-electrolyte interface. Electrochim Acta 33:461-476

84.   Chazalviel J (1990) Impedance studies at semiconductor electrodes: classical and more exotic techniques. Electrochim Acta 35:1545-1552

85.   Marcus RA (1956) On the theory of oxidation reduction reaction involving electron transfer. J Chem Phys 24:966-978

86.   Sutin N (1983) Theory of electron transfer reactions: Insights and hindsights. In: Prog Inorg Chem, Lippard SJ (Ed) 30:441-448, John Wiley & Sons, New York

87.   Peter LM (1991) Dynamic aspects of semiconductor photoelectrochemistry. Chem Rev 90:753-769

88.   Miller RDJ, Mclendon G, Nojik AJ, Schmickler W, Willing F (1995) Surface electron transfer Processes, VCH Publishers, New York

89.   Lewis NS (1991) An analysis of charge transfer rate constant for semiconductor-liquid interfaces. Annu Rev Phys Chem 42:541

90.   Lewis NS (1997) Progress in understanding electron transfer reaction at semiconductor/liquid interfaces. J Phys Chem B 102:4843-4855

91.   Gerischer H (1991) Electron transfer kinetics of redox reactions at semiconductor/electrolyte contact: a new approach. J Phys Chem

92.   Gao YQ, Gerogievskii Y, Marcus RA (2000) On the theory of electron transfer reactions at semiconductor/liquid interfaces. J Chem Phys 112:3358-3369

93.   Smith BB, Nozik AJ (1996) Study of electron transfer at semiconductor-liquid interfaces addressing the full system electronic structure. Chem Phys 205:47-72

94.   Smith BB, Halley JW, Nozik AJ (1996) On the Marcus model of electron transfer at immiscible liquid interface and its application to the semiconductor liquid interface. Chem Phys 205:245-267

95.   Boroda YG, Voth GA (1996) A theory of adiabatic electron transfer processes across the semiconductor-electrolyte interface. J Chem Phys 106:6168-6183

96.   Nishida M (1980) A Theoretical treatment of charge transfer via surface states at the semiconductor electrolyte interface. Analysis of water electrolysis process.

97.   Tauc J (1970) Absorption edge and internal electric field in amorphous semiconductors. Mater Res Bull 5:721-729

98.   Pankove JL (1971) Optical process of semiconductors. Prentice Hall, Englewood Cliffs, NJ, USA

99.   Rajeshwar K (1993) Spectroscopy 8:16

100.   Salvador P (2001) Semiconductor photoelectrochemistry: A kinetic and thermodynamic analysis in the light of

equilibrium and non-equilibrium models. J Phys Chem 105:6128–6141

101. Bird RE, Hulstrom RL, Lewis LJ (1983) Terrestrial solar spectral data sets. Solar energy 30:563-573

102. Bolton JR (1996) Solar photoproduction of hydrogen: A review. Solar Energy 57:37-50

103. Bolton JR, Strickler SJ, Connolly JS (1985) Limiting and realizable efficiencies of solar photolysis of water. Nature 316:495-500

104. Bolton JR (1978) Solar Fuels. Science 202:705-711

105. Archer MD, Bolton JR (1990) Requirements for ideal performance of photochemical and photovoltaic solar energy converters. J Phys Chem 94:8028-8036

106. Grimes DM; Grimes, CA (2006) A unique electromagnetic photon field using Feynman's electron characteristics and Maxwell's equations. J Computational and Theoretical Nanoscience 3:649-663

107. Shockley W, Queisser HJ (1961) Detailed balance limit of efficiency of p-n junction solar cells. J Appl Phys 32:510-519

108. Hanna MC, Nozik AJ (2006) Solar conversion efficiency of photovoltaic and photoelectrolysis cells with carrier multiplication absorbers. J Appl Phys 100: 074510 (8 pages)

109. Murphy AB, Barnes PRF, Randeniya LK, Plumb IC, Grey IE, Horne MD, Glasscock JA (2006) Efficiency of solar water splitting using semiconductor electrodes. Int J Hydrogen Energy 31:1999-2017

110. Gerischer H (1981) The principles of photoelectrochemical energy conversion. In: Cardon F, Gomes WP, Dekeyser W (Eds) Photovoltaic and photoelectrochemical solar energy conversion, Plenum, New York, pp. 199-245

111. Weber MF, Dignam MJ (1984) Efficiency of splitting water with semiconducting photoelectrodes. J Electrochem Soc 131:1258-1265

112. Heller A (1981) Conversion of sunlight into electrical power and photoassisted electrolysis of water in photoelectro-chemical cells. Acc Chem Res 14:154-162

113.   Fujishima A, Kohayakawa K, Honda K (1975) Hydrogen production under sunlight with an electrochemical photocell. J Electrochem Soc 122:1487-1489

114.   Bard AJ, Faulkner LR (2001) Electrochemical methods: Fundamentals and applications. John Wiley & Sons, New Jersey

115.   Parkinson B (1984) On the efficiency and stability of photoelectrochemical devices. Acc Chem Res 17:431-437

116.   Ang PGP, Sammells AF (1984) Hydrogen evolution at p-InP photocathodes in alkaline electrolyte. J Electrochem Soc 131:1462-1464

117.   Dohrmann JK, Schaaf NS (1992) Energy conversion by photoelectrolysis of water: determination of efficiency by in situ photocalorimetry. J Phys Chem 96:4558-4563

118.   Heller A (1982) Electrochemical solar cells. Solar energy 29:153-162

119.   Aharon-Shalom E, Heller A (1982) Efficient p-InP (Rh-H alloy) and p-InP (Re-H alloy) hydrogen evolving photocathodes. J Electrochem Soc 129:2865-2866

120.   Heller A, Vadimsky RG (1981) Efficient solar to chemical conversion: 12% efficient photoassisted electrolysis in the [p-type InP(Ru)]/HCl-KCl/Pt(Rh) cell. Phys Rev Lett 46:1153-1156

121.   Khan SUM, Al-shahry M, Ingler Jr. WB (2002) Efficient photochemical water splitting by a chemically modified n-TiO_2. Science 297:2243–2245

122.   Lackner KS (2003) Comment on "Efficient photochemical water splitting by a chemically modified n-TiO_2" - (III). Science 301:1673c

123.   Hagglund C, Gratzel M, Kasemo B (2003) Comment on "Efficient photochemical water splitting by a chemically modified n-TiO_2" - (II). Science 301:1673b

124.   Fujishima A (2003) Comment on "Efficient photochemical water splitting by a chemically modified n-TiO_2" - (II). Science 301:1673a

125.   Raja KS, Mahajan VK, Misra M (2006) Determination of photoconversion efficiency of nanotubular titanium oxide

photo-electrochemical cell for solar hydrogen generation. J Power Sources 159:1258-1265

126. Khan SUM, Akikusa J (1999) Photoelectrochemical splitting of water at nanocrystalline n-Fe_2O_3 thin-film electrodes. J Phys Chem B 103:7184-7189

127. Tomkiewicz M, Woodall JM (1977) Photoelectrolysis of water with semiconductor materials. J Electrochem Soc 124:1436-1440

128. Butler MA, Ginley DS (1980) Principles of photoelectrochemical, solar energy conversion. J Mater Sci 15:1-19

129. Nozik AJ (1975) Photoelectrolysis of water using semiconducting TiO_2 crystals. Nature 257:383-386

130. Wrighton MS, Ginley DS, Wolczanski PT, Ellis AB, Morse DL, Linz A (1975) Photoassisted electrolysis of water by irradiation of a titanium dioxide electrode. Proc Nat Acad Sci 72:1518-1522

131. Bockris JOM, Murphy OJ (1982-1983) The two efficiency expressions used in evaluating photo-assisted electrolysis. Appl Phys Commun 2:203-207.

132. Varghese, OK; Grimes, CA (2007) Appropriate Strategies For Determining The Photoconversion Efficiency Of Water Photoelectrolysis Cells: A Review With Examples Using Titania Nanotube Array Photoanodes. Solar Energy Materials and Solar Cells, in press.

133. Ghosh AK, Maruska HP (1977) Photoelectrolysis of water in sunlight with sensitized semiconductor electrodes. J Electrochem Soc 124:1516-1522

134. Bezman R, Fujishima A, Kohayakawa K, Honda K (1976) Hydrogen production under sunlight with an electro-chemical photocell. J Electrochem Soc 123:842-843

135. Kainthla RC, Zelenay B, Bockris JOM (1987) Significant efficiency increase in self-driven photoelectrochemical cell for water photoelectrolysis. J Electrochem Soc 134:841-845

136. Kay A, Cesar I, Gratzel M (2006) New Benchmark for water photooxidation by nanostructured α-Fe_2O_3 films. J Am Chem Soc 128:15714-15721

137. Bard AJ, Memming R, Miller B (1991) Terminology in semiconductor electrochemistry and photoelectrochemical energy conversion. Pure Appl Chem 63:569-596

138. Nazeeruddin MK, Kay A, Rodicio I, Humphry-Baker R, Muller E, Liska P, Vlachopoulos N, Gratzel M (1993) Conversion of light to electricity by cis-X_2Bis(2,2'-bipyridyl-4,4'-dicarboxylate)ruthenium(II) charge-transfer sensitizers (X=Cl⁻, Br⁻, I⁻, CN⁻, and SCN⁻) on nanocrystalline TiO_2 electrodes. J Am Chem Soc 115:6382-6390

Chapter 4

OXIDE SEMICONDUCTING
MATERIALS AS PHOTOANODES

4.1 Introduction

The standing objective in water photolysis research is development of a robust, efficient, reliable, cost-effective, and stable photoelectrochemical system for splitting water using only sunlight as the energy input. During the last three decades various metal oxide (TiO_2, $SrTiO_3$, Fe_2O_3, etc.) and non-oxide (GaAs, CdS, InP, etc.) semiconductors have been employed as photoanodes in photoelectrochemical cells. The photocorrosion stability of the photoanode and/or photocathode, its wavelength response, and current-voltage behavior are key factors underlying the ability to achieve a useful, i.e. cost effective, phoelectrochemical device. However to date the materials investigated have been found lacking. For example, lower bandgap materials respond more fully to visible spectrum light but are less stable, i.e. more prone to corrosion, under operation in a photoelectrochemical cell than their higher bandgap counterparts. In contrast higher bandgap materials, while quite stable under operation in a photoelectrochemical cell, respond to a significantly smaller fraction of the incident light energy, e.g. that of the UV range with $\lambda < 400$ nm.

   This chapter considers photoanodes comprised of metal oxide semiconductors, which are of relatively low cost and relatively greater stability than their non-oxide counterparts. In 1972 Fujishima and Honda [1] first used a crystal wafer of n-type TiO_2 (rutile) as a photoanode. A photoelectrochemical cell was constructed for the decomposition of water in which the TiO_2 photoanode was connected with a Pt cathode through an external circuit. With illumination of the TiO_2 current flowed from the Pt electrode to the

TiO$_2$ electrode through the external circuit. The direction of the current revealed that oxygen evolution occurred at TiO$_2$ photoanode and hydrogen evolution at Pt metal cathode.

Since the groundbreaking report of Fujishima and Honda many metal oxide photoanode materials, and material architectures, have been investigated however oxides demonstrating suitable stability and charge transfer properties, e.g. TiO$_2$, have an energy gap sufficient for capturing only a small, approximately 5%, fraction of the incident solar energy that reaches earth. Efforts have been made to improve the efficiency of oxide semiconductors, with studies performed on single and polycrystalline materials in either thin film form or as sintered specimens. Of particular interest is recent work on the use of nanocrystalline, nanoporous, and nanotubular semiconducting metal oxides in their application to hydrogen photoproduction. In this chapter we detail some methods and strategies used for improving device efficiencies.

4.2 Photoanode Reaction Mechanisms

The solar photoelectrochemical process is one of the most attractive methods for conversion of solar to chemical energy fuel, with hydrogen as the energy carrier, by means of water splitting. There are three possible ways to configure the electrodes in such systems: {1} Photoanode made of n-type semiconductor and cathode made of metal. {2} Photoanode made of n-type semiconductor and cathode made of p-type semiconductor. {3} Photocathode made of p-type semiconductor and anode made of metal. The simplest photoelectrochemical cell designed for such purpose consists of a semiconductor photoelectrode, or photoanode, and a metal counter electrode, or cathode, immersed in an electrolyte solution, see **Fig. 4.1**. With light incident upon the photoanode, the photoanode absorbs part of the light generating an electron-hole pair which are then used for water electrolysis.

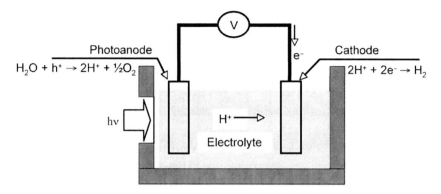

Fig. 4.1: Skeleton structure of a photoelectrochemical cell (PEC) comprised of a photoanode and cathode. Potentials of both are measured with reference to a third electrode, the standard calomel electrode.

Photoelectrolysis of water involves several processes within a photoanode, and at the photoanode-electrolyte interface. These are: (1) Absorption of light at or above the band gap. (2) Excitation of electrons from the valance band to the conduction band leaving holes behind in the valance band; that is to say, photo-generation of charge carriers (electron-hole pairs) due to light induced intrinsic ionization of the photoanode. (3) Charge separation and migration, at the same time, of electrons passing through the photoanode to the back-side electrical contact, and holes to the interface between the photoanode and electrolyte. (4) Oxidation of water at the photoanode by holes. (5) The transport of H^+ ions from the photoanode to the cathode through the electrolyte. (6) The transport of electrons from photoanode to cathode through the external circuit, leading to the reduction of H^+ ions at the cathode by these electrons.

The overall reaction of the photoelectrochemical cell (PEC), $H_2O + hv \rightarrow H_2 + 1/2O_2$, takes place when the energy of the photon absorbed by the photoanode is equal to or larger than the threshold energy of 1.23 eV. At standard conditions water can be reversibly electrolyzed at a potential of 1.23 V, but sustained electrolysis generally requires ~1.5 V to overcome the impedance of the PEC. Ideally, a photoelectrochemical cell should operate with no external bias so as to maximize efficiency and ease of construction. When an n-type photoanode is placed in the electrolyte charge distribution occurs, in both the semiconductor and at the semiconductor-electrolyte interface, leading to the formation of both a depletion layer and an accumulation layer, see **Fig. 4.2**. The photoelectrode zero charge potential can be observed when there is no excess charge on the semiconductor, which forms the basis for the flat band potential, V_{FB}. An accumulation layer forms when negative excess charge associated with the semiconductor electrons accumulate at the interface, which are compensated by positive ions from the electrolyte. Similarly a depletion layer forms where the positive charge is depleted of electrons, and thus negative counter ions compensate the charge at the interface from ions in the electrolyte.

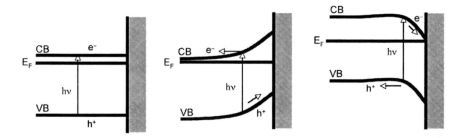

Fig. 4.2: For an n-type bulk semiconductor in the presence of an electrolyte illustrated is: (left) no space charge layer, (center) a space charge layer in a depletion region, (right) a space charge layer in an accumulation region.

The flat band potentials of a semiconductor can be determined from the photocurrent-potential relationship for small band bending [equation (4.2.1)], or derived from the intercept of Mott-Schottky plot [equation (4.2.2)] using following equations

$$V - V_{FB} = BJ^2_{PC}(\lambda) \tag{4.2.1}$$

$B = \left(\dfrac{1}{\alpha} \cdot W \cdot e \cdot I_O\right)^2$, W is the width of depletion layer for a band bending of 1.0 V, I_O is the flux of incident photon for any given wavelength λ, and J_{pc} is the photocurrent density at any given wavelength λ.

$$\frac{1}{C_{SC}^2} = \frac{2}{e\varepsilon\varepsilon_0 N_D}\left(V - V_{FB} - \frac{kT}{e}\right)$$ (4.2.2)

where C_{SC} is space charge capacitance, ε is dielectric constant of the semiconductor, ε_0 is permittivity of the vacuum, N_D is the donor density, V is the applied potential (external bias) and $\frac{kT}{e}$ is the temperature dependence term in Mott-Schottky equation which is 0.03V at 298K.

The absolute placement of the valence and conduction bands effectively controls the degree of band bending E_B, since the Fermi level E_F for heavily doped n-type metal oxide semiconductors coincides with the conduction band level. The value of E_B is generally taken as the difference between E_F and the Fermi level of electrolyte, which is taken as the redox potential, $E_{F,redox}$, for the reactions occurring at the photoanode. For n-type semiconductors $E_{F,redox}$ is the O_2/OH^- potentials. The maximum open circuit photopotential is E_B, and if it exceeds 1.23 V it is possible to run the water splitting reaction without addition of external energy. In the case that E_B falls short of 1.23 V, external bias is needed. This means that the conduction band position must be $\geq$ 1.23 V above the O_2/OH^-. For rapid and energetically feasible electron transfer the valence band must be slightly below the O_2/OH^- level. The ideal situation is depicted in **Fig. 4.3.**

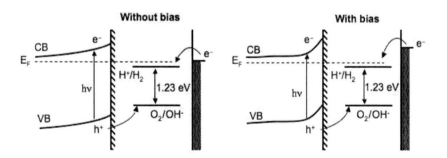

Fig. 4.3: Illustration of the effect of bias on water photoelectrolysis at photoanode/cathode.

Semiconductors suitable to serve as photoanodes for solar water photoelectrolysis must have the following general traits: (a) Chemical stability both under illumination and dark. (b) A band gap of approximately 2.0 eV to absorb maximum solar radiation. (c) Absence of charge recombination centers to prevent recombination of the photogenerated charge carriers. (d) Moderate conductivity; too high and the series resistance degrades the efficiency, too low and the electrolyte electrically shorts to the photoelectrochemical cell. (e) Suitable band edge positioning with respect to the H^+/H_2 reduction potential and O_2/OH^- oxidation potentials.

4.3 General Description of Oxide Semiconductor Photoanodes

The most frequently studied semiconducting photoanode materials are TiO_2, $SrTiO_3$, WO_3, SnO_2 and Fe_2O_3, see **Table 4.1**. The strengths and weakness of each material vary; for example, the maximum value obtained for the photo-voltage of a photoelectrochemical cell using a TiO_2 photoanode is ~0.9 V. Hence, for TiO_2 (rutile, bandgap ≈ 3.0 eV) photoanodes the conduction band edge potential is insufficiently negative to generate hydrogen at a useful rate. Consequently it requires a bias in order to decompose water by use of an externally applied bias voltage, or by imposing an internal bias voltage through the use of different hydrogen ion concentrations (pH). Moreover, such large band gaps (< 420 nm) do not permit one to effectively capture the suns' energy which is predominately in the visible region. Hematite, α-Fe_2O_3, has a bandgap of approximately 2.2 eV hence is well suited for collecting solar energy, however the charge transport properties of

Table 4.1: Properties of single crystal metal oxide photoanodes

Photoanode	E_{BG}[a]	Donor Concn.,[b] cm^{-3}	E_{VB}[c]	E_{CB} $\approx V_{FB}$[d]	E_B[e]
TiO_2	3.0	3×10^{20}	1.95	−1.05	1.0
SnO_2	3.5	2×10^{18}	2.8	−0.7	0.6
$SrTiO_3$	3.2	3×10^{20}	2.8	−1.3	1.3
$KTaO_3$	3.4	3×10^{20}	1.8	−1.4	1.2
Fe_2O_3	2.2	2.5×10^{17}	2.5	−0.3	-
Fe_2O_3 + 1% TiO_2	2.2	6.4×10^{19}	-	-	-
Fe_2O_3 + 1% SnO_2	2.2	7.8×10^{18}	-	-	-

[a]E_{BG} represents band gap; [b]derived from the slope of Mott-schottky plots using equnation $s = 2(q\epsilon\epsilon_0 A^2 N)^{-1}$; [c]Vs SCE at pH = 13; [d] V_{FB} is from the intercept of Mott-Schottky plot; [e]E_B is independent of pH.

the material are generally quite low resulting in low photoconversion efficiencies. There are several approaches to improving the efficiency of metal oxide based PECs. Doping with substitutional elements is one such approach which, in general, modestly improves the photo-response of oxide semiconductors. Since most of the PEC configurations involve n-type oxide semiconductor photoanodes and metal cathodes, use of two similar and/or different, n-type and p-type, oxide and/or non-oxide semiconductors as electrodes are an approach to harvesting a large part (visible region) of the solar spectrum energy. If the electron affinity of the n-type electrode is greater than that of p-type electrode then the available electron-hole potential for driving a chemical reaction within the electrolyte is greatly enhanced by the simultaneous irradiation of the (n-type and p-type) electrodes. Due to enhanced electron-hole potentials, various combinations of n-type and p-type semiconductors such as n-TiO_2/p-GaP, n-$SrTiO_3$/p-CdTe, n-$SrTiO_3$/p-GaP, n-Fe_2O_3/p-Fe_2O_3 have been used to eliminate the bias needed for semiconductor electrodes when the flat band potential is below the reduction potential of hydrogen. However a truly useful architecture of such combinations, involving an n-type semiconductor as a photoanode and a p-type semiconductor as a photocathode, has yet to be achieved. The main reason for this is the lack of a suitable p-type oxide semiconductor with suitable electrical properties. One can modify various oxide semiconductors at room temperature to p-type, but obtaining electrical properties to match the need for efficient water photoelectrolysis is still elusive. For example, while Fe_2O_3 exhibits fairly good p-type properties with magnesium doping charge transport properties have still been so low as to prevent useful application of the material [2].

For heavily doped n-type semiconductors, the flat band is nearly coincident with the conduction band, while for heavily doped p-type semiconductors the flat band lies very close to the valence band edge. A necessary thermodynamic condition for the photoproduction of hydrogen and oxygen is that the p-type conduction band must be at or above the H^+/H_2 half cell potential, while n-type valence band must lie below the O_2/OH^- half cell potential.

Cells with TiO_2 photoanodes of various types (e.g., single crystal, polycrystalline, thin film) have been heavily investigated, largely because TiO_2 resists photocorrosion, and demonstrates excellent charge transport properties. The stability of the semiconductor in contact with a liquid while under irradiation is a critical factor. A necessary requirement of a photoanode is resistance to corrosion reactions at the electrolyte interface. Photocorrosion refers to chemical reactions between the electrolyte and semiconductor in the presence of light generated charge carriers. In liquid environments the semiconductor may suffer electrochemical corrosion, that is react with the electrolyte even in the absence of light generated charge carriers. Some oxide materials, such as n-TiO_2, n-SnO_2, n-WO_3, n-$SrTiO_3$ are resistant to electrochemical corrosion, while ZnO is stable only as a photocathode (p-ZnO). Cuprous oxide, with a band gap excellent for capturing solar spectrum energy, 2.2 eV, generally suffers from electrochemical corrosion.

Rapid charge transfer across the electrolye-semiconductor photoanode interface can occur when the appropriate bands overlap the electrolyte redox level. Thus the competition between recombination of the photo-generated electron-hole pairs and charge transfer across the electrolyte-photoanode interface influences the quantum efficiencies. To generate oxygen, rather energetic photogenerated holes are required and these tend to cause decomposition of the semiconductor. Thus a key requirement in single junction PECs cells is the discovery of a semiconductor with an appropriate band gap for effectively capturing solar spectrum energy (generally less than 2.5 eV), with the conduction band sufficiently negative for hydrogen evolution and the valence band sufficiently positive for oxygen evolution, so that it remains stable under irradiation. This efficient and stable single-junction semiconductor electrode has yet to be realized. **Table 4.2** summarizes experimental data on photoelectrochemical cell efficiencies involving some of the representative single crystals and polycrystalline photoanodes.

Table 4.2: Experimental data on efficiencies of photoelectrochemical cell involving representative single crystals and polycrystalline photoanodes

Photoelectrochemical cells			Efficiencies			Light source	Ref.
Photoanode	Cathode	Electrolyte	Bias (V)	η_{qe} (%)	η_c (%)		
Single crystal							
TiO_2	Pt	Fe^{3+}	-	10	-	Xe Lamp, 500 W	1
TiO_2	Pt	1) 0.2 M H_2SO_4	0.6		2.4	UV light, 26 mW/cm²	5
			0.8		4.1		
		2) 0.1 M KOH (A) + 0.2 M H_2SO_4 (C)	0.0		9.5		
			0.4		10.1		
TiO_2	Pt	0.5 M K_2SO_4 + 0.05 M CH_3COOH + 0.05 M CH_3COONa	-	0.067	0.029	Xe Lamp, 500 W	6
TiO_2	Pt	5.0 M KOH	-	-	0.4	Sunlight, 105 mW/cm²	8
TiO_2:Cr					0.44		
TiO_2:Al					1.3		
		5.0 M KOH + 2 M H_2SO_4			0.6		
TiO_2-SiO_2	Pt	1.0 M NaOH	-	40	-	Xe lamp, 500 W	12
TiO_2-SiO_2-				82			
Al_2O_3							
TiO_2-SiO_2-				62.5			
In_2O_3							
$SrTiO_3$	Pt	9.5 M NaOH	0.0	-	20.0	Hg Arc laser, 200 W	13
$SrTiO_3$	Pt	10.0 M NaOH	0.0	10	5.0	3.8 eV photon	15
			0.7	~75	25.0		
Fe_2O_3	Pt	0.1 M NaOH	0.4	0.5	-	150 W Xe lamp	20
Polycrystalline							
TiO_2	Pt	1.0 M Na_2SO_4	0.5		0.5	Hg lamp, 400 W	28
TiO_2:Pt			0.5		2.7		
TiO_2:W	Pt	0.1 M KCl	0.65		0.20	Xe Lamp, 450W	32
WO_3:Ti	Pt	0.1 M KCl	0.65	4.2	~0.025	Xe Lamp, 450W	33
TiO_2	Pt	4.0 M KOH (A) + 4.0 M HCl (C)	0.86*		0.4	Sunlight	34
TiO_2:Cr	Pt	1.0 M KCl (A) + 3.0 M HCl (C)	0.44*	-		Sunlight	34

*Chemically biased. A stands for anode, and C for cathode.

4.4 Single Crystal Materials as Photoanodes

4.4.1 Crystalline n-TiO_2

The electrochemical behavior of n-type titanium oxide (n-TiO_2) photoanodes was first studied in conjunction with a Pt counter electrode connected via an external circuit [1,3]; the modern era of semiconductor electrodes and interest in these photoelectrochemical devices for energy conversion, especially via the water-splitting reaction, can be traced from these efforts. In those early experiments n-TiO_2 single crystals were used after a reduction treatment to improve electric conductivity. Flow of anodic current is observed at a potential more positive than −0.5 V (*vs* SCE) in an electrolyte (Na$_2$SO$_4$) solution of pH 4.7 upon irradiation of light having a wavelength less than 415 nm, i.e. 3.0 eV, which corresponds to the band gap of rutile TiO$_2$, while oxygen evolution seen at the surface of the TiO$_2$ photoanode due to the photo-induced holes produced in the TiO$_2$ valence band [3]. The direction of current revealed that oxidation (oxygen evolution) occurs at the TiO$_2$ electrode and reduction (hydrogen evolution) at the Pt electrode. An observed quantum efficiency of approximately 10% has been achieved when more reducible species, such as Fe^{3+} ions, were added in the Pt electrode compartment with light from a 500 W Xe lamp used to irradiate TiO_2 surface [1]. In the absence of bias voltage low overall solar conversion efficiency of photoelectrolysis is observed, since the band bending at the TiO_2 surface is not sufficient to cause efficient separation of the electron-hole pairs [4]. Nozik reported the photoelectrolysis of water using a photoanode comprised of semiconducting TiO_2 crystals in the presence of various electrolytes such as 1.0 M phosphate buffer, 0.1 M KOH as well as 0.1M KOH in anodic and 0.2 M H_2SO_4 in cathodic compartments [5]. In these experiments the TiO_2 crystals were illuminated (UV light, 300-400 nm) with an intensity of 26 mW cm^{-2}. A homogeneous cell containing 0.2 M H_2SO_4 electrolyte shows photoconversion efficiencies of 2.1% and 4.1 % at 0.6 and 0.8 bias voltage, while the differential pH cell containing 0.1 M KOH in anodic and 0.2 M H_2SO_4 in cathodic compartments exhibits 9.5% efficiency at zero bias voltage and 10.1% at a bias voltage of 0.4 V [5]. Ohnishi and coworkers have developed a similar homogeneous photoelectro-chemical cell containing mixed-electrolyte solutions, see **Table 4.1** [6]. We now consider reactions on the surface of a TiO_2 electrode and Pt Cathode in neutral and alkaline solutions [7]:

Reaction on TiO₂ photoanode in neutral medium:

$$TiO_2 + h\nu \rightarrow e^-(TiO_2) + h^+(TiO_2) \qquad (4.4.1)$$
$$h^+(TiO_2) + H_2O \rightarrow OH + H^+_{(aq)} \qquad (4.4.2)$$
$$OH + OH \rightarrow H_2O_2 \ (k = 6 \times 10^9 \ M^{-1}S^{-1}) \qquad (4.4.3)$$
$$OH + H_2O_2 \rightarrow HO_2 + H_2O \ (k = 1.3 \times 10^{10} \ M^{-1}S^{-1}) \quad (4.4.4)$$
$$HO_2 \rightarrow O_2^-{}_{(aq)} + H^+_{(aq)} \ (pK = 4.8) \qquad (4.4.5)$$
$$O_2^-{}_{(aq)} + h^+(TiO_2) \rightarrow TiO_2 + O_2 \uparrow \qquad (4.4.6)$$
$$O_2^-{}_{(aq)} + O_2^-{}_{(aq)} \rightarrow O_2^{2-} + O_2 \uparrow \ (pH \leq 6) \qquad (4.4.7)$$
$$HO_2 + O_2^- \rightarrow HO_2^- + O_2 \uparrow \qquad (4.4.8)$$
$$HO_2 + HO_2 \rightarrow H_2O_2 + O_2 \uparrow \ (pH \leq 3) \qquad (4.4.9)$$

Reaction on TiO₂ photoanode in alkaline medium:

$$h^+(TiO_2) + OH^- \rightarrow TiO_2 + OH \qquad (4.4.10)$$
$$OH \rightarrow O^-{}_{(aq)} + H^+_{(aq)} \ (pK = 11.9) \qquad (4.4.11)$$
$$h^+(TiO_2) + O^-{}_{(aq)} \rightarrow TiO_2 + O_{(ads)} \qquad (4.4.12)$$
$$O_{(ads)} + O_{(ads)} \rightarrow O_2 \uparrow \qquad (4.4.13)$$

Reaction on Pt cathode:

$$e^-(TiO_2) \rightarrow e^-(Pt) + H^+_{(aq)} \rightarrow H_{(ads)} \qquad (4.4.14)$$
$$e^-(Pt) + H^+_{(aq)} \rightarrow H_{(ads)} \qquad (4.4.15)$$
$$H_{(ads)} + H_{(ads)} \rightarrow H_2\uparrow \qquad (4.4.16)$$
$$e^- \rightarrow e^-_{(sol)} \text{ (Solvated electron at interface)} \qquad (4.4.17)$$
$$e^-_{(sol)} + H^+_{(aq)} \rightarrow H \text{ } (k = 2.3 \times 10^{10} \text{ M}^{-1}\text{s}^{-1}) \qquad (4.4.18)$$
$$H + H \rightarrow H_2\uparrow \qquad (4.4.19)$$

Inhibition of H_2 formation can be seen when the anode-volume is saturated with O_2; application of an external bias up to 0.7 V can usually prevent this effect. increased yield of H_2 when Pt is present as a cathode has been rationalized in terms of three factors: (a) the removal of conduction band electron from TiO_2 to Pt [equation (4.4.14)], (b) The ease of reactions (4.4.15) and (4.4.16) because of a low overpotential for H_2 evolution from water at the Pt cathode, and (c) H atom migration to the Pt cathode.

Marusuka and coworkers showed that the spectral response of TiO_2 can be extended into the visible portion of the spectrum by impurity doping [8]. Photoanodes made up of single crystals of TiO_2 doped with Al^{3+} (0.05 wt%) and Cr^{3+} (0.0004-0.4%) were illuminated with solar spectrum light at an intensity of 105 mWcm^{-2}. The presence of chromium in the rutile lattice has been found to give rise to optical absorption extending from fundamental band edge of 415 nm to 550 nm. The magnitude of the photoresponse depends upon the chromium concentration; a maximum response was found for a 0.4 wt% (10^{20} Cr atom cm^{-3}) dopant concentration. A 10% increase in sunlight conversion efficiency is observed for TiO_2:Cr under visible light irradiation as compared to undoped TiO_2 ($\eta \approx 0.4\%$). Although Al doped titania (TiO_2:Al) has not shown significant photoresponse in the visible region, it does enhance performance in the UV region. Doping of Al increases the minority (hole) carrier diffusion length therefore permitting an increase in the collection of photogenerated charge. In addition, doping can introduce or modify the surface states that are essential for oxidation of OH$^-$. Marusuka and coworkers [8] report a solar conversion efficiency of 1.3% for TiO_2:Al ($\lambda = 375$ nm) in a 5.0 M KOH electrolyte. In the case of Al doped TiO_2 crystals, where Al is believed to have entered at either interstitial or substitutional sites, it appears that the electrochemical stability of a photoanode is closely connected with these defects [9]. By introducing pentavalent impurities such as Nb^{5+} to TiO_2, followed by adequate reduction, n-type semiconductors can be obtained for better PEC performance with well-defined electron (donor) concentrations ($N_D \approx 10^{19}$ cm^{-3}) and no indication of (interstitial, vacancy) defects [9-11]. The size of Nb^{5+} (0.69 Å) and Ti^{4+} (0.64 Å) are close to those needed for isotropic dissolution and thus allows niobium to act as an electron donor to reduce Ti^{4+} to Ti^{3+}. This leads to an increase in the concentration of the majority (electron) carriers and therefore the conductivity. The photocurrent densities sometimes vary due to incomplete dissolution as well as large dopant concentrations [10]. Raising the sintering temperature of TiO_2:Nb to 1350°C and maintaining the doping concentration $N_D/[Ti^{3+}]$ in the range of 0.05 at%, leads to an increase in photocurrent by a factor three [11].

Doping TiO_2 with various metal oxides, for example: (a) TiO_2-SiO_2 (3.4 % SiO_2), (b) TiO_2-SiO_2-Al_2O_3 (3.4 % SiO_2, 0.2% Al_2O_3), (c) TiO_2-SiO_2-In_2O_3 (3.4 % SiO_2, 0.2% Al_2O_3), and (d) TiO_2-SiO_2-RuO_2 (3.4 % SiO_2, 0.2% Al_2O_3) are reported for photoelectrolysis of water [12]. As revealed by SEM and XRD analysis, Al_2O_3, In_2O_3 and RuO_2 introduce a surface-active state suitable for the oxidation of OH^-. High photocurrent densities have been found in RuO_2 doped oxides due to the catalytic nature of Ru. A considerable increase in the extent of water photoelectrolysis, performed at bias of 1.0 V with 500 W Xe lamp, is observed for photoanodes (b)-(d) in the presence of 1.0 M NaOH electrolyte solution.   At a wavelength of 333 nm, nearly 82% quantum efficiency is observed for the TiO_2-SiO_2-Al_2O_3 (3.4 % SiO_2, 0.2% Al_2O_3), phonoanode, compared to 40% for the TiO_2-SiO_2 (3.4 % SiO_2) photoanode; apparently the presence of Al_2O_3 enhances the carrier diffusion length and facilitates charge carriers separation.

4.4.2 Crystalline n-SrTiO$_3$

$SrTiO_3$ is a mechanically and chemically robust material with a simple perovskite structure. N-type $SrTiO_3$ can be achieved by reducing it with H_2 for 4h at 1050-1100°C; the resulting single crystal thus obtained is blue-black in appearance. Wrighton and coworkers [13,14] first introduced the use of single crystal n-SrTiO$_3$ as a photoanode and Pt as cathode in an electrochemical cell, resulting in a sustained conversion of water to O$_2$ and H$_2$ with no external bias. The photoelectrochemical properties of SrTiO$_3$ electrodes are similar to those of TiO$_2$, although the onset potential of photocurrent shifts negative by about 0.3-0.35 V. In 9.5 M NaOH electrolyte solution, O$_2$ evolution is observed at the photoanode at a potential more positive than -1.3 V (*vs* SCE) while H$_2$ evolution is seen at the Pt cathode. With a slight increase in external bias the photocurrent increases more rapidly and sharply than that in comparison to TiO$_2$ or SnO$_2$. Even at zero bias a reasonable amount of photocurrent is observed that could initiate stiochiometric water splitting. The anodic photocurrent, which is proportional to the intensity of light on the photoanode, begins to flow for wavelengths less than 390 nm (4.2 eV) and reaches a maximum for a wavelength of 330 nm. The response is then nearly constant with increasing excitation energy [13]. Mavroides and coworkers illuminated a SrTiO$_3$ photoanode with a light intensity of 3.8 eV and observed water splitting without bias in presence of 10 M NaOH [15]. The maximum external quantum efficiency is found to be 10%, about an order of magnitude greater than the highest value obtained for photoelectrolysis in cells with crystalline TiO$_2$ electrodes in the absence of a bias voltage. This increase in quantum efficiency is due to the increased band bending of SrTiO$_3$, which is about 0.2 eV for a cell without a bias voltage. The efficiency of electron hole pair separation increases by the application of a bias voltage to make the anode Fermi level more positive than the cathode Fermi level, which is equivalent to $E_{red}(H^+/H_2)$ in accordance with increasing amount of band bending [15].

   To extend the photoresponse of SrTiO$_3$ electrodes into the visible region, single crystals of SrTiO$_3$ photo-electrodes have been doped with various transition metal ions (M^{n+}). The response to visible light for water decomposition has been found to decrease in the following order; $Cr^{3+} > Co^{2+} > Ni^{2+} > Mg^{2+} > Rh^{3+}$ [16]. As observed in the absorption spectrum, chromium doping of SrTiO$_3$ single crystals extends the fundamental band edge of 390 nm to about 600 nm. Absorption in the visible region is attributed to

$Cr^{3+} \rightarrow Ti^{4+}$ charge transfer. Such photoresponse can only be expected if the energy level of the M^{n+} is near the top of the valence band, otherwise a generated hole will not reach the semiconductor electrolyte interface [16, 17], the reason why several other dopant metal ions such as V^{3+}, Mn^{3+}, Fe^{2+}, Fe^{3+}, Co^{3+} and oxides of Nb, Mo and W show no sensitization in the visible region. Their corresponding energy levels are too far above the valence band so water cannot be oxidized [16]. Single crystal photoanodes made of Niobium-doped strontium titanate, $SrTiO_3$:Nb (Nb = 0.07, 0.69 mol%) are reported to show relatively efficient water photoelectrolysis under irradiation from a 500 W Xe lamp in the presence of 0.1 M Na_2SO_4 (pH =5.92) [18]; under monochromatic light irradiation of 298.2 nm and 448.2 nm, the corresponding IPCEs values are 15.67% and 0.26% (for 0.07 mol% Nb doped $SrTiO_3$), 4.32% and 0.19% (for 0.69 mol% Nb doped $SrTiO_3$), respectively. The IPCE value for a pure $SrTiO_3$ anode is 0.92% at 298.2 nm. Nb doping results in two optical absorption bands, one near 387.5 nm (3.2 eV) and another at 516 nm (2.4 eV). The existence of intermediate defect states created by oxygen vacancies within the band gap appears responsible for the conversion of visible light. Although a further increase of Nb-doping increases the concentration of intermediate defect states a sharp decrease in photon conversion efficiency is observed [18]. It is suggested that the existence of oxygen vacancies creates weakly bonded electrons that can be ionized, which in turn scatter the incident photons [18].

4.4.3 Crystalline α-Fe_2O_3

Ferric oxide (Fe_2O_3) is one of the most interesting semiconducting materials that can be used as photoanode because its band gap (2.0 – 2.2 eV) is well suited for capturing the majority of the solar spectrum energy. It can absorb all UV light and most of the visible light, up to 565 nm, which comprises nearly 38% of sunlight photons at air mass (AM) 1.5. Single crystal α-Fe_2O_3 can be obtained by flux growth from a Li_2MoO_4 flux melt [19] and by hydrothermal growth with 8M NaOH [20]. Three electrode systems comprising α-Fe_2O_3, Pt metal and SCE as photoanode, cathode and reference electrode, respectively, have been employed for water photoelectrolysis. An external bias is needed when this n-type semiconductor is used as photoanode, connected with a metal cathode through external circuit. The photoresponse studied in the absorption range of 280-580 nm through various electrolytes in the range of pH 6.5-13.6 does not show any sign of electrode degradation at a bias voltage of +0.5 V [21,22]. Pentavalent impurity doping such as Nb improves the n-type behavior of α-Fe_2O_3 single crystals for a quantum efficiency of 30% at 370 nm and 0 V (*vs* SCE) [23]. Photopotential, capacitance and transient photocurrent measurements suggest that the performance of α-Fe_2O_3 photoanodes is hampered by low charge carrier mobilities and slow charge transfer across the interface [20, 23]. The major factor determining the behavior of α-Fe_2O_3 in aqueous solutions is the observation that photogenerated holes are apparently located in 'd' orbitals that form narrow bands. When the photogenerated electron-hole pairs are located deep in the semiconductor bulk the low mobility and small diffusion length of the holes leads to a high probability of recombination. The hopping process in the narrow 'd' bands of Fe_2O_3 leads to low carrier mobility because strong coupling with the lattice phonons must occur for the carriers to migrate. If the mobility is low, the diffusion length of the carrier is smaller than the photon penetration depth. Hence the probability of these carriers reaching the space charge layer is low; the few that reach the space charge layer are accelerated to the surface encountering kinetic barriers at the interface. Thus low quantum efficiency at longer wavelengths is observed. At short wavelengths the photogenerated electron-hole pairs are produced in the near-surface region where a space charge layer exists. The slow charge transfer kinetics at the semiconductor-electrolyte interface appear due to a mismatch in energy between the acceptor 'd' orbitals of Fe_2O_3 and the donor 'p' orbital of the oxygen or hydroxide redox couple in solution [23].

4.4.4 Other Crystalline Oxides

Crystalline SnO_2 and WO_3 photoanodes for water photoelectrolysis have been reported [24-26]. The band gap of SnO_2 (E_{BG} = 3.5 eV) makes single crystals of this material of little interest with respect to solar induced optical properties. Wrighton and coworkers observed that n-type single crystals of Sb doped SnO_2 can serve as photoanodes to electrolyze water to H_2 and O_2 [24]; substantial photocurrent (~3 mA) response has been seen at a bias voltage of + 0.5 V upon illumination of such a photoanode (immersed in 12.0 M NaOH) with UV light (313 nm) from a 200 W high pressure Hg lamp. The positive current corresponds to O_2 evolution at the SnO_2:Sb anode. Compared to TiO_2, SnO_2 requires a slightly larger potential to achieve photocurrent onset.

Butler and coworkers introduced the use of tungsten trioxide, WO_3, as a photoanode material [25,26]. Its band gap of 2.7 eV results in the potential utilization of nearly 12% of the AM 1 solar spectrum as compared to 4% for TiO_2. However in various aqueous solutions, for example 1 M NH_4OH + NH_4NO_3 (pH 7) solution, WO_3 yields tungstic acid. With proper electrolyte selection WO_3 could be a useful photoanode oxide [25]. For example, in a 1.0 M sodium acetate solution no visible sample deterioration, and no measurable photocurrent decay were observed for tests over extended durations. A detailed analysis of WO_3 photoanode behavior has provided a number of important insights into the general behavior of photosensitive semiconducting electrodes [26]. It is seen that the minority carrier (hole) diffusion length plays a limiting role in determining the photoresponse of n-WO_3 photoanodes; to obtain moderate quantum efficiencies the diffusion length should be comparable to the optical absorption depth.

4.5 Polycrystalline Photoanode Materials

Polycrystalline oxide materials, both undoped and doped, have been extensively examined for use as photoanodes. TiO_2 electrodes have been prepared by thermal oxidation of a Ti plate in an electric furnace in air at 300-800°C (15-60 min) and in a flame at 1300°C (20 min) [27-30]. XRD analysis of thermally oxidized samples indicates the formation of metallic sub-oxide interstitial compounds, i.e. TiO_{0+x} (x < 0.33) or Ti_2O_{1-y} (0 < y < 0.33) and Ti_3O together with rutile TiO_2 [27]. The characteristic reflection of metallic titanium decreases in intensity after prolonged oxidation (60 min) at 800°C indicating the presence of a fairly thick oxide layer (10-15 μm). Oxidation at 900°C leads to poor adhesion of the oxide film to the metal underlayer (used as the back electrical contact). The flame oxidized samples are predominately rutile with trace amounts of sub-oxides. A maximum photoelectrochemical conversion efficiency of 0.65% is found for the flame-oxidized samples at an external bias voltage of +0.2 V under UV light irradiation using 400 W Hg lamp [27]. In order to modify the rutile electrode, a 700°C air oxidized Ti thin plate was repeatedly impregnated with aqueous solutions containing equimolar (1 M) amounts of $TiCl_3$ and nitrates or chlorides of various metals, followed by flame oxidation at 1300°C until a cohesive layer formed [28]. Impregnation of Pt (in this case, H_2PtCl_6 was used) was found to be the most effective in raising the photoconversion efficiency (out of 32 metal cations studied); using a 400 W medium pressure Hg lamp light source a 1 sun photoconversion efficiency of 2.7% was obtained in comparison to 0.5% for the pure TiO_2 (0.5 V bias in presence of 1.0 M Na_2SO_4 electrolyte). The thermally oxidized Ti plate (500-800°C, one hour) doped with Ru, Rh, Pt and Au via aqueous metal salt solutions followed by 400°C heating exhibit a cathodic photocurrent as well as a visible light response [29]. The response of these electrodes to the visible region is thought due to the formation of an impurity band near the π^* conduction band in the case of Ru, Rh and Pt doping, and near the π valence band for Au doped TiO_2.

Stable $CuWO_4$ and Cu_3WO_6 photoanodes were prepared by solid-state reaction [30]. These polycrystalline compounds were doped with indium oxide to enhance the n-type properties of the final composition, $Cu_{0.99}In_{0.01}WO_4$ and $Cu_{2.90}In_{0.10}WO_6$. The band gaps of these two compounds are close to 2.0 eV. The flat band potential in 0.1 M Na_2HPO_4 at pH = 9.4 is ~ −0.2 V, suggesting that external bias is needed for PEC operation [30] due to strong overlap between the Cu-3d and O-2p orbitals.

The use of polycrystalline La^{3+} doped $SrTiO_3$ as a photoanode has also been investigated [31]. La is considered an appropriate dopant because of its stability in trivalent state and the similarity of the ionic radii for Sr^{2+} (1.40 Å) and La^{3+} (1.32 Å), which ensures its incorporation at Sr^{2+} sites. Doping leads to the formation of a new energetic deep donor sub-level (Ti^{3+}_{3d}) in the band gap, thereby shifting its response towards the visible region. Similarly W^{6+} doping of polycrystalline TiO_2 has shown an increase in photoconversion efficiency from 0.007% (undoped, bias = 0.3 V) to 0.20% (0.1 at% W, bias = 0.65 V) [32]. Other efforts in this field have shown Ti doping of WO_3 shifts the flat band potential towards more positive values, with a maximum IPCE of 12% at 400 nm observed at 0.4 at.% Ti [33]. A positive value for V_{FB} (+0.288 V for 0.4 at% Ti) indicates that the Ti-doped WO_3 photoanode PE cell requires external bias for water photoelectrolysis; a maximum solar conversion efficiency of about 0.025% has been observed at a bias voltage of 0.65 V (in 0.1 M KCl).

Bak and coworkers [34] used a polycrystalline TiO_2 photoanode made by isostatic cold pressing (100 Mpa) and then sintered at 1175°C in a reductive atmosphere (Ar + 1% H_2) for 10 hours. 1% Cr-doped TiO_2 was also prepared by an impregnation method using an aqueous $Cr(NO_3)_2$ solution, then calcined at 727°C and annealed at 1000°C. The photoelectrochemical cell used the described TiO_2 anode and a Pt cathode placed in two different chambers, containing electrolytes of 4.0 M KOH and 4.0 M HCl separated by agar-agar solution while maintaining these electrodes at pHs of 14.3 and 0.3, respectively. For the Cr-doped TiO_2 photoanode 1.0 M KCl was used as an electrolyte while the Pt cathode was immersed in 3.0 M HCl. Water photoelectrolysis was performed directly in sunlight at an angle of Sun's radiation close to 48°. The initial light-off voltage of 0.35 V of a chemically biased cell (ΔpH =14.6) results in an increase of photovoltage to 1.6 V by the exposure of direct sunlight achieved within 30 min. Termination of the light results in a decrease of the photovoltage to 0.4 V over 30 min, while re-exposure of the photoanode to sunlight results in a 30 min increase of photovoltage to 1.3 V. It appears that polarization-related phenomena stemming from the chemical bias are responsible for the lack of reproducibility in the voltage dynamics [34].

Polycrystalline Fe_2O_3 was obtained by thermal decomposition of ferric oxalate at 800°C, finely ground and then compacted into discs, with the addition of metal oxides as desired to prepare doped Fe_2O_3 samples [35]. Improved characteristics such as saturation current, flat band potential, and minority charge carrier diffusion length were found for TiO_2-doped Fe_2O_3 samples relative to undoped, and SnO_2 as well as ZrO_2 doped samples. The highest saturation current, observed with 1% $TiO_2:Fe_2O_3$, 45 μA at $10mW/cm^2$, was approximately 3x greater than that achieved in the undoped samples. The flat band potential, although improved upon doping, remains below the H^+/H_2 level and requires the 0.5-0.6 V reverse or anodic bias to trigger water photoelectrolysis. Ta_2O_5: Fe_2O_3 does not measurably respond to solar light, possibly due to the incipient reduction of Fe^{3+} to metallic iron [35]. In combination with a n-type photoanode, using Mg-doped Fe_2O_3 as a p-type cathode results in a self-biasing PEC [36]. In the n-type photoanode, the holes driven toward the electrode/electrolyte interface mediate the conversion of OH^- ions in solution to O_2 gas, while electrons driven toward the surface of the p-type electrode reduce H^+ to H_2. At both electrodes the majority carriers diffuse away from the surface to contribute to the observed photocurrents.

4.6 Thin Film Photoanode Materials

Chemical vapor deposition [37,38], and thermal or anodic oxidation of Ti substrates [39,40,41] have been used to prepare polycrystalline thin films of TiO_2. For example, thin films of TiO_2 prepared by anodic oxidation of Ti, followed by electrodeposition of In_2O_3 from 0.5 M $In_2(SO_4)_3$ show enhanced optical absorption up to 500 nm [42] with the In_2O_3 modified electrode showing enhanced photocurrent and photovoltage partially due to the low electrical resistance (10^{-2} Ω) and reduced overvoltage of the photoanode.

4.6.1 General Synthesis Techniques of Semiconducting Thin-Films

The photoelectrochemical behavior of a given photoanode is dependent on its method of synthesis. Various methods, some of which we now briefly consider, such as anodic oxidation, spray pyrolysis, reactive sputtering and vapor deposition are commonly employed to make polycrystalline thin films.

Anodic oxidation

Before anodizing a metal sample careful cleaning of the substrate is necessary for impurity removal. Several methods are available; one such method [41] involves ultrasonic washing in D. I. Water, methyl alcohol and trichloroethylene followed by chemical etching in a mixture of 20% nitric acid, 30% hydrofluoric acid and 50% water. Use of dilute sulfuric acid, nitric acid or hydrochloric acid followed by ethanol and water rinse [5,43], or using NaOH [44] are a few common methods. Oxidation of Ti can be carried out either in acidic or alkaline mediums applying sufficient power (30 V and 50-70 mA/cm^2) [43] until a noticeable decrease in the current is observed due to oxide formation [45].

Thermal oxidation

The thermal oxidation of Ti metal to form n-type TiO_2 is one of most convenient ways to make a photosensitive electrode. This method involves careful cleaning and etching of the metal sample, typically a foil or sheet, as described previously followed by heating at temperatures above 400°C in oxygen or air. Use of an oven [46] as well as natural gas burner [39] have been described. Film formation is dependent upon oxidation temperature, duration, and oxygen partial pressure. The oxidation of Ti metal in pure oxygen atmosphere, resulting in a light-gray colored TiO_2 film, at 600°C for 10-30 min provides the best reported TiO_2 photoanodes with respect to corrosion stability and photoconversion efficiency [47].

Chemical Vapor Deposition

Chemical vapor deposition (CVD) is a process by which gaseous molecules are transformed into a solid thin film or powder on the substrate surface [48]. Plasma enhanced chemical vapor deposition (PECVD) [48,49] provides deposition of various layers at relatively low substrate temperatures, typically 200-300°C. Film growth, and resulting film properties are affected by deposition variables as power density, gas pressure, frequency, gas composition, and substrate temperature. TiO_2 thin film photoanodes deposited on a Ti substrate by PECVD using $TiCl_4$ vapor and O_2 feed gas has shown superior quantum efficiencies in comparison to thermally grown thin films [50]. Polycrystalline anatase films have been prepared by metal organic chemical vapor deposition (MOCVD); polarization at about 0.5 V below the flat band potential leads to a dramatic improvement in the donor density from 2×10^{17} to 10^{19} cm^{-3} as determined by Mott-Schottky plots [51].

Sputtering

Sputtering has proven to be a successful method of coating a variety of substrates with thin films of electrically conductive or non-conductive materials. The process occurs by bombardment of a sputtering target surface with plasma induced ions, with the momentum of the ion resulting in material ejection from the target that becomes the deposited film.   Reactive sputtering refers to deposition of a film formed by chemical reaction between the target material and a gas introduced into the sputtering chamber. One example of preparing a thin film photoelectrode by reactive sputtering is $Sn_{1-x}Pb_xO_2$, resulting in polycrystalline films of amber to black color as dependent upon the Pb concentration [52].

Sol-Gel Processes

Sol-gel involves the transition of a system from a liquid (the colloidal sol) into a solid (the gel) phase. This process allows fabrication of thin-films by dip coating or spin coating on various substrates. Gel modification can be carried out by the addition of suitable dopants in order to have materials of desired properties. Sol-gel-derived thin films have several advantages in comparison to other film fabrication techniques, including low cost and high film uniformity. Thermodynamically meta-stable films of high porosity, hence high surface area, can be readily obtained using sol gel, for example thin films of anatase TiO_2 [53-55].

TiO_2 sol can be prepared by slowly adding anhydrous ethanol or water-ethanol to titanium isopropoxide placed in an ice-chilled bath with, generally, a small amount of HCl used as a catalyst. A uniform TiO_2 gel film can be made on glass substrate by dipping it in a TiO_2 sol and subsequently pulling it up at a constant speed of reasonable rate. The obtained gel film can then be annealed for drying and crystallization, with the steps repeated for increased film thickness [53]. Studies on sol-gel derived TiO_2 electrodes show significant dependence on the solvent used which in turn affects film thickness, crystallinity, surface area and Ti^{3+} ion density. The photoelectrochemical properties of titania thin films synthesized from three different solvents are summarized in **Table 4.3** [54]. Sol-gel methods have been used to prepare mixed oxide films; one such example is the synthesis of $Ti_{1x}VO_2$ films that, used as a photoanode, were found to respond to visible spectrum light [55]. A solution of water, vanadium(IV) oxyacetylacetonate, nitric acid and ethanol is added dropwise to the solution of titanium isopropoxide, followed by dip coating on desired substrate; the $Ti_{1-x}V_xO_2$ (x = 0.15) showed a 2.12 eV bandgap.

Table 4.3: Photocurrents, flat-band potentials and donor densities of 10-times deposited, as prepared and optimized TiO_2 film electrodes

Electrodes	Photocurrent (mA/cm²)	Flat band potential (V vs SCE)			Donor density (cm⁻³)	Band gap (eV)
		Dark	Light	diff		
A	28	−1.00	−0.50	0.50	0.66×10^{20}	3.10
B	52	−1.10	−0.58	0.52	1.24×10^{20}	3.03
C	50	−1.07	−0.60	0.47	1.79×10^{20}	3.03
A (800°C, 10 min)	50	−1.00	−0.56	0.44	1.80×10^{20}	2.98
B (600°C, 30 min)	54	−1.00	−0.60	0.40	4.82×10^{20}	2.98
C (500°C, 20 min)	56	−1.00	−0.58	0.42	3.28×10^{20}	3.03

A: obtained from ethanol; **B:** obtained from ethanol mixed with isopropanol; **C:** obtained from ethanol mixed with 2-ethoxyethanol

4.6.2 Thin Film Characterization

Optical microscopy, scanning electron microscopy and transmission electron microscopy are commonly used to characterize surface morphologies. Different surface structures obtained from the various preparative methods provide insight into how surface features can affect surface reactions. For example, thermal oxidation of Ti films in oxygen result in an island-like TiO_2 topology [46,56] while vapor deposition generally results in smooth films [43,45,56] of smaller geometric surface area.

The crystalline structure, size and shape of the unit cell and the crystallite size of a material can be determined using X-ray diffraction spectroscopy (XRD). Rutile TiO_2 is the only component obtained from thermal oxidation of titanium, with samples heated above 700°C showing a splitting of the (210) peak indicating a transfer from orthorhombic to tetragonal symmetry. In general anodic oxidation produces anatase, but rutile can be obtained by various oxidation techniques. Rutile TiO_2 appears to be more desirable for application in photoelectrochemical cells. Due to large

Pilling-Bedworth ratio of 1:96, anatase phase films can show cracks and fissures with, consequently, a loss of mechanical stability, however a hydrogen reduction treatment above 600°C leads to phase transition from anatase (101) to rutile (110) [43] with XRD detecting TiH$_2$ upon prolonged hydrogen treatment of titania. As shown in **Fig. 4.4**, introduction of vanadium increases the intensity of the anatase TiO$_2$ peak; above 700°C disappearance of the vanadium (001) peak and the simultaneous appearance of the rutile (110) peak are observed, but anatase continues to dominate even after heat treatment at 800°C. A sharp vanadium (001) peak is observed for heat treatments carried out in air, while no vanadium peak has been seen in the case of heat treatment at 600°C in presence of Ar/H$_2$.

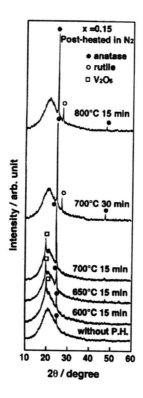

Fig. 4.4: XRD patterns of Ti$_{1-x}$V$_x$O$_2$ samples with x=0.15 post-heated at various temp (Reprinted with permission from Ref. [55])

Table 4.4: Photocurrent-Potential data of s ol-gel derived TiO$_2$ thin film photoanode at different film thicknesses [57].

T (µm)	0.3	1.0	1.8	2.3
V	A	A	A	A
-1.0	-0.425	-0.392	-0.070	-0.237
-0.8	-0.291	0.237	0.127	-0.196
-0.6	-0.189	0.905	0.213	-0.091
-0.4	-0.106	1.255	0.254	-0.023
-0.2	-0.011	0.287	0.391	0.042
0.0	0.118	0.317	0.463	0.126
0.2	0.219	0.395	0.517	0.248
0.4	0.493	0.691	0.602	0.391
0.6	0.290	0.523	0.819	0.569
0.8	0.493	0.691	1.212	0.851
1.0	0.716	0.925	1.504	1.260
1.2	0.754	1.378	1.828	1.285
1.4	0.823	1.960	2.516	1.577
1.6	0.912	2.213	2.816	1.563
1.8	0.980	2.341	3.183	1.563
2.0	1.020	2.477	3.231	1.568

T film thickness, µm
V potential, V
A photocurrent, mA

Jung and coworkers [57] employed sol-gel to prepare porous n-TiO$_2$ thin film electrodes on glass. XRD measurements suggest anatase crystals begin to be formed around 450°C. The amount of photocatalyst deposited on the substrate during a dip coating process depends on several variables, including the suspension viscosity, the number of coating steps, and the speed with which the electrode is withdrawn from the suspension. In general, increasing the particle concentration produces a suspension of higher viscosity, which in turns lead to greater coating thickness. There exists a strong relationship between the photocurrent and film thickness of a photoanode, as summarized in **Table 4.4** [57]. An optimal film thickness exists where the resistivity of the semiconductor and the light absorption reach optimum values. According to Yazawa et al. [41] the photoresponse in a semiconductor electrode is expected to show a maximum value by the following equation:

$$\chi_T \approx \frac{1}{\alpha} \approx W \qquad (4.6.1)$$

where χ_T, α, and W are, respectively, the film thickness, absorption coefficient of light, and depletion layer width. From (4.6.1), most of the incident light should be absorbed inside the depletion layer and film thickness should not exceed the width of the depletion layer [58]. Assuming $\alpha = 4 \times 10^4$ cm^{-1} (345 nm) in TiO$_2$, the optimum value of film thickness and donor density (N_D) from equation (4.6.1) are, respectively, about 0.2 μm and 10^{17} cm^{-3}. TiO$_2$ film thickness produced by controlled thermal oxidation is found in the range of 0.1 – 2.0 μm. If the film thickness is of the order of the width of space charge region (depletion layer) most of the photogenerated carriers formed inside the film can reach the n-TiO$_2$-electrolyte interface. Takahashi et al. correlated sol-gel prepared 10 nm – 540 nm TiO$_2$ thin films with measured photocurrent amplitudes under 500 W Xe lamp illumination [59]. A 70 nm electrode showed maximum photocurrent, nearly six times higher than those of the 180–540 nm electrodes. The authors suggest that films having thicknesses smaller than the space charge layer show larger photocurrents due to more effective electron-hole separation. Films below 70 nm exhibited poor photocurrent amplitudes, a behavior that the authors attributed to poor film crystallinity.

4.6.3 Thin Film Photoanodes

Photoanodes comprised of materials of low carrier mobility suffer recombination losses in the space charge region, bulk, and surface. However if the electrode is thin enough, so that it is entirely occupied by the space charge region, then recombination can be reduced. To compensate for the lack of absorption in a given thin layer, a device comprised of several sequentially placed layers [60] can be used. The example illustrates how different measures can be taken to address different issues. In addition to physical geometry, PEC efficiency can be enhanced and photocorrosion minimized by electrolyte modification. For example, addition of thiourea and 2-aminopyridine to an electrolyte solution of 0.1 M NaOH buffered with potassium hydrogen phthalate at pH 4.7, containing n-TiO$_2$ anodic thin film as the photoanode, has been found to suppress the rate of recombination and photocorrosion [61]. In the case of thiourea addition, the onset potential is ~0.1 V more negative, while it is completely absent in the case of 2-aminopyridine, suggesting that the rate of recombination can be suppressed [61]. **Table 4.5** summarizes experimental data on photoelectrochemical cell efficiencies involving thin film photoanodes.

Table 4.5: Experimental data on efficiencies of photoelectrochemical cell involving representative photoanode in the form of thin film

Photoelectrochemical cells			Efficiencies			Light source	Ref.
Photoanode	Cathode	Electrolyte	Bias (V)	η_w (%)	η_c (%)		
Thin films							
TiO_2-In_2O_3	Pt	1.0 M NaOH	-	100	-	Xe lamp, 500 W	42
TiO_2	Pt	5.0 M NaOH (a) + 3.0 M H_2SO_4 (c)	-	-	1.6	Xe arc lamp, 150 W	47
CM-TiO_2	Pt	5.0 M KOH	0.3		8.35, 11	Xe arc lamp, 150 W, 40 mA/cm²	65
CM-TiO_2	Pt	5.0 M KOH	0.3		1.4[a]	Xe arc lamp, 150 W, 40 mA/cm²	66
$Zn_{0.958}Co_{0.042}O$	Pt	0.5 M KNO_3	0.0	-	0.6-1.0	1.0 kW Xe	77
Fe_2O_3:Au (nc)	Pt	Buffer solution, pH =7		20.0[f]		Xe lamp, 500W	87
Fe_2O_3:Si (ns)	Pt	1 M NaOH		42.0[f]		Xe lamp, 450 W, 1000 mW/cm²	88
TiO_2 (nc)	Pt	0.1M HCl + 50% (v/v) MeOH·H_2O	0.0	30.0[f]		UVA Lamp, 12.8 V	110
Fe_2O_3 (nc)	Pt	1.0 M NaOH	0.6	22.5	4.92	Xe arc lamp, 150 W	111

[a]Calculated value under similar reaction condition. [nc] stands for nanocrystalline. [f]IPCE.

The photoelectrochemical behavior of thin film polycrystalline anatase TiO_2 films has been extensively studied. The PEC behavior of such films has been found to vary with time, in most cases decreasing during water photoelectrolysis [62,63]. Grätzel and coworkers found that the relative lack of oxygen during photolysis within a closed system is due to photo-uptake of O_2 by the TiO_2 particles [63], a behavior particularly evident in highly alkaline solutions due to formation of linear or μ-peroxo-bridged Ti species at the surface. Mn_2O_3 covered n-TiO_2 thin film electrodes (bandgap = 2.85 eV) have been found stable with time, showing significant improvement in oxygen versus H_2O_2 evolution at the photoanode [64].

Thermally oxidized n-TiO_2 polycrystalline thin film photoanodes [47] have been examined in a three electrode system comprising a Pt cathode and SCE counter electrode in a self-driven two compartment cell having a pH gradient of 15.5 and structure n-TiO_2|5M KOH|3M H_2SO_4|Pt. A maximum solar to H_2 conversion efficiency of 1.6% at zero bias is obtained. For the films synthesized at 850°C, quantum efficiencies of 60% and 54% were calculated for the two absorption peaks 330 nm and 370 nm. The intercepts of the Mott-Schottky plots for the films prepared at three temperatures in **Fig. 4.5** meet at −1.10 V (*vs* SCE). This gives a flatband potential of −1.13 V (*vs* SCE) when a temperature dependent factor of 0.03 V is included, indicating that the flatband potential of n-TiO_2 films do not depend on the oxidation temperature. The flatband potential of −1.13 V (*vs* SCE) from the Mott- Schottky plot is 0.16 V away from the flatband potential of −0.97 V (*vs* SCE) obtained from equation (4.2.1), J^2_{pc} *vs* applied potential V. which is the correct onset potential and not the flatband potential in the Schottky junction model. The measured open circuit potential under illumination has been found to be equal to the onset potential. Hence, from the difference between the open circuit potential (−0.97 V) and the flatband potential (−1.13 V), one determines the built-in band bending of 0.16 V under a light intensity of 50 mW cm^{-2}.

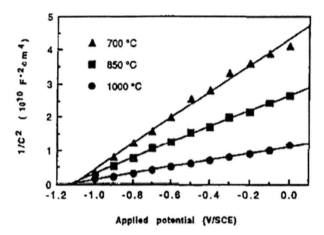

Fig. 4.5: Mott-Schottky plot of n-TiO$_2$ prepared at different temperatures. AC frequency: 1000 Hz. Reprinted with permission from Ref. [47].

A photoanode comprised of flame oxidized carbon doped n-TiO$_2$ films have been reported to perform water splitting with high photoconversion efficiencies [65]. While chemically modified n-TiO$_2$ can be prepared by the controlled combustion of Ti metal in a natural gas flame the authors, in investigating this technique [66], have found reproducibility to be a challenge. Various authors [67,68,69] have discussed in considerable depth issues surrounding the stated photoconversion efficiencies of [65].

Cr-doped TiO$_2$ has been widely studied for photoanode application since addition of Cr^{3+} shifts the optical absorption spectrum towards the visible range, with the caveat that excess chromium doping leads to higher recombination rates. Various synthesis methods have been employed to date. For example Radecka an coworkers used r.f. sputtering to prepare up to 16 at% Cr-doped TiO$_2$ [70]. XRD reveals that at low Cr concentration the films are predominately anatase. As the Cr concentration increases the rutile phase prevails; the rutile-to-anatase ratio increases from 35% for 1 at% Cr to 58% for 16 at% Cr. As calculated from Mott-Schottky plots [70] the effective donor density concentration increases with increasing Cr concentration, reaching an optimal value at 7.6 at.% Cr. The increase in Cr concentration shifts the entire absorption spectrum towards the visible region by lowering the oxide band gap to ~2.4 eV. Polycrystalline thermally oxidized WO_3 thin films (E_{BG} = 2.8 eV) are intrinsically unstable when used as photoanodes [71] due to dissolution. Efforts have been made to stabilize WO_3 thin film anodes [72], with the greatest success coming from modification of the redox couple ($Fe^{2+/3+}$) used in an aqueous electrolyte solution.

Thermally oxidized iron oxide thin films synthesized using iron laminae in an electric furnace in the presence of air at temperatures between 200 to 1000°C for 15 min have been used as photoanodes for water photoelectrolysis [73]. XRD measurements suggests that oxidation proceeds through the simultaneous formation of several phases, namely, FeO, Fe_3O_4, $Fe_{3-x}O_4$, α- and γ-Fe_2O_3. Samples prepared below 400°C primarily show Fe_3O_4 phase, while α-Fe_2O_3 phase is detected for samples calcined at 400°C. Calcination above 500°C and 700°C exhibited a diffraction band of non-stoichiometric $Fe_{3-x}O_4$ and FeO respectively. Thin film thickness gradually increases for samples prepared from 400°C to 700°C, with the presence of thick α-Fe_2O_3 overlayers seen between 500 and 800°C greatly diminishing the photocurrent density. Therefore, maximum photocurrents are found for electrodes prepared and activated at low temperatures (200–400°C) [73]. Thin films of Fe_2O_3 deposited on tin oxide, as well as RuO_2 - Fe_2O_3 on tin oxide substrates, by spray pyrolysis have been investigated as photoanodes [74]. A maximum photocurrent density of 0.7 mA/cm^2 has been found for the Fe_2O_3 electrodes. Although there was no improvement in overall current density for the RuO_2-Fe_2O_3 electrodes, the onset potential shifted in the cathodic direction 120 mV using electrolyte solutions of 0.2 M NaOH and 0.5 M Na_2SO_4. This shift of the onset potential signifies that one will need to apply

120 mV less to split water as compared to that of the n-Fe_2O_3 electrode [74].

Spray pyrolysis of ethanolic solutions of Fe(acetylacetone)$_3$ or $FeCl_3$ between 370°C and 450°C onto a glass substrate are reported for the fabrication of α-Fe_2O_3 thin-film photoanodes [75]. Upon illumination by a 150 W Xe lamp samples consistently demonstrate photocurrents of 0.9 mAcm^{-2}, IPCE values up to 15%, and robust mechanical stability with no signs of photocorrosion for the undoped samples. With simultaneous multiple doping of 1% Al and 5% Ti, an IPCE of 25% can be reached at 400 nm. Zn doping is known to induce p-type character in Fe_2O_3 thin film electrodes [76].

An automated rapid serial electrochemical deposition system has been used to synthesize cobalt-doped ZnO thin films [77]. $Zn_{0.956}Co_{0.044}O$ exhibits a four-fold increase in photoelectrochemical properties over pure ZnO, with no external bias. XRD and XPS measurements show the wurtzite structure typical of pure ZnO, while the addition of Co^{2+} lowers the band gap of the film from 4.2 eV to 2.75 eV. At a bias of 1.1 V, photocurrent generation improves by a factor of 2.5 as compared to zero applied bias. The measured flat band potentials of −0.21 V(*vs* Ag/AgCl) have been found to change a negligible amount as a function of composition, which indicates that the conduction band lies at approximately the same energy regardless of $Zn_{1-x}Co_xO$ composition. Given the change in the absorption band edge of these materials, the valence band edge must be raised with increasing Co concentration.

4.7 Nanocrystalline and Nanoporous Thin Film Materials as Photoanodes

Nanocrystalline semiconductor thin film photoanodes, commonly comprised of a three dimensional network of inter-connected nanoparticles, are an active area of photoelectrochemistry research [78–82] demonstrating novel optical and electrical properties compared with that of a bulk, thick or thin film semiconductor [79,80]. In a thin film semiconductor electrode a space charge layer (depletion layer) forms at the semiconductor-electrolyte interface; charge carrier separation occurs as a result of the internal electric field formed at the depletion layer. However in a nanocrystalline semiconductor electrode the interfacial kinetics are considered more important than the internal electric field because the individual particles are too small to form a depletion layer [81,82], hence charge separation of the photo-excited electron-hole pairs is dominated by diffusion rather than electric field assisted drift [77-83].

Lee and coworkers [84] have extensively studied the photoelectrochemical behavior of nanocrystalline TiO_2 electrodes using a three-electrode cell comprising a Pt cathode, SCE reference electrode, and TiO_2-coated ITO glass photoanode; the electrolyte pH was adjusted by use of 0.1 M KOH and HCl solutions. Photoanodes made up of TiO_2 nanoparticles of three different diameters of ~ 5 nm, ~ 11 nm and 25 nm were employed. Sol-gel was employed in the synthesis of the TiO_2 nanoparticle 'sol' and then spin-coated for preparation of the photoanodes. Annealing at 450°C for 2 hr gives the nanostructured thin films an anatase phase (101) character as observed by XRD. Capacitance and cyclic voltammetry measurements infer the formation of a depletion layer in the semiconductor film, showing that the flat band potential (V_{FB}) of the TiO_2 films depends upon the particle size. Band gap excitation of the electrode in aqueous electrolyte solutions generated anodic photocurrents for water oxidation, with photocapacitance measurements showing that the band edges of the electrodes shift anodically under illumination [84]. An advantage of nanocrystalline anatase is that it can be sintered at a comparatively lower temperature of 450-500°C without significant transformation to rutile [84,85]. Experiments performed on TiO_2 nanocrystalline photoanodes of several geometric areas, namely 0.21, 0.50, 0.72, 1.47 and 1.85 cm^2 show the highest photocurrent density corresponds to the smallest electrode area, decreasing for increasing electrode area [82]. Such scaling issues are, obviously, of great importance regarding scale implementation of photoelectrochemical cells for solar hydrogen production.

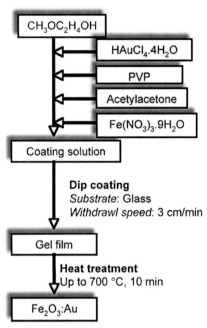

Fig. 4.6: Flow chart showing preparation of Fe_2O_3 thin film containing dispersed Au particles.

Sol–gel techniques have been applied to the fabrication of a variety of metal oxide nanocrystalline thin films. For example, the synthesis of sol-gel-derived iron oxide thin films with embedded gold catalytic nanoparticles is shown in **Fig. 4.6**. The addition of Au nanoparticles has been found to enhance the quantum efficiency of α-Fe_2O_3 thin films. For 40 nm thick α-Fe_2O_3 thin films the IPCE at $\lambda = 400$ nm increased a factor of eight, to 4%, with addition of Au particles 5–20 nm in size (the α-Fe_2O_3:Au photoanode was immersed in an aqueous buffer solution containing 0.2 M $Na_2B_2O_4$, 0.14 M H_2SO_4, 0.3 M Na_2SO_4)[86]. It is hypothesized that the quantum efficiency can be significantly increased if the desired nanocrystalline thin film can be made sufficiently porous, thereby providing a large number of photoelectrochemical sites for water oxidation, and sufficiently thin- i.e. thinner than the space charge layer [60,87]; see nanotube array discussion in Chapter 5.

Chemical vapor deposition has been found an efficient route to deposit nanostructured thin films on a suitable substrate. Grätzel and coworkers have used this technique to deposit Si doped α-Fe_2O_3 on FTO coated glass at 415°C using $Fe(CO)_5$ and tetraethoxy silane as precursors [88]. $Fe(CO)_5$ yields high purity α-Fe_2O_3, while other precursors such as $Fe(acetylacetone)_3$ or $FeCl_3$ leave carbon or chlorine residues that can act as an electron donor. The observed (110) reflection of Fe_2O_3:Si in XRD patterns corresponds to pure hematite, while FESEM imaging reveals that the iron oxide films have a dendritic morphology with a highly branched nanostructure 'growing' towards the electrolyte. Under AM 1.5 illumination (450 W Xe lamp, 1000 mW cm^{-2}) a α-Fe_2O_3:Si (Si = 1.5 at%) thin film photoanode has recorded an IPCE of 42% at 370 nm at an electrode potential of 1.23 V (vs RHE, reversible hydrogen electrode) [88]. This remarkable IPCE value, the largest yet obtained for a Fe_2O_3 film is due partly to the short hole diffusion length, a consequence of the film architecture, and due partly to deposition of a thin insulating SiO_2 layer below and a cobalt monolayer atop the α-Fe_2O_3 film.

The photoinduced super-hydrophilic properties of photocatalytically active mesoporous titania thin films, composed of small monodisperse spherical particles about 15 nm in diameter, are being used as antifogging and self-cleaning materials [89]. Mesoporous thin films, possessing large surface areas, show higher photocatalytic activity and better light-induced hydrophilicity than TiO_2 thin films. Wang et al. observed that the Ti^{3+} ion is closely associated with the hydrophilicity of TiO_2 [90], carrying out several experiments to prove the existence of this ion on the surface of titania electrodes in presence of aqueous electrolytes. It is well established that the dissociated electrons and holes produced by UV illumination are trapped, respectively, by surface Ti^{4+} ions and bridging site O^{2-} ions to form Ti^{3+} ions and oxygen defects where water molecules are dissociatively adsorbed [91] resulting in a hydrophilic surface. In this process electroactive Ti^{3+} ions can be reoxidized to Ti^{4+} ions by anodic polarization, underlying the presence of oxidative peaks in cyclic voltammograms of the samples. A possible reaction mechanism is:

$$TiO_2 + h\nu \rightarrow e^-(TiO_2) + h^+(TiO_2) \qquad (4.7.1)$$
$$h^+(TiO_2) + H_2O \rightarrow 2H^+ + \tfrac{1}{2}\,O_2 \qquad (4.7.2)$$
$$e^-(TiO_2) + 4H^+ + TiO_2 \rightarrow Ti^{3+} + 2H_2O \qquad (4.7.3)$$
$$Ti^{3+} + 2H_2O \rightarrow TiO_2.4H^+ + e^-(\text{on the anode}) \qquad (4.7.4)$$

It has been noted that extended exposure to UV light is necessary for generation and accumulation of the electroactive Ti^{3+} ions, hence the oxidative peaks are difficult to regain in the second positive cycle once Ti^{3+} ions are exhausted in the anodic polarization process (4.7.4).

Nanocrystalline and nanoporous semiconductors have opened new directions in semiconductor photoelectrochemistry [92,93]. In an experiment carried out by Tang and co-workers [94], highly crystalline mesoporous TiO_2 thin films were synthesized, then the pores filled with carbon, with the motivation that the carbon would act as a filler to prevent pore collapse due to the stress induced by a 500°C anneal to induce crystallization. A study of the resulting $C-TiO_2$ crystallites shows weak, broad and small angle XRD peaks, suggesting the partial destruction of long-range mesostructure. Despite the disruption of the long-range order, the C-TiO_2 thin film photoanode at zero bias exhibits a solar spectrum photoconversion efficiency of 2.5% (40 mW/cm^2 illumination), a value approximately 4-10 times higher than a control sample of comparable crystallinity. The coupling of the TiO_2 layer with a thin layer of $C-TiO_2$ facilitates light capture while maintaining the excellent charge transfer properties of TiO_2. The use of bi-layers to improve photoconversion efficiency is an appealing one, with difficulties inherent in obtaining uniform coverage of the underlying material architecture [95,96,97,98].

A sol-gel dip-coating fabricated multi-layer thin film architecture demonstrating a spinodal phase separation structure (SPSS) was prepared using a colloidal sol containing polyoxyethylene(20) nenonylphenyl ether (NPE20); the resulting films demonstrated significantly higher anodic photocurrents than dense TiO_2 thin film electrodes [99]. It is suggested that the origin of this improved photocurrent is due to the large specific surface area inherent in the SPSS, and the higher concentration of Ti^{3+} species obtained by the reducing action of the polymer upon sintering. **Figure 4.7** shows the confocal scanning microscope observation of the D5 SPSS TiO_2 thin film on ITO coated glass, indicating that it has a well-defined spinodal phase separation structure. The horizontal lines in the bottom cross-section profile show the height of each layer from the substrate.

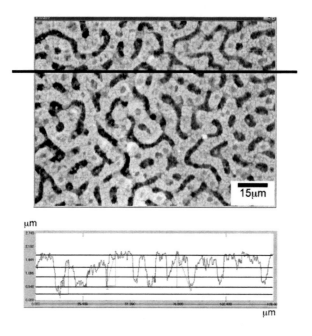

Fig. 4.7: Confocal scanning microscope image of (top) a spinodal phase separation structure TiO$_2$ thin film, and (bottom) the cross-sectional height profile of the above picture across the solid line. Reproduced with permission from Ref. [100].

The photoelectrochemical characteristics of nanoporous WO$_3$ thin films are described by Santanto and coworkers [100]. A 0.4 mol dm^{-3} tungstic acid and 25% w/w of poly(ethylene glycol) solution was applied to the substrate (F-doped SnO$_2$ glass) and then annealed in flowing oxygen at 550°C for 30 min. The WO$_3$ films, comprised of particles 20 nm – 50 nm in diameter, were obtained with multiple coating steps with each coating step followed by a heat treatment, resulting in an increase of film thickness with each coating step of 0.2–0.4 μm. The band-gap energy of 2.5 eV for the resulting films corresponds to a photoresponse of up to 500 nm, and is identical with that reported for bulk WO$_3$ films formed by thermal oxidation of the metal at 700–750°C. A maximum IPCE is obtained of 75% at approximately 410 nm for a 2 μm thick WO$_3$ film in 1.0 M HClO$_4$ solution, while AM 1.5 spectrum illumination results in photocurrents of approximately 2.5 mA/cm^2.

Various techniques have been investigated for the synthesis of nanostructured hematite, α-Fe_2O_3, with an aim towards overcoming low charge carrier mobility and high bulk recombination losses [101] while offering large surface areas. Duret and co-workers demonstrated photocurrents in the mA range under 1.5 AM solar illumination from mesoporous hematite α-Fe_2O_3 thin films consisting of 5–10 nm thick nanoleaflets synthesized via ultrasonic spray pyrolysis [102]. Other efforts have included the preparation of silicon doped α-Fe_2O_3 photoanodes [88,103], and α-Fe_2O_3 based self-biased photoelectrochemical cells using n-type and p-type α-Fe_2O_3 films [104]. Grimes and co-workers [105] have reported synthesis of self-organized nanoporous α-Fe_2O_3 films via potentiostatic anodization of Fe foil in an electrolyte consisting of 1 vol. % HF + 0.5 wt. % ammonium fluoride (NH_4F) + 0.2 vol. % 0.1M nitric acid (HNO_3) in glycerol (pH 3) at 10°C, see **Fig. 4.8**. The as-anodized sample, initially amorphous, is annealed at 400°C for 30 minutes to form crystalline hematite; longer annealing periods and higher temperatures resulted in a greater fraction of magnetite and pore filling. From **Fig. 4.9(a)** one can see a crystalline nanoporous structure of excellent uniformity. The cross sectional image of **Fig. 4.9(b)** shows the extent of the pore depth (~ 380 nm) and the barrier layer thickness (~ 600 nm) of the annealed sample. The annealed nanoporous films exhibit significant absorbance in the visible range. The current-potential response of an illustrative sample, 115 nm pore diameter, 400 nm pore depth, with potential measured relative to Ag/AgCl standard electrode under dark and 1.5 A.M simulated solar illumination is shown in **Fig. 4.10** (1M NaOH electrolyte). The dark current increases from zero at 0.46 V to 0.020mA at 0.6V, beyond which it rapidly increases; the photocurrent at 0.6V is 0.26 mA, see **Fig. 4.10(a)**. The photoelectrochemical response of a sample in an electrolyte containing 0.5 M H_2O_2 (50%) and 1 M NaOH (50%) are given in **Fig. 4.10(b)**; the onset potential is –0.37 mV, several hundred mV more negative than in the 1 M NaOH solution. The addition of H_2O_2 enhances the reaction kinetics, as the photo oxidation rate of H_2O_2 is much larger than that of water. Thus by introducing 50% of 0.5 M H_2O_2 the nanoporous structure demonstrated a net, light illuminated less dark, photocurrent of 0.51 mA at 0.6 V Vs Ag/AgCl. It is expected that crystallization and retaining the nanoporous structure in the thermal annealing step would significantly improve the photoelectrochemical properties of the resulting films.

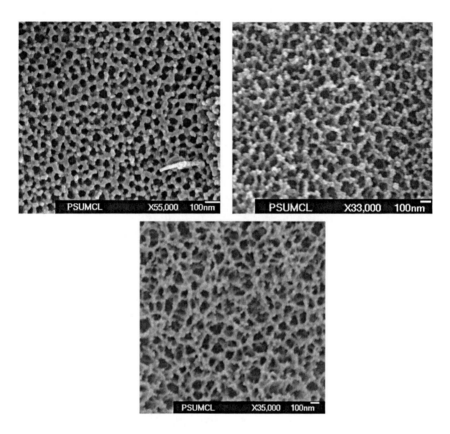

Fig. 4.8: FESEM image of samples anodized in 1% HF + 0.3 wt % NH₄F + 0.2 % 0.1 M HNO in glycerol at 10°C **(a)** 40V, **(b)** 60V, and **(c)** 90 V.

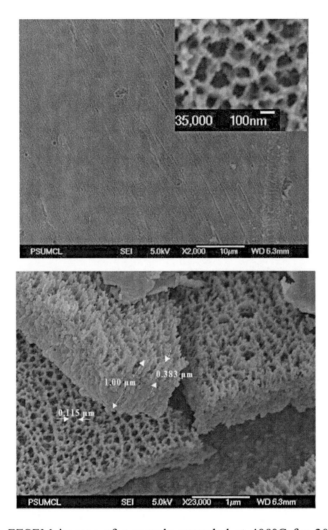

Fig. 4.9: FESEM images of a sample annealed at 400°C for 30 minutes showing: **(a)** the uniformity of the pore formation across the sample (insert shows the pore diameter after annealing); **(b)** cross sectional image showing the pore depth as 383 nm, pore diameter as about 100 nm, and the barrier oxide thickness as approximately 600 nm.

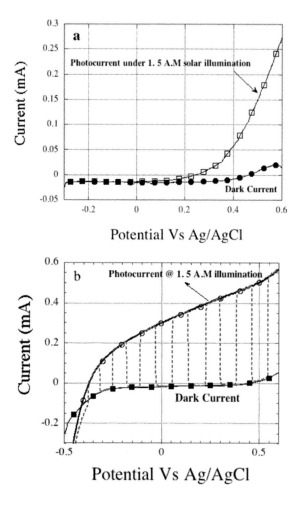

Fig. 4.10: Photocurrent as a function of measured potential for the Iron (III) Oxide photoanode (anodized in 0.2 % 0.1M HNO_3, 1%HF, and 0.3 wt % NH_4F in glycerol at 10°C) recorded in: (a) 1M NaOH solution, and (b) 0.5 M H_2O_2 + 1M NaOH solution under 1.5 A.M simulated solar illumination.

4.7.1 Water Photoelectrolysis using Nanocrystalline Photoanodes

Various approaches are being pursued to circumvent the poor solar light response of photocorrosion-stable wide band gap semiconductors. For example, in an experiment carried out by Karn et al. [106] nanostructured TiO_2, and nanostructured mixed In_2O_3-TiO_2 were employed as photoanodes in a three electrode PE cell using 1.0 M NaOH electrolyte. Under 0.5 W cm^{-2} illumination from a 1000 W Xe-Hg lamp the measured H_2 gas evolution rate is increased from 0.12 mL/hr for conventional planar-film TiO_2, to 0.2 mL/hr for nanostructured TiO_2, which further increases to 0.4 mL/hr for nanostructured In_2O_3 - TiO_2. For the In_2O_3 - TiO_2 mixed oxide this enhancement is due to the decreased energy band gap and slight shift in the flat band potential.   The presence of photoactive islands of In_2O_3 on the nanostructured TiO_2 further increases the surface area and hence the photoresponse increasing the photocurrent and photovoltage from 4 mA/cm^2 and 0.6 V respectively to 14 mA/cm^2 and 0.9 V for In_2O_3–TiO_2.

Wang et al. [107] studied the photoelectrochemical properties of various metal ion doped TiO_2 nanocrystalline electrodes; doped quantum sized TiO_2 nanoparticles (2-3 nm) prepared in the presence of metal salts of $M(NO_3)_3$ (M = La^{3+}, Nd^{3+}, Pr^{3+}, Sm^{3+}, Fe^{3+}, Eu^{3+} and Cr^{3+}) and $M(NO_3)_2$ (M = Co^{2+}, Zn^{2+}, Cd^{3+}). The photocurrents generated by these metal-doped TiO_2 electrodes, except for Fe^{3+}, Cr^{3+}, Co^{3+}, Cd^{3+}-doped TiO_2, were about 6 times higher than that of the undoped TiO_2 electrode (4.6 mA/cm^2) in aqueous thiocyanide solution (pH 4) under 200 W Xe lamp (λ = 350 nm; intensity 305.1 mW/cm^2) and at an applied electrode potential of +0.3 V. Doped Fe^{3+}-TiO_2 electrodes yielded lower photocurrent densities than the undoped n-type TiO_2 electrode showing the coexistence of p-n characteristics, with pure p-type characteristics observed at electrode potential of +0.6 V. This behavior is most likely due to the replacement of Ti^{4+} ions by Fe^{3+} thereby forming p-type surfaces, which in turn form a p-n junction with the core TiO_2 [107]. In the case of lanathanide and Zn^{2+}, Cd^{2+} ions, interstitial substitution into TiO_2 lattice occur, which lead to increase the electron density in the conduction band. It has also been observed that hydrothermal processing carried out at high temperature (170 °C) facilitates better lattice doping as compared to low temperature (65°C) sol-gel process. Amount of metal ion doping and film thickness affected the nanocrystalline properties of the photoelectrode and the IPCE values were found maximum at 0.5 mol% Zn^{2+} doping and ~2 mm of film thickness (in the case Zn^{2+}-doped TiO_2 electrode).

Fretwell and co-workers [108] developed a modified electrochemical cell using nanocrystalline TiO_2 thin film photoanodes made by dip-coating from TiO_2 sols. A metal cathode and anode were deposited as parallel strips on FTO coated glass. In the absence of a cathode, irradiation of this film in contact with 0.1 M HCl in 50% (v/v) $MeOH$-H_2O generates hydrogen with the quantum yield of 4.6%. In the presence of a Pt cathode the rate of H_2 production increases nearly four times to a quantum yield of 19.3%. The geometry and size of the Pt cathodes did not appear to affect the hydrogen evolution quantum yield. In presence of 0.1 N HCl, hydrogen evolution can be seen both at cathode (70%) and photoanode (30%). In acidic solution, at pH = 0, the reduction potential of anatase TiO_2 conduction band is 0.2 V more negative than that for the H_2/H^+. In presence of aqueous methanol solution, following reactions occur:

$$CH_3OH \rightarrow CH_3O^- \text{ (ads)} + H^+ \text{ (ads)} \tag{4.7.1}$$
$$e^-(CB) + H^+(\text{ads on } TiO_2) \rightarrow H(\text{ads}) \rightarrow \tfrac{1}{2} H_2 \text{ (ads.)} \rightarrow \tfrac{1}{2} H_2 \text{ (aq.)} \tag{4.7.2}$$
$$e^-(CB) \rightarrow e^-(Pt) \tag{4.7.3}$$
$$e^-(Pt) + H^+(\text{ads on Pt}) \rightarrow \tfrac{1}{2} H_2 \text{ (ads.)} \rightarrow \tfrac{1}{2} H_2 \text{ (aq.)} \tag{4.7.4}$$

In 0.1 N NaOH electrolyte solutions, hydrogen evolution is seen exclusively at the TiO_2 photoanode in the presence of a sacrificial electron donor, CH_3OH, by the following reaction;

$$CH_3O^- + h^+ \text{ (VB)} \rightarrow CH_2O + H(\text{ads}) \rightarrow \tfrac{1}{2} H_2 \text{ (ads.)} \rightarrow \tfrac{1}{2} H_2 \text{ (aq.)} \tag{4.7.5}$$

Using a spray pyrolysis technique, nanocrystalline n-Fe_2O_3 thin-films have been deposited on a SnO_2-coated glass substrate and their use as photoanodes reported [109]. Using these films for photoelectrochemical water splitting a (light + electrical to chemical) energy conversion efficiency of 4.92% is achieved at an applied potential of 0.2 V (vs SCE) pH 14 electrolyte [109]. Film architecture was optimized by manipulating spray velocity, dwell time, solvent, and spray solution concentration in order to obtain high quality and efficient nanocrystalline thin-films used as water-splitting photoanodes. A maximum photocurrent density of 3.7 mA/cm^2 at 0.7 V bias, AM 1.5 light intensity of 50 mW/cm^2, was obtained from a hematite n-Fe_2O_3 film, of estimated 150 nm thickness using 60 s spray time at a substrate temperature of 350°C and a spray solution of 0.11 M $FeCl_3$ in absolute ethanol. A maximum quantum efficiency of 22.5% is obtained for the film at 370 nm. Thicker films show a decreased photocurrent amplitude attributed to increased electrical resistance and a higher recombination rate of the photo-generated charge carriers.

4.8 Quantum-size Effects in Nanocrystalline Semiconductors

A unique aspect of nanotechnology is the prospect of designing and fabricating semiconducting nanoscale materials leading to substantial increases in the surface area to volume ratios. Bottom-up approaches involve the self-assembly of smaller components via synthetic chemistry, while top-down approaches generally involve externally controlled engineering strategies for transforming larger entities, e.g. silicon wafers, into smaller architectures via lithographic, printing or various deposition techniques. As the size of the system decreases a number of physical properties, including optical and electronic behavior, reflect quantized motion of electrons and holes in a confined state [110]. So called quantum size effects can be observed when the size of the particle, or structural feature, is comparable to the size of electron-hole pair [111,112]. Consequently the electronic properties of solids change dramatically with a reduction in architectural feature sizes, nominally 1 nm to 10 nm, owing to quantum mechanical effects where the relatively well behaved continous electronic band strucuture of a bulk semiconductor changes to one comprised of discrete energy levels [113]. With an aim towards using semiconducting nanoparticles as light sensitive components in a highly effcient photocatalytic system, various attempts have been made to analyze the distinctive features of nanostructured semiconductor materials demonstrating quantum size effects. A number of oxide and non-oxide semiconductor nanoparticles have shown a blue shift in their absorption spectra as a consequence of exciton confinement with decreasing particle size [114]. Various oxide semiconductors display quantum size effects in the range $2R \leq 10$ nm (R = radius of the particles), and their absorption and luminescence have been properties investigated for this reason [115-123]. For example, Kormann et al. [116] demonstrated quantum size effects attributed to exciton confinement, a 0.15 eV blue shift of the UV absorption edge, in particles of 2R $\approx$ 2.4 nm prepared by the hydrolysis of TiCl$_4$. A smilar 0.1 eV blue shift of the optical absorption edge has been observed by Kavan and co-workers [117] for a film consisting of 2R $\approx$ 2.0 nm TiO$_2$ particles prepared by anodic oxidative hydrolysis of TiCl$_4$. Photophysical and photochemical behavior of colloidal ZnO particles (2R < 5.0 nm) have been investigated by absorption and luminescence measurements [118-122], with measurements suggesting a blue shift in the optical absorption edge due to quantum size effects. As the ZnO particles, which are nearly spherical in shape, grow in size comparable to 2000-3000 molecules they begin to exhibit the photophysical properties of bulk ZnO [119]. Luminescence spectra of the ZnO sols suggest that adsorbed electron relays are necessary for the transfer of electrons from the conduction band into lower lying traps [120,121].

Anpo and co-workers [121] reported size quantization effects in rutile (2R = 5.5 nm) and anatase (2R = 3.8 nm) particles that displayed an increase in the band gap by ~0.093 eV and ~0.156 eV, respectively, relative to the bulk values of 3.0 and 3.2 eV for rutile and anatase. Improved photocatalytic activity has been observed for both rutile and anatse with the blue shift in the band gap [121] due, apparently, to modification of the electronic properties of the smaller particles. However the general results per size quantification appear to be strongly influenced by the nuances of the topological and structural features. For example, the absorption and luminescence behavior of TiO$_2$ colloidal particles with sizes 2R = 2.1 nm, 13.3 nm and 26.7 nm were studied by Serpone and co-workers [122], who found no significant differences in the absorption edges. Monticone et al. [123] measured a 0.17 eV blue shift for anatase particles of 2R = 1.0 nm, with no shift in band gap energy observed for sizes 2R $\geq$ 1.5 nm.

4.8.1 Quantum Size Effects: A Theoretical Overview

A theoretical description of the above-noted size-dependent behavior has been proposed by L. E. Brus [113,114] and Y. Kayanuma [124].

As the size of a semiconductor crystal becomes 'small' a regime is entered in which the electronic properties, e.g. ionization potential and electron affinity, are determined by size and shape of the crystals [113]. When a quantum of light (hv) with energy exceeding the band gap falls on the surface of a semiconductor crystal there appears a bounded electron-hole pair known as an exciton ($e^-...h^+$):

$$Semiconductor + hv \rightarrow \left(e^-...h^+\right) \rightarrow e^-_{CB} + h^+_{VB} \qquad (4.8.1)$$

Coulombic interaction, $E(R) = -\dfrac{e^2}{\varepsilon R}$, between the electron and hole binds the exciton which usually remains delocalized over a length much greater than the lattice constant. Bohr's model of the hydrogen atom has been used to describe the movement of the exciton (bound electron-hole pair) [125,126] by which the region of the electron-hole pair delocalization can be appraised by the Bohr radius a_B of the exciton. The energy levels of an exciton created in a direct band gap semiconductor are shown in **Fig. 4.11**.

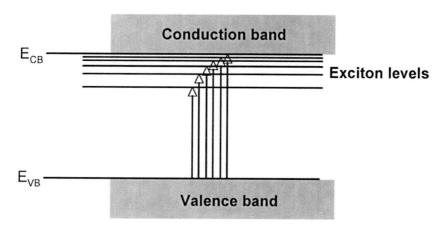

Fig. 4.11: Representative diagram for optical transitions showing the formation of excitons from the top of the valence band.

As the radius of a semiconductor crystallite approaches the exciton-Bohr-radius its electronic properties begin to change, whereupon quantum size effects can be expected. The Bohr radius a_B of an exciton is given by

$$a_B = \frac{\hbar^2 \varepsilon}{e^2}\left(\frac{1}{m_e^*} + \frac{1}{m_h^*}\right) \qquad (4.8.2)$$

where $\hbar = \dfrac{h}{2\pi}$, h = planck's constant, 6.626×10^{-34} J.s or 4.135×10^{-14} eV•s), m_e^* and m_h^* are the effective masses of the electron and the hole, ε is the dielectric constant of the semiconductor, and e is the electron charge. The exciton radius (R_{exc}) can be calculated [127]:

$$R_{exc} = \left(\frac{\varepsilon}{m_e^* / m_e}\right) \cdot a_B \qquad (4.8.3)$$

The quantitative size-dependence of the minimum electronic excitation energy of a nanoparticle can be approximated taking into account the uncertainty in position (Δx) and momentum (Δp) of the exciton:

$$\Delta x \times \Delta p \geq \frac{\hbar}{2} \qquad (4.8.4)$$

In a single particle approximation there is a filled valence band, an empty conduction band, and a range of forbidden energies (band gaps). The delocalized exciton is defined by its energy and momentum, having no fixed position in the periodic potential of the bulk crystals. At low energies delocalized electron waves follow a quadratic relationship between wave vector k and energy E given by:

$$E = \frac{\hbar^2 k^2}{2m^*} \qquad (4.8.5)$$

Near k = 0 the electron and hole effective masses must be isotropic and constant. For a nanoparticle, the uncertainty in the exciton position depends upon its size, $\Delta x \approx 2R$. If the relation between energy and momentum is independent of particle size the exciton energy will be inversely proportional to the square of the nanoparticle size. With respect to the bandgap of the bulk material, the size-dependent shift in the nanoparticle exciton energy is:

$$\Delta E = \frac{\hbar^2 \pi^2}{2R^2 \mu} - \frac{1.8e^2}{\varepsilon R} - 0.248R_y \qquad (4.8.6)$$

where R is the nanoparticle radius, μ the reduced mass of the exciton $\mu^{-1} = (m_e^*)^{-1} + (m_h^*)^{-1}$, and R_y the effective Rydberg energy

$$= \frac{e^4}{2\varepsilon^2 \hbar^2 \mu}.$$

The first two terms in equation (4.8.6) have been given by Brus [115,116,131]. The first term in equation (4.8.6) with the $1/R^2$ dependence represents the quantum localization energy, and is the sum of the confinement energy for the electron and holes. The second term, with $1/R$ dependence, accounts for the coulombic interaction between the electron and hole. The third term in equation (4.8.6) was introduced by Kayanuma [126] and is a result of spatial correlation in the limit as $\frac{R}{R_{exc}} \to 0$ [126]. The last size-independent term is usually small but can become significant for a semiconductor with a small dielectric constant.

Equation (4.8.6) is used to estimate the shift in band gap energy ΔE due to size quantization. One can derive a shift in the band gap of 0.15 eV, in good agreement with experimental observations for TiO$_2$, taking R = 1.2 nm and ε = 184 (one finds m_e^* = 1.63 m_e as calculated from the absorption spectrum). Using equation (4.8.3) with Bohr radius a_B = 0.053 nm (the Bohr radius of hydrogen) and electron effective mass values between 5 m_e and 13 m_e, exciton radii values between 0.75 nm and 1.9 nm have been estimated for TiO$_2$ [118], hence per the described model particles consisting of a few hundred molecules can be anticipated to exhibit quantum-size effects. Precise values of electron and hole effective masses in anisotropic semiconductors remain elusive with values dependent upon the direction of motion, whereupon equation (4.8.6) becomes significantly more complex. Comparison of experimental results with theoretical predictions for oxide semiconductors

[118-122, 125] and non-oxide semiconductors [127,128] indicate the model works effectively only for weakly confined relatively large particles. The observed exciton energy of dense nanoparticle films appears lower than those predicted from equation (4.8.6) resulting in overestimation of nanoparticle size.

Another indicator of quantum size effects is the increase in the volume-normalized oscillator strength of excitonic absorption with decreasing nanoparticle size. For restricted geometries the binding energy and oscillator strength can increase due to the coherent motion of the exciton, and the enhanced spatial overlap between the electron and hole wave functions [127]. The oscillator strength of an exciton is given by

$$f = \frac{2m}{\hbar^2} \Delta E \left| \mu_D^2 \right| \left| U\left(0\right)^2 \right| \tag{4.8.7}$$

Where $\left| U(0)^2 \right|$ is the probability of finding the electron and hole at the same location and μ_D is the transition dipole moment.

The overlap integral of the electron and hole wave functions are independent of the particle size for large spherical particles (R > a_B), and the oscillator strength of excitonic absorption then determined by the bulk dipole moment of the transition. Assuming an exciton moves coherently and is delocalized throughout the particles, the corresponding exciton energy band has discrete eigenvalues near k = 0 of the form:

$$\frac{\hbar^2 \pi^2 n^2}{2\left(m_h^* + m_e^*\right) R} \tag{4.8.8}$$

where n varies from 1 to $(R/a_B)^3$. In the low temperature limit the oscillator strength, f, of all $(R/a_B)^3$ levels are concentrated in the lowest excitonic transition. The oscillator strength of a particle is proportional to, and increases linearly with the nanoparticle volume.

$$f\alpha \left(\frac{R}{a_B}\right)^3 \tag{4.8.9}$$

For small particles ($R < a_B$) the overlap integral $|U(0)^2|$ of the electron and hole wave functions increases with decreasing nanoparticle volume; as a result f is only weakly dependent on the particle size [113]. However, volume normalized oscillator strength (f/V) increases with decreasing nanoparticle size and can be estimated as [126]:

$$\frac{f}{V} = \left(\frac{a_B}{R}\right)^3 \qquad (4.8.10)$$

While f determines the radiative lifetime of an exciton at low temperature limit, f/V determines the light absorption coefficient, hence the exciton absorption band will get stronger with decreasing nanoparticle size in the $R < a_B$ size regime.

Effects due to spatial confinement of photogenerated charges appear in architectures of dimensions either comparable to or smaller than the exciton delocalization region $(a_B/R)^3$. Assuming spherical particles, the extent of the resultant changes are determined by the ratio of the nanoparticle radius R and a_B. For $a_B \leq R$ the electronic structure of the nanoparticle nominally remains the same as in the bulk crystal. In a strong quantum confinement regime, $a_B > R$, a noticeable increase in the exciton energy is observed due to a gradual change of the semiconductor energy bands into a set of discrete electronic levels.

4.8.2 Surface Activities

A series of competing processes arise from photogenerated charge carriers in a semiconductor nanoparticle since a large percentage of the atoms are at the surface and behave differently than those in the bulk. An exciton at the surface is rapidly (picosecond) trapped due to surface defects, with the electron-hole pair subsequently participating in transfer between the semiconductor nanoparticle and the electrolyte adsorbed on its surface:

$$h_{VB}^+ \text{ (or } e_{CB}^-) \rightarrow e_{tr}^- \text{ (or } h_{tr}^+) \qquad (4.8.11)$$

$$h_{VB}^+ + e_{CB}^- \rightarrow h\nu \qquad (4.8.12)$$

$$e_{tr}^- \text{ (or } h_{tr}^+) + h_{VB}^+ \text{ (or } e_{CB}^-) \rightarrow h\nu' \qquad (4.8.13)$$

with hv′ defect luminescence [129] due to the structural inhomogeneity of the intrinsic nanoparticle crystals [131].

One of the most distinguishing features of semiconductor nanoparticles for use in photoelectrocatalysis is the absence of band bending at the semiconductor-electrolyte interface, see **Fig. 4.2**. In contrast to bulk behavior, for a colloidal semiconductor or a semiconductor comprised of a nanocrystalline network in contact with an electrolyte the difference in potentials between the center ($r = 0$) of the particle and a distance r from the center can be expressed [83]:

$$\Delta\varphi_{SC} = \frac{kT}{6e}\left[\frac{r - (r_0 - W)}{L_D}\right]^2\left[1 + \frac{2(r_0 - W)}{r}\right] \quad (4.8.14)$$

$$L_D = \left(\frac{\varepsilon\varepsilon_0 kT}{2e^2 N_D}\right)^{1/2} \quad (4.8.15)$$

where $\Delta\varphi_{sc}$ is the potential drop within the layer, r_0 is the radius of the semiconductor particle, W is the width of band bending (or space charge layer) and L_D is the Debye length determined by the concentration of ionized dopant impurities, N_D. (ε and ε_0 are the dielectric constants of the semiconductor and medium respectively).

Figure 4.12 is an illustration of the potential distribution for n-type semiconductor particles at the semiconductor-electrolyte interface. There are two limiting cases of equation (4.8.11) for photo-induced electron transfer in semiconductors. For large particles the potential drop within the semiconductor is defined by:

$$\Delta\varphi = \frac{kT}{2e}\left[\frac{W}{L_D}\right]^2 \quad (4.8.16)$$

For small particles ($r_0 \ll L_D$), generally 1 nm to 10 nm, the potential drop within the semiconductor is given by:

$$\Delta\varphi = \frac{kT}{6e}\left[\frac{r_0}{L_D}\right]^2 \quad (4.8.17)$$

A 50 mV drop in the potential of a r_0 = 6 nm colloidal TiO_2 particle correlates to an ionized donor impurity concentration of 5×10^{19} cm^{-3}, correlating to a Debye length L_D on the order of 10^{-7} m thus satisfying the condition $r_0 \ll L_D$.

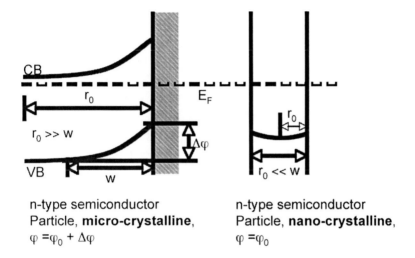

n-type semiconductor
Particle, **micro-crystalline**,
$\varphi = \varphi_0 + \Delta\varphi$

n-type semiconductor
Particle, **nano-crystalline**,
$\varphi = \varphi_0$

Fig. 4.12: Diagram illustrating space charge layer formation in microcrystalline and nanocrystalline particles in equilibrium in a semiconductor-electrolyte interface. The nanoparticles are almost completely depleted of charge carriers with negligibly small band bending.

As indicated by equation (4.8.17), for nanoscale crystals the difference in the potential is constant throughout the semiconductor volume and its surface; the particles are too small to develop a space charge layer. Hence there is no barrier, due to band bending, to electron transfer across the semiconductor nanoparticle-solution interface. Furthermore band-to-band electron hole recombination, which reduces the effectiveness of the photoelectrochemical pathways, is suppressed. Charge carrier dynamics in illuminated semiconductor nanocrystals can be described as localized charges in surface traps [130]. The depletion layer at the interface between a bulk semiconductor and a liquid medium plays an important role in light induced charge separation. The local electrostatic field present in the space charge layer serves to separate the electron-hole pair. In the case of a colloidal or nanocrystalline semiconductor, band bending is small and thus charge separation occurs via diffusion. A characteristic feature of semiconductor nanoparticles is the extremely short diffusion lifetime (τ_d), bulk to surface, of the photogenerated charge in comparison to the electron-hole recombination time (τ_r) [83,131]. For 10 nm TiO_2 nanoparticles a lifetime of 10 ps (τ_d) is recorded for an electron traveling to the surface, whereas the lifetime of direct electron-hole recombination τ_r is on the order of 100 ns [133]. As the particle size increases, τ_d approaches τ_r becoming essentially equivalent, $\tau_d \approx \tau_r$, for a 1 μm particle at which point electron-hole recombination behavior starts dominating over surface trap phenomena. Since charge carrier diffusion from the interior of a particle to the surface can occur more rapidly with charge localized in the surface traps (e_{tr}^- and h_{tr}^+) than electron-hole recombination (e_{CB}^- and h_{VB}^+) it is possible to obtain quantum yields approaching unity for photo-redox processes [83] if at least one of the charge carriers are rapidly removed from the surface-interface.

4.8.3 Particle Size Effects on the Photoelectrochemical Properties

Nanocrystalline TiO_2 electrodes prepared by reactive sputtering with an average particle size below 22 nm exhibit quantum size effects manifested by a blue shift in the band gap energy [99]. A maximum photocurrent activity has been observed for a 50 nm particle size and film thickness of 1.2 μm. It is observed that when the mean particle size decreases from 51 nm to 32 nm the film orientation changes from (112) to (101) simultaneously resulting in a lower photocurrent activity; the columnar structured (112)-oriented anatase provides a lower grain contact resistance than the (101)-oriented anatase of irregular structure. Sol-gel synthesized nanocrystalline TiO_2 and V-doped TiO_2 of average particle size 5 nm to 10 nm demonstrate a 0.19 eV blue shift in the band gap energy as compared to bulk TiO_2, a behavior manifest by a negative shift in the flat band potential [132] and dependent on the electron and hole effective masses [133]. Saki and co-workers reported a nanocrystalline TiO_2 multilayer thin film with a flat band potential approximately 0.1 V more negative, and band gap approximately 0.6 eV higher than bulk anatase TiO_2 ($E_{fb} = -1.15$ V and $E_{BG} = 3.2$ eV) [134].

A similar blue shift in the absorption edge of various sol-gel synthesized nanocomposite nanoparticles such as TiO_2-SnO_2 [135], TiO_2-SiO_2 [136], TiO_2-WO_3[137] and TiO_2-In_2O_3[138] have been observed and are assigned to quantum size effects. Various binary composite nanostructured film electrodes have demonstrated extended photoresponses and improved photocurrents [137-139]. However binary nanocomposite TiO_2-In_2O_3 nanocrystalline film electrodes show a sharp drop in photocurrent generation (UV illumination) in TiO_2 rich (78% TiO_2, 2 $\mu A \cdot cm^{-2}$) or In_2O_3 rich (73% In_2O_3, 30 $\mu A \cdot cm^{-2}$) formulations as compared to pure TiO_2 (60 $\mu A \cdot cm^{-2}$) and pure In_2O_3 (40 $\mu A \cdot cm^{-2}$) [140]. Although the photocurrent onset potential for the TiO_2-In_2O_3 (78% TiO_2) is 0.30 V more negative than that for the TiO_2 electrode, indicating favorable electron separation, a comparative decrease in photocurrent is observed [140]. While the measured photocurrent amplitudes are low enough that the reported variations may not be significant we note that in nanocrystalline films photogenerated electrons and holes can be independently involved in surface oxidation and reduction processes. Since nanostructured films have extremely high surface areas and the nanocrystals are surrounded by the electrolyte, a key factor resulting in a decrease in the photocurrent efficiency is due to the recombination of electron-hole pairs through surface states or through photoelectrochemical cell reaction products.

A potential advantage of nanostructured electrodes over conventional semiconductor electrodes is the relatively large density of surface states that may facilitate rapid separation of photogenerated electron-hole pairs. Boschloo and co-workers [139] have observed the presence of surface states at nanostructured TiO_2/aqueous electrolyte interfaces. The electrons trapped in these states have an absorption spectrum significantly different from that of conduction band electrons. The induced optical changes in nanostructured TiO_2 appear mainly due to electron accumulation in conduction band states. The accumulation of excess charge on semiconductor nanoparticles substantially alters the photophysical and photoelectrochemical characteristics and also affects the kinetics of electron phase transfer, forming the basis of the *Burstein-Moss* effect [140]. The accumulation of excess negative charge localized in the surface traps results in displacement of the Fermi level towards the edge of the conduction band, thereby increasing band gap energy; the Burstein-Moss effect is manifest in a shift of the spectral limit of light absorption to higher energies because of electron-filled lower energy states in the conduction band [131,141].

Fujishima and co-workers examined the photoelectro-chemical behavior of Nb-doped TiO_2 ($Ti_{1-x}Nb_xO_2$; $x = 0.01$, 0.03, 0.06, 0.1) photoelectrodes prepared by pulse laser deposition on $LaAlO_3$ and $SrTiO_3$ [142]. Extensive Nb doping turned an otherwise n-type TiO_2 semiconductor into a degenerate semiconductor with metallic-like conductivity with a blue shift of the band gap due to the Burstein-Moss effect resulting in decreased light absorption. The Nb doping also led to the formation of recombination centers that drastically reduced the photoelectrochemical efficiency of the $Ti_{1-x}Nb_xO_2$ thin film electrodes. Since high electron concentrations increase band-to-band recombination the Nb doping caused a severe decay of photocurrent amplitudes despite the increase in the conductivity. Miyagi and co-workers [143] observed similar phenomena, with Nb doping of titania leading to the formation of defects and deep electron traps that serve as recombination centers responsible for a dramatic loss in photocatalytic activity.

References

1.   Fujishima A, Honda K (1972) Electrochemical photolysis of water at a semiconductor electrode. Nature 238:37–38
2.   Sieber KD, Sanchez, C, Turner JE, Somorjai GA (1985) Preparation, electrical and photoelectrochemical properties of magnesium doped iron oxide sintered discs. Mat Res Bull 20:153–162
3.   Fujishima A, Honda K (1971) Electrochemical evidence for the mechanism of the primary stage of photosynthesis. Bull Chem Soc Jpn 44:1148–50
4.   Mavroides JG, Tchernev DI, Kafalas JA, Kolesar DF (1975) Photoelectrolysis of water in cells with TiO_2 anodes. Mater Res Bull 10:1023–1030

5. Nozik AJ (1975) Photoelectrolysis of water using semiconducting TiO_2 crystals. Nature 257:383–386

6. Ohnishi T, Nakato Y, Tsubumura H (1975) Quantum yield of photolysis of water on titanium oxide. Ber Bunsenges Phys Chem 79:523–525

7. Getoff N (1990) Photoelectrochemical and photocatalytic hydrogen production. Int J Hydrogen Energy 15:407–417

8. Ghosh AK, Muruska HP (1977) Photoelectrolysis of water in sunlight with sensitized semiconductor electrodes. J Electrochem Soc 128:1516–1522

9. Gautron J, Marucco JF, Lemasson P (1981) Reduction and doping of semiconducting rutile. Mater Res Bull 16:575–580

10. Salvador P (1980) The influence of Niobium doping on the efficiency of n-TiO_2 electrode in water photoelectrolysis. Sol Energy Mater 2:413–421

11. Wang MH, Guo RJ, Tso TL, Perng TP (1995) Effects of sintering on the photoelectrochemical properties of Nb-doped TiO_2 electrode. Int J Hydrogen Energy 20:555–560

12. Nair MP, Rao KVC, Nair CGR (1991) Investigation of the mixed-oxide materials TiO_2-SiO_2, TiO_2-SiO_2-Al_2O_3, TiO_2-SiO_2-In_2O_3 and TiO_2-SiO_2-RuO_2 in regard to the photoelectrolysis of water. Int J Hydrogen Energy 16: 449–459

13. Wrighton MS, Ellis AB, Wolczanski PT, Morse DL, Abrahamson HB, Ginley DS (1976) Strontium titanate photoelectrodes. Efficient photoassisted electrolysis of water at zero applied potential. J Am Chem Soc 98:2774–2779

14. Bolts JM, Wrighton MS (1976) Correlation of photocurrent-voltage curves with flat-band potential for stable photoelectrodes for the electrolysis of water. J Phys Chem 80:2641–2646

15. Mavroides JG, Kafalas JA, Kolesar DF (1976) Photoelectrolysis of water in cells with $SrTiO_3$ anodes. Apl Phys Lett 28:241–243

16. 't Lam Rue, de Haart LGJ, Wiersma AW, Blasse G, Tinnemans AHA, Mackor A (1981) The sensitization of

SrTiO$_3$ photoanodes by doping with various transition metal ions. Mater Res Bull 16:1593–1600

17. Mackor A, Blasse G (1981). Visible light induced photocurrents in SrTiO$_3$-LaCrO$_3$ single-crystalline electrode. Chem Phys Lett 77:6-8

18. Yin J, Ye J, Zou Z (2004) Enhanced photoelectrolysis of water with photoanode Nb:SrTiO$_3$. Appl Phys Lett 85: 689-691

19. Redon AM, Vigneron J, Heindl R, Sella C, Martin C, Dalbera JP (1981). Differences in the optical and photoelectrochemical behaviors of single crystals and amorphous ferric oxide. Solar Cells 3:179–186

20. Kung HH, Jarrett HS, Sleight AW, Ferretti A (1977) Semiconducting oxide anodes in photoassisted electrolysis of water. J Appl Phys 48:2463–2469

21. Quin RK, Nashby RD, Baughman RJ (1976) Photoassisted electrolysis of water using single crystals of α-Fe$_2$O$_3$ anode. Mater Res Bull 11:1011–1017

22. Gartner WW (1959) Depletion layer photoeffects in semiconductors. Phys Rev 116:84–87

23. Sanchez C, Sieber KD, Somorjai GA (1988) The photoelectrochemistry of α-Fe$_2$O$_3$. J Electroanal Chem 252:269–290

24. Wrighton MS, Morse DL, Ellis AB, Ginley DS, Abrahamson HB (1976) Photoassisted electrolysis of water by ultraviolet irradiation of an antimony doped stannic oxide electrode. J Am Chem Soc 98:44-48

25. Butler MA, Nasby RD, Quinn RK (1976) Tungsten trioxide as an electrode for photoelectrolysis of water. Sol State Commun 19:1011-1014

26. Butler MA (1977) Photoelectrolysis and physical properties of semiconducting anode. J Appl Phys 48:1914-1920

27. Antonucci V, Giordano N, Bart JCJ (1982) Structure and photoelectrochemical efficiency of oxidized titanium electrodes. Int J Hydrogen Energy 7:769-774

28. Giordano N, Antonucci V, Cavallaro S, Lembo R, Bart JCJ (1982) Photoassisted decomposition of water over modified rutile electrodes. Int J Hydrogen Energy 7:867–72

29. Matsumoto Y, Shimizu T, Sato E (1982) Photoelectrochemical properties of thermally oxidized TiO_2. Electrochem Acta 27:419–424

30. Benko FA, MacLaurin CL, Koffyberg FP (1982). $CuWO_4$ and Cu_3WO_6 as anodes for the photoelectrolysis of water. Mater Res Bull 17:133–136

31. Campet G, Claverie J, Hagenmuller P, Chang BT (1984) Influence of lanthanum-doping on the photoelectrochemical properties of $SrTiO_3$ polycrystalline anodes. Mater Lett 3:5–10

32. Radecka M, Sobas P, Trenczek A, Rekas M (2004) Photorespose of undoped and W-doped TiO_2. Polish J Chem 78:1925–1934

33. Radecka M, Sobas, Wierzbicka PM, Rekas M (2005) Photoelectrochemical properties of undoped and Ti-doped WO_3. Physica B 364:85–92

34. Bak T, Nowotny J, Rekas M, Sorrell CC (2002) Photoelectrochemical properties of TiO_2-Pt system in aqueous solutions. Int J Hydrogen Energy 27:19–26

35. Sastri MVC, Nagasubramanian G (1982) Studies of ferric oxide electrodes for the photo-assisted electrolysis of water. Int J Hydrogen Energy 11:873–876

36. Turner JE, Hendewerk M, Somorjai GA (1981) The photodissociation of water by doped iron oxides: the unbiased p/n assembly. Chem Phys Lett 105:581–585

37. Hardee KL, Bard AJ (1975) Semiconductor electrodes I. J Electrochem Soc 122:739–742

38. Gerischer H (1977) On the stability of semiconductor electrodes against photodecomposition. J Electroanal Chem 82:133–143

39. Fujishima A, Kohayakawa K, Honda K (1975). Hydrogen production under sunlight with an electrochemical photocell. J electrochem Soc 122:1487–1489

40. Keeny J, Weinstein DH, Haas GM (1975) Electricity from photosensitization of Ti. Nature 253:719–720

41. Yazawa K, Kamugawa H, Morisaki H. (1979) Semiconducting TiO$_2$ films for photoelectrolysis of water. Int J Hydrogen Energy 4:205–209

42. Babu KSC, Srivastava ON (1989) Structural and photoelectrochemical studies of In$_2$O$_3$ modified TiO$_2$ in regard to hydrogen-production through photoelectrolysis. Int J Hydrogen Energy 14:529–535

43. Hartig KJ, Getoff N, Kotschev KD, Kanev ST (1983) Influence of hydrogen reduction on photoelectrochemical behavior of anodic oxidized n-TiO$_2$ layer. Sol Energy Mater 9:167–195

44. Wilson RH (1977) A model for the current-voltage curve of photoexcited semiconductor electrodes. J Appl Phys 48:4292–4297

45. Bicelli LP, Razzini G (1985) Photoelectrochemical performance of anodic n-TiO$_2$ films submitted to hydrogen reduction. Int J Hydrogen Energy 10:645–649

46. Lindquist SE, Lindgren A, Ning ZY (1985) On the origin of the bandshifts in the action spectra of polycrystalline TiO$_2$ electrode prepared by thermal oxidation of titanium. J Electochem Soc 132:623-631

47. Akikusa J, Khan SUM (1997) Photo response and AC impedance characterization of n-TiO$_2$ during hydrogen and oxygen evolution in an electrochemical cell. Int J Hydrogen Energy 22:875–882

48. Dobkin DM, Zurao MK (2003) Priniciples of chemical vapor deposition. Kluwer, Dordrecht

49. Kern W, Vossen J (1991) Thin film processes II. Academic press, New York.

50. Williams LM, Hess DW (1984) Phtoelectrochemical properties of plasma deposited TiO$_2$ thin film. Thin Solid Films 115:13-18

51. Boschloo GK, Goossens A, Schoonman J (1997) J Photoelectrochemical study of thin film of anatase TiO$_2$ films prepared by metal organic chemical vapor deposition. J Electrochem Soc 144:1311–1317

52. Levi-Clement C, Schleich DM, Gorochov O, Czapla A (1983) $Sn_{1-x}Pb_xO_2$ sputtered thin film as photoanode for photoelectrochemical cells. Mater Res Bull 18:1471–1476

53. Yoko T, Yuasa A, Kamiya K, Sakka S (1991) Sol-gel-derived TiO_2 thin film semiconductor electrode for photocleavage of water. J Electrochem Soc 138:2279–2784

54. Yoko T, L. Hu L, Kozuka H, Sakka S (1996) Photoelectrochemical properties of TiO_2 coating films prepared using different solvent by the sol-gel method. Thin Solid Films 283:188–195

55. Zhao G, Kozuka H, Lin H, Yoko T (1999) Sol-gel preparation of $Ti_{1-x}V_xO_2$ solid solution film electrodes with conspicuous photoresponse in the visible region. Thin Solid Films 339:123–128

56. Hartig KJ, Getoff N, Nauer G (1983) Comparison of photoelectrochemical properties of n-TiO_2 films obtained by different production methods. Int J Hydrogen Energy 8:603–607

57. Jung HC, Kim KS, Yoon DH, Nam SS, Sun KH (1991) The stability of PEC electrodes (TiO_2 anode and Pt cathode) and Cell for H_2 production. Int J Hydrogen Energy 16:379–386

58. Dyer CK, Leech JSL (1978) Reversible optical changes within anodic oxide films on titanium and niobium. J Electrochem Soc 125:23–29

59. Takahashi M, Tsukigi K, Uchino T, Yoko T (2001) Enhanced photocurrent in thin film TiO_2 electrodes prepared by sol-gel method. Thin Solid Films 388, 231–236

60. Bockris JOM, Itoh K (1984) Stacked thin film electrode from iron oxide. J Appl Phys 56:874–876

61. Prasad G, Rao NN, Srivastava ON (1988) On the photoelectrodes TiO_2 and Wse_2 for hydrogen production through photoelectrolysis. Int J Hydrogen Energy 13:399–405

62. Augustynski J (1993) The role of the surface intermediates in the photoelectrochemical behavior of anatase and rutile TiO_2. Electochim Acta 38:43–46

63. Yeshodharan E, Grätzel M (1983) Photodecomposition of liquid water with TiO_2 supported noble metal cluster. Helv Chim Acta 66:2145–2153

64. Khan SUM, Akikusa J (1998) Stability and photoresponse of nanocrystalline n-TiO_2 and n-TiO_2/Mn_2O_3 thin film electrodes during water splitting reactions. J Electrochem Soc 145:89–93

65. Khan SUM, Al-shahry M, Ingler Jr. WB (2002) Efficient photochemical water splitting by a chemically modified n-TiO_2. Science 297:2243–2245

66. Shankar K, Paulose M, Mor GK, Varghese OK, Grimes CA (2005) A study on the spectral photoresponse and photoelectrochemical properties of flame annealed titania nanotube arrays. J Phys D: Appl Phys 38:3543–3549.

67. Hagglund C, Gratzel M, Kasemo B (2003) Comments on "Efficient photochemical water splitting by a chemically modified n-TiO_2" (II). Science 301:1673b

68. Aroutiounian VM, Arakelyan VM, Shahnazaryan GE (2005) Metal oxides photoelectrode for hydrogen generation using solar water radiation driven water splitting. Sol Energy 78:581–592

69. Noworyta K, Augustynski J (2004) Spectral photorespnses of carbon-doped TiO_2 film electrode. Electrochem Solid-State Lett 7:E31–E33

70. Radecka M, Zakrzewska Wierzbicka KM, Gorzkowska A, Komornicki S (2003) Study of the TiO_2-Cr_2O_3 system for photoelectrolytic decomposition of water. Solid state Ionics 157:379–386

71. Hodes G, Cahen D, Mannasen J. (1976) Tungsten trioxide as a photoanode for photoelectrochemical cell (PEC). Nature 260:312–313

72. Quarto FD, Paola AD, Sunseri C (1981) Semiconducting properties of anodic WO_3 amorphous film. Electrochim Acta 26:1177–1184

73. Giordano N, Passalacqua E, Antonucci V, Bart JCJ (1983) Iron oxide electrodes for photoelectrolysis of water. Int J Hydrogen Energy 10:763–766

74.   Majumder SA, Khan SUM (1994) Photoelectrolysis of water at bare and electrocatalyst covered thin film Fe_2O_3 Int J Hydrogen Energy 19:881–887

75.   Sartoretti CJ, Alexander BD, Solarska R, Rutkowaska IA, Augustynski J (2005) Photoelectrochemical oxidation of water at transparent ferric oxide film electrode Phys Chem B 109:13685-13692

76.   Ingler WB, Baltrus JP, Khan SUM (2004) Photoresponse of p-type zinc-doped iron(III) oxide films. J Am Chem Soc 126:10238–10239

77.   Jaramillo TF, Baeck SH, Shwarsctein AK, Choi KS, Stucky, McFarland EW (2005) Automatated electrochemical synthesis and photoelectrochemical characterization of $Zn_{1-x}Co_xO$ thin film for solar hydrogen production. J Comb Chem 7:264–271

78.   Regan BO, Grätzel M (1991) A low-cost high-efficiency solar cell based on dye-sensitized colloidal TiO2 thin film Nature 353:737–740

79.   Katoh R, Furabe A, Barzykin, Arakawa H, and Tachiya H (2004) The kinetics and mechanism in electron injection and charge recombination in dye-sensitized nanocrystalline semiconductor. Coord Chem Rev 248:1195-1213

80.   Frank AJ, N Kopidakis, J Lagemaat (2004) Electron in nanostructured TiO_2 solar cell:transport, recombination and photovoltaic properties. Coord Chem Rev 248:1195-1213

81.   Gregg BA (2004) Interfacial processes in the dye sensitized solar cell. Coord Chem Rev 248:1512-1224

82.   Lewis NS (2005) Chemical control of charge transfer and recombination at semiconductor photoelectrode surfaces. Inorg Chem 44:6900–6911

83.   Hagfeldt A, Gratzel M (1995) Light-Induced Redox Reactions in Nanocrystalline Systems Chem Rev 95:49-68

84.   Lee MS, Cheon IC, Kim YI (2003) Photoelectrochemical studies of nanocrystalline TiO2 film electrodes. Bull Kor Chem Soc 24:1155–1162

85.   Mishra PR, Shukla PK, Singh AK, Srivastava ON (2003) Investigation and optimization of nanostructured TiO_2 photoelectrode in regard to hydrogen production through

photoelectrochemical process. Int J Hydrogen Energy 28:1089–1094

86.   Watanabe A, Kozuka H (2003) Photoanodic properties of sol-gel derived Fe_2O_3 thin films containing dispersed gold and silver particles. J Phys Chem 107:12713–12720

87.   Hida Y, Kozuka H (2005) Photo anodic properties of sol-gel-derived iron oxide thin films with embedded gold nanoparticles: Effects of polyvinylpyrrolidone in coating solutions. Thin Solid Films 476:264–271

88.   Kay A, Cesar I, Grätzel M (2006) New Benchmark for water photooxidation by nanostructured α-Fe_2O_3 films. J Am Chem Soc 128:15714–15721

89.   Fujishima A, Hashimoto K, Watanabe T (1999) TiO_2 Photocatalysis: Fandamentals and Applications. BKC Inc, Tokyo, pp. 1–176

90.   Wang R, Hashimoto K, Fujishima A, Chikuni M, Kojima E, Kitamura A, Shimohigoshi M, Watanabe T (1997) Light induced amiphilic surfaces. Nature 388:431

91.   Sun RD, Akira N, Fujishima A, Watanabe T, Hashimoto K (2001) Photoinduced surface wettability conversion of ZnO and TiO_2 thin film. J Phys Chem B 105: 1984–1989

92.   Lewis N (2001) Frontiers of research of in photoelectrochemical solar energy conversion. J Electroanal Chem 508:1–10

93.   Tomkievich M (2000) Scaling properties in photocatalysis. Catal Today 58:115–123

94.   Tang J, Wu Y, McFarland EW, Stucky GD (2004) Synthesis and photocatalytic properties of highly crystalline and ordered mesoporous thin film. Chem Commun 14:1670–1671

95.   Vigil E, Gonzalez B, Zumeta I, Domingo C, Domenech X, Ayllon JA (2005) Preparation of photoelectrode with spectral response in the visible without applied bias based on photochemically deposited copper oxide inside a porous titanium dioxide thin film. Thin Solid Films 14:489:50–55

96.   Guo B, Liu Z, Hong L, Jiang H, Lee JY (2005) Photocatalytic effect of sol-gel derived nanoporous TiO_2 transparent thin films. Thin Solid Films 479:310–315

97.   Bei Z, Ren D, Cui X, Shen J, Yang X, Zhang Z (2004) Photoelectrochemical properties and crystalline structure change of Sb-doped TiO_2 thin films prepared by the sol-gel method. J Mater Res 19:3189–3195

98.   Takahashi M, Tsukigi K, Dorajpalam E, Tokuda Y, Yoko T (2003) Efficient photogeneration in $TiO_2/VO_2/TiO_2$ multilayer thin film electrodes prepared by sputtering method. J Phys Chem B 107:13455–13458

99.   Mori R, Takahashi M, Yoko T (2005) Photoelectrochemical and photocatalytic properties of multilayered TiO_2 thin films with a spinodal phase separation structure prepared by a sol-gel process. J Mater Res 20:121–127

100.   Santato C, Ulman M, Augustynski J (2001) Photoelectrochemical properties of Nanostructured WO_3 thin films. J Phys Chem B 105:936–940

101.   Beermann N, Vayssieres L, Lindquist SE, Hagfeldt A (2000) Phtotelectrochemical studies of oriented nanorod thin film of hematite. J Electrochem Soc 147:2456–2461

102.   Duret A, Gratzel M (2005) Visible light induced water oxidation on mesoscopic α-Fe_2O_3 films made by ultrasonic spray pyrolysis. J Phys Chem B 109:17184–17191

103.   Cesar I, Kay A, Martinez, JAG, Gratzel M (2006) Translucent thin film Fe_2O_3 photoanode for efficient water splitting by sun light: nanostrucutre directed effect of Si doping. J Am Chem Soc 128:4582–4583

104.   Ingler WB, Khan SUM (2006) A self-driven p/n-Fe_2O_3 tandem photoelectrochemical cell for water splitting. Electrochem Solid-State Lett 9:G144–G146

105.   Prakasam HE, Varghese OK, Paulose M, Mor GK, Grimes CA (2006) Synthesis and photoelectrochemical properties of nanoporous Iron (III) oxide by potentiostatic anodization. Nanotechnology 17:4285–4291

106.   Karn RK, Srivastava ON (1998) On the structural and photochemical studies of In_2O_3-admixed nanostructured TiO_2 with regard to hydrogen production through photoelectrolysis. Int J Hydrogen Energy 23:439–444

107.   Wang Y, Cheng H, Hao Y, Ma J, Li W, Cai S (1999) Photoelectrochemical properties of metal-ion-doped TiO_2 nanocrystalline electrode. Thin Solid Films 349:120–125

108.   Fretwell R, Douglas P (2002) Nanocrystalline TiO_2-Pt photoelectrochemical cells-UV induced hydrogen evolution from aqueous solution of ethanol. Photochem Photobiol Sci 1:793–798

109.   Khan SUM, Akikusa J (1999) Photoelectrochemical splitting of water at nanocrystalline n-Fe_2O_3 thin-film electrode. J Phys Chem B 103:7184–7189

110.   Sargeev, GB (2006) Nanochemistry. Elsevier, Amsterdam

111.   Brus LE (1983) A simple model for ionization potential, electron affinity and aqueous redox potential of small semiconductor electrolytes. J Chem Phys 79:5566–5571

112.   Brus LE (1984) Electron-electron and electron-hole interactions in small semiconductor crystallites: the size dependence of the lowest electronic excited state. J Chem Phys 80:4403–4409

113.   Alivisatos AP (1996) Semiconductor clusters, nanocrystals, and quantum dots. Science 271:933–937

114.   Henglin A (1989) Small-particle research:Physicochemical properties of extremely small colloidal metal and semiconductor particle. Chem Rev 89:1861-1873

115.   Kamat PV (1993) Photochemistry on non reactive and reactive (semiconductor) surfaces. Chem Rev 93:267-300

116.   Kormann C, Bahnemann DW, Hofmann MR (1988) Preparation and characterization of quantum-size titanium oxide J Phys Chem 92:5196–5201

117.   Kavan L, Stoto T, Gratzel M, Fitzmaurice D, Shklover V (1993) Quantum size effects in nanocrystalline semiconducting TiO_2 layers prepared by anodic hydrolysis of $TiCl_3$. J Phys Chem 97:9493–9498

118.   Koch U, Fojtik A, Weller H, Henglein A (1985) Photochemistry of semiconductor colloids: preparation of extremely small ZnO particles, fluorescence phenomena and size quantization effects. Chem Phys Lett 122:507–510

119. Bahnemann DW, Kormann C, Hoffmann MR (1987) Preparation and characterization of quantum size zinc oxide: a detailed spectroscopic study. J Phys Chem 91:3789–3798

120. Kamat PV, Patrick B (1992) Photophysics and photochemistry of quantized ZnO colloids. J Phys Chem 96:6829–6834

121. Anpo M, Shima T, Kodama S Kubukawa Y (1987) Phtotcatalytic hydrogenation of propyne with water on small-particle titania: size quantization effects and reaction intermediates. J Phys Chem 91:4305

122. Serpone N, Lawless D and Khairutdinov R (1995) Size effects on the photophysical properties of colloidal anatase TiO₂ particles: size quantization or direct transition in this indirect semiconductor. J Phys Chem 99:16646–16654

123. Monticone S, Tufeu R, Kanaev AV, Scolan E, Sanchez C (2000) Quantum size effect in TiO₂ nanoparticles: does it exist? Appl Surf Sci 162-163:565–570

124. Kayanuma Y (1988) Quantum size effects of interacting electrons and holes in semiconductor microcrystals with spherical shape. Phys Rev B 38:9797–9805

125. Wang Y, Herron N (1991) Nano-meter sized semiconductor clusters, materials synthesis and quantum size effects, and photophysical properties. J Phys Chem 95:525-532

126. Gaponenko SV (1998) Optical properties of semiconductor nanocrystals. University Press, Cambridge

127. Brus LE (1986) Electronic wave functions in semiconductor clusters: experiments and theory J Phys Chem 90:2555–2561

128. Lippens PE, Lanno M (1990) Comparison between calculated and experimental values of the lowest excited electronic state of the small CdSe crystallites. Phys Rev B 41:6079-6081

129. Stroyuk AL, Kryukov AI, Kuchmii SY, Pokhodenko VD (2005) Quantum size effect in semiconductor photocatalysis. Theoretical and Experimental Chemistry 41:207–228

130. Nojik AJ and Memming RJ (1996) Physical chemistry of semiconductor-liquid interfaces. J Phys Chem 100: 13061–13078

131. Jortner J, Rao CNR (2002) Nanostructure advanced materials: perspectives and directions. Pure Appl Chem 74:1491–1506

132. Sene JJ, Zeltner WA, Anderson MA (2003) Fundamental photoelectrocatalytic and electrophoretic mobility studies of TiO_2 and V-doped TiO_2 thin film doped materials. J Phys Chem B107:1597–1603

133. Enright B, Fitzmaurice D (1996) Spectroscopic determination of electron and hole effective masses in a nanocrystalline semiconductor film J Phys Chem 100:1027–1035

134. Sakai N, Ebina Y, Takada K, Sasaki T (2004) Electronic band structure of titania semiconductor nanosheet revealed by electrochemical and thotoelectrochemical studies. J Am Chem Soc 126:5851–5858

135. Bedja I, Kamat PV (1999) Capped semiconductor colloids: synthesis and photoelectrochemical behavior of TiO_2 capped SnO_2 electrolyte. J Phys Chem 99:9182–9188

136. Davis RJ, Liu Z (1997) Titania silica: a model binary oxide system. Chem Mater 9:2311–2324

137. Shiyanovskaya I, Hepel M (1999) Bicomponent WO_3/TiO_2 films as photoelectrodes. J Electrochem Soc 146:243–249

138. Poznyak SK, Talpin DV, Kulak AI (2001) Structural, optical and photoelectrochemical properties of nanocrystalline TiO_2-In_2O_3 composite solids and films prepared by sol-gel method. J Phys Chem B 105:4816–4823

139. Boschloo GK, Fitzmaurice D (1999) Spectroelectrochemical investigation of surface states in nanostructured TiO_2 electrodes. J Phys Chem B 103:2228–2231

140. Burstein E (1954) Anamolous optical absorption in InSb. Phys Rev 93:632–633

141. Liu CY, Bard AJ (1989) Effect of excess charge on band energetics (optical absorption length and carrier redox potential) in small semiconductor particles. Phys Chem 93:3232–3237.

142. Emelin AV, Furubayashi YV, Zhang X, Jin M, Murakami T, Fujishima A (2005) Photoelectrochemical behavior of Nb-doped TiO_2 electrode. J Phys Chem B 109:24441–24444

143. Miyagi T, Kamei M, Sakaguchi I, Mitsuhashi T, Yamazaki A (2004) Photocatalytic properties and deep level of Nb-doped anatase TiO_2 film grown by metalorganic chemical vapor deposition. Jpn J Appl Phys 43:775–776

Chapter 5

OXIDE SEMICONDUCTORS: NANO-CRYSTALLINE, TUBULAR AND POROUS SYSTEMS

5.1 Introduction

Issues of photocorrosion stability and material availability favor the application of certain metal oxides, in particular TiO_2 and iron oxide to water photoelectrolysis. Tubular nano-structures have generated considerable interest for their potential use in the conversion of solar energy into H_2 via photoelectrolysis. For example, several recent studies have indicated that titania nanotubes have improved properties compared to that of colloidal or any other form of titania for application in photocatalysis [1,2], photoelectrolysis [3-5], and photovoltaics [6-9]. Titania nanotubes and arrays thereof have been produced by variety of methods; depositing into free standing nanoporous alumina used as a template [10-13], sol–gel transcription processes using organo-gelators as templates [14,15], a seeded growth mechanism [16], and hydrothermal processes [17-19]. While each of has advantages, none of them are capable of providing better control over the nanotube dimensions than the anodization of titanium in fluoride based bath [20-27]. In particular highly ordered nanotube arrays, vertically oriented from the surface, appear ideal for water photolysis allowing efficient charge transfer in combination with a high surface area readily accessible to electrolyte percolation.

The thermodynamic potential for splitting water into H_2 and O_2 at 25°C is 1.23 V. Considering overvoltage losses the actual voltage required for water dissociation is in the range of 1.6 to 1.8 V. A 650 nm wavelength, which is in the lower-energy red portion of the visible spectrum, corresponds to an energy of 1.9 eV. Hence essentially the entire visible spectrum of light has sufficient energy to split water into H_2 and O_2. The key is to find a light-harvesting catalytic material architecture that can efficiently collect the energy and direct it with minimal loss to the water-splitting reaction. Since the bulk of the terrestrial solar energy is in the visible spectrum, any shift in the optical response of titania, or its high-bandgap energy metal oxide counterparts, from the UV towards the visible spectrum offers the potential for a significant, positive impact on the photocatalytic and photoelectrochemical utility of the material(s). Hence in addition to issues associated with achieving and characterizing a specific material architecture, considerable efforts have focused on band gap engineering of metal oxide materials so that they more fully respond to full spectrum light while maintaining needed charge transfer and chemical stability properties.

Of the materials being developed for photocatalytic applications, titania remains the most promising because of its high efficiency, low cost, chemical inertness, and photostability [28-30]. Consequently titania has been the most widely investigated photoelectrode material [31]. The choice of photoanode is crucial in a photoelectrochemical cell and involves at least three criteria; photocorrosion stability, the existence of a band gap suitable for capturing as much energy as possible from incident light, and minimization of charge carrier recombination. The geometry of titania nanotube arrays grown vertically from a substrate is, apparently, ideal for water photolysis with demonstrated IPCE values of over 85%, and 320 nm – 400 nm spectrum photoconversion efficiencies to date, from unoptimized systems, of over 16.25%. However, the widespread technological use of titania is impaired by its wide band gap to 3.2 eV; any shift in the optical response of titania from the UV to the visible spectral range, while maintaining their excellent charge transfer properties, would have a profound positive effect on the photocatalytic and photoelectrochemical efficiency of the material. There are two factors to be considered when employing dopants to change the optical response of titania. First it is desirable to maintain the integrity of crystal structure of the host material while producing significant changes in the electronic structure. The crystal structure of the material is directly related to the ratio of cation and anion size in the crystal lattice. While it appears easier to replace Ti^{4+} in titania with any cation, it is more difficult to substitute O^{2-} with any other anion due to the difference in the charge states and ionic radii. Doping through the use of various transition metal cations has been intensively attempted [32-39]. Except for a few cases [37,38], the photoactivity of the cation-doped titania have shown a noticeable diminution of the photoactivity due to an enhancement of the recombination mechanism of the photoexcited electron-hole pairs and/or higher thermal instability [39]. Recently, some groups have demonstrated the substitution of a nonmetal atom such as nitrogen [40-44] and fluorine [45-47] for oxygen. However, insertional doping should be considered as a possibility where the inherent lattice strain in nanometer-sized material provides an opportunity to dope titania to a larger extent.

This chapter considers the fabrication of oxide semiconductor photoanode materials possessing tubular-form geometries and their application to water photoelectrolysis; due to their demonstrated excellent photo-conversion efficiencies particular emphasis is given in this chapter to highly-ordered TiO_2 nanotube arrays made by anodic oxidation of titanium in fluoride based electrolytes. Since photoconversion efficiencies are intricately tied to surface area and architectural features, the ability to fabricate nanotube arrays of different pore size, length, wall thickness, and composition are considered, with fabrication and crystallization variables discussed in relationship to a nanotube-array growth model.

5.2 Synthesis of Nanotubular Oxide Semiconductors

Metal oxide nanotubes have been synthesized by a diverse variety of fabrication routes. For example titania nanotubes, and nanotube arrays, have been produced by deposition into a nanoporous alumina template [48-51], sol–gel transcription using organo-gelators as templates [52,53], seeded growth [54], hydrothermal processes [55-57] and anodic oxidation [58-65].

5.2.1 Template Synthesis

Various strategies of template synthesis are reported which involve nanoporous anodic alumina, organogelators, hydrogels, nanotubular cholesterol, carbon nanotubes and crystalline nanowires as templates [48-53,66-88]. P. Hoyer [48] synthesized titania nanotubes using a polymer [poly(methyl methacrylate), PMMA] mold with a negatype (or replicated) structure of an anodic alumina porous membrane. Titanium dioxide was electrochemically deposited onto the polymer mold; dissolution of polymer resulted in a nanotubular structure. The tube diameters of were in the range ~70-100 nm, being controlled by the membrane pore size. The titania phase obtained was amorphous, and calcination to induce crystallization led to deformation of the tubular structure. Liu and co-workers [66] have synthesized single-crystalline nanotubes of TiO_2 by hydrolyzing TiF_4 under acidic condition at 60°C in presence of an anodic aluminum oxide (AAO) membrane template. Dipping AAO membranes in titania sols prepared from the Ti-isoproxides in alcohols is another route to prepare well aligned TiO_2 nanotubes [67].

Nanotubes of In_2O_3 and Ga_2O_3 have also been synthesized by employing sol–gel chemistry and porous alumina templates [68]. Aqueous ammonia is added to In (or Ga) containing sols to obtain precipitates, which are then peptized with nitric acid to produce stable sols. The alumina membrane is immersed in the sol and then air-dried followed by annealing in air at elevated temperatures for 12 h to obtain the oxides. The alumina template is then dissolved in alkaline solution to yield the free tubes. The hollow nanotubes so obtained have lengths of up to 10 μm. The positively charged sol particles adhere to the negatively charged pore walls of the templating membrane leading to the formation of composite semiconductor nanotubes. SnO_2 nanotube arrays have successfully been fabricated using colloidal infiltration employing porous anodic alumina (AAO) membranes as the template [69,70]. Urea is used as a ligand in the alcohol medium. Its hydrolysis at elevated temperature is utilized to tune the pH value to accelerate the formation of sol followed by ordered SnO_2 nanotube arrays through sol-gel template route.

Sol-gel chemistry that employs gels derived from low to moderate molecular weight organic compounds has been extended for the synthesis of metal oxide nanotubes. The driving forces behind of such physical gelation involve hydrogen bonding, van der Waals force, π-π interaction and electrostatic interaction through cooperating noncovalent interactions of gelator molecules. The outstanding feature responsible for gelation is thought to rely on the formation of three-dimensional networks composed of highly intertwined fibers. The first stage of physical gelation is the self-assembly of gelator molecules, with the self-assembled bundle organogelator acting as a template in the sol-gel polymerization process. Oxide nanotubes are thus synthesized by coating these templates with oxide precursors followed by gel dissolution in a suitable solvent and subsequent calcination, **Fig. 5.1**. By this reconstructive synthesis hierarchical morphologies can be obtained, with the gelators acting as the structure-directing agents.

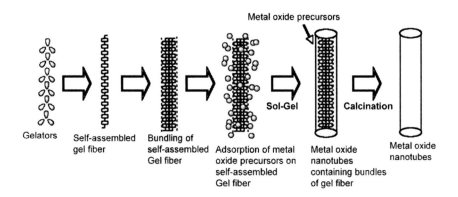

Fig. 5.1: Template (organogelator) synthesis of metal oxide nanotubes.

Nakamura and Matsui [71] prepared silica nanotubes as a spin-off product of sol–gel synthesis wherein tetraethylorthosilicate (TEOS) was hydrolyzed in the presence of ammonia and D, L-tartaric acid. Ono et al. [72] showed that certain cholesterol derivatives can gelate tetraethyl orthosilicate (TEOS) to obtain tubular silica structures. Using cholesterol based gelators nanotubes of transition metal (Ti, V and Ta) oxides can also be prepared. The organogelators used in these processes are chiral diamino cyclohexane derivatives [73], sugar-appended porphyrin [74], trans-(1R, 2R)- 1,2-cyclohexanedi(11-aminocarbonylundecylpyridiniu -m) hexaflourophosphate [75], and N-Carbobenzyloxy-L-valylaminooctadecane [76], see **Fig 5.2**. In all these cases metal alkoxides have been used as oxidic precursors. In order to use simple water-soluble metal salts as precursors, tripodal cholamine has successfully been used as a hydrogelator [77] enabling fabrication of a wide range of oxide nanotubes including SiO_2, TiO_2, ZrO_2, ZnO and WO_3. In aqueous solutions the aggregation process is predominantly driven by the hydrophobic surface of the hydrogelator.

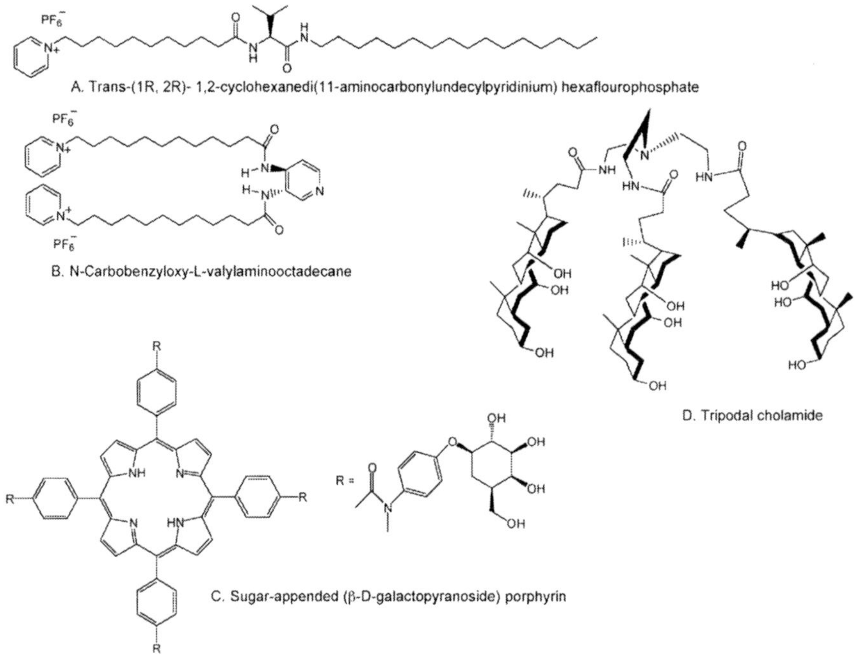

A. Trans-(1R, 2R)- 1,2-cyclohexanedi(11-aminocarbonylundecylpyridinium) hexaflourophosphate

B. N-Carbobenzyloxy-L-valylaminooctadecane

C. Sugar-appended (β-D-galactopyranoside) porphyrin

D. Tripodal cholamide

Fig. 5.2: Structure of gelators employed for template synthesis of various nanotubes.

The advantage of template synthesis is that organo or hydro-gelator templates can direct the shape-controlled synthesis of oxide nanotubes. Recent reports describe the use of carbon nanofibers as a template for the shape-controlled synthesis of zirconia, alumina and silica nanotubes [78]. The shape of vapor grown carbon nanofiber

(VGCF) is found to be straight having diameters in the range 100-200 nm, with the nanotube shape similar to that of the VGCF template. For synthesis, dropwise addition of oxidic precursors such as $Zr(^nOPr)_4$ in ethanol, $Al(OBu)_3$ in CCl_4, and $SiCl_4$ in CCl_4 onto the template placed in a suction filtration unit, which immediately infiltrates into its fibrous structure followed by drying, hydrolysis and then calcination in air at 750°C yields ZrO_2, Al_2O_3 and SiO_2 nanotubes, with the nanotube wall thickness controlled by the number of coating steps.

Adachi and coworkers [79] were the first to report the synthesis of silica nanotubes by employing surfactant-assisted growth. Laurylamine hydrochloride was used as the surfactant template around which TEOS was hydrolyzed. Tube formation was followed by trisilylation treatment. Trimethylsilylation inactivated the silanol groups on the surface of the tube, thus inhibiting the condensation of silanol groups between the different bundles, and yielding long individual silica tubes. Another advantage of the trisilylation treatment was that the surfactant was removed without calcination. The sol–gel method in the presence of citric acid as the structure modifier has also been used to prepare individual silica nanotubes [80].

Carbon nanotubes (CNTs) have been successfully used as removable templates for the synthesis of a variety of oxide nanotubes. Ajayan and coworkers [81] reported the preparation of V_2O_5 nanotubes by using partially oxidized carbon nanotubes as templates. Surface tension induces the growth of uniform, thin metal oxide films, sometimes a monolayer thick, on the walls of the CNTs [81]. Annealing a mixture of partially oxidized carbon nanotubes and V_2O_5 at 750°C, and subsequent preferential removal of carbon nanotubes templates by oxidation at 650-675°C yielded nanotubular V_2O_5. Rao and coworkers [82,83] have prepared a variety of oxide nanotubes including SiO_2, Al_2O_3, V_2O_5, MoO_3 and RuO_2 employing CNT templates. In these preparations acid-treated multi-wall CNT bundles were coated with a suitable metal oxide precursor, with subsequent heating of the coated composites to remove the carbon template.

A few synthetic methodologies have been reported for 1D-nanostructured α-Fe_2O_3 materials. For example, Mann et al. [84]

used biomacromolecules as templates to synthesize Fe_2O_3 nanotubular materials. Chen and co-workers [85] used AAO templates to synthesize α-Fe_2O_3 nanotubes, first filling an AAO template with 4.0 M solution of $Fe(NO_3)_3$ followed by drying and then annealing at 400°C for 5 h. The alumina template was then dissolved in 6 M NaOH to obtain the nanotubes [85]. When employing CNTs as templates [86], the CNTs were first coated with iron oxide nanoparticles via the thermal decomposition of $Fe(NO_3)_3$ in a supercritical (SC) CO_2 and ethanol solution at 150°C. The resulting Fe_2O_3/CNT composites were then heated in an oxygen atmosphere at 500°C for 6 hr to remove the CNTs obtaining α-Fe_2O_3 nanotubes [86]. Liu et al. reported an efficient and convenient method to synthesize single-crystalline α-Fe_2O_3 nanotubes at relatively lower temperatures, in which polyisobutylene bissuccinimide surfactant was used as a template with aqueous butanol as a solvent and carbamide(NH_2CONH_2) as a base [87]. At a critical concentration value, polyisobutylene bissuccinimide surfactant molecules spontaneously organize into rod-shaped micelles. Carbamide in an aqueous butanol solution (above 90°C) decomposes to form NH_4OH and thus provides OH^-, which in turn reacts with Fe^{3+} ion in solution to form crystalline FeOOH nucleus at the coordination site. These FeOOH molecules are interconnected through hydrogen bonds around a molecule of the surfactant or along exterior surface of rod-shaped micelles. The dehydration between the molecules at 150°C and surfactant removal by organic solvent yields crystalline 1D tubular nanostructured Fe_2O_3.

Zhou and his coworkers [88] used a three-step process for the preparation of single-crystalline Fe_3O_4 nanotubes using a nanowire template. Single crystalline MgO nanowires were first grown on Si/SiO_2 substrates. A conformal layer of Fe_3O_4 was then deposited onto the nanowires using pulsed laser deposition (PLD) to obtain MgO/Fe_3O_4 core-shell nanowires. The MgO inner cores of the MgO/Fe_3O_4 core-shell nanowires were then selectively etched in $(NH_4)_2SO_4$ solution (10 wt %, pH ~ 6.0) at 80°C.

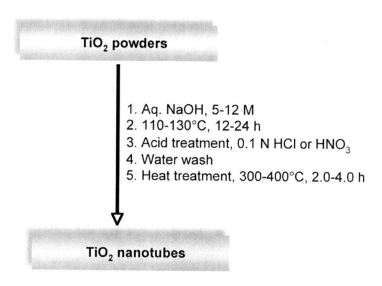

Fig. 5.3: Wet chemical synthesis route for TiO_2 nanotubes.

5.2.2 Hydrothermal Synthesis

Metal oxide nanotube formation via hydrothermal synthesis commonly involves the use of a sealed teflon-lined autoclave, or direct reflux of the alkaline solution under a nitrogen ambient. The synthesis scheme for titania nanotubes is shown in **Fig. 5.3**. Several groups have studied the effects of hydrothermal conditions on the resulting metal oxide nanotubes [55-58,89-96]. **Table 5.1** summarizes the synthesis conditions for preparation of various metal oxide nanotubes, with essential parameters including temperature, concentration of the alkaline or basic medium, pH value, and the reaction time. Highly concentrated NaOH solution has been found most appropriate for the synthesis of TiO_2 nanotubes. Seo et al. [89] observed that reflux reactions carried out at 150°C in N_2 atmosphere lead to the formation of uniform titania nanotubes 10-15 nm in diameter and 150 nm in length. Below this temperature uneven formation of nanotubes occurred with a large amount of sphere-like nanoparticles. Above this temperature the nanotubes grows longer with reduced diameter. Calcination at 400°C and above exhibits anatase phase [89], with the crystallinity of the anatase phase increasing as the heat treatment temperature increases. The transformation of the anatase phase to the rutile phase occurs around 700°C, with only the rutile phase observed above 800°C. Similar observations have been made for those reactions that were carried out in a sealed Teflon-lined autoclave. Tsai and Teng [93] observed that titania nanotubes are most readily synthesized from 10 M NaOH solution in the temperature range 120-130°C; above or below this temperature range the nanotube quality significantly degrades. In other words, hydrothermal treatment temperatures and conditions determine the extent of the precursor conversions, which influences the geometrical features of the resulting nanotubes. As the synthesis temperatures increases from 100°C to 130°C the surface area of the resulting titania nanotubes increases, reaching a maximum of 400 m²/g at 130°C, and then starts decreasing with further rises in temperature. The increase in surface area up to 130°C can be attributed to the enhanced stretching of the Ti–O–Ti bonds to form Ti–O–Na and Ti–OH bonds [55,91-93], leading to the formation of lamellar sheets due to the electrostatic repulsion of the charge on the sodium [55]. Above 130°C, the destruction of lamellar titania could be the possible cause of low surface area and low pore volume [93]. Heat treatment of the as-synthesized nanotubes above 400°C results in sintering leading to the formation of rod-like structures with an abrupt decrease in surface area. At about 800°C the agglomerated TiO_2 forms cylindrical particles of larger size.

Table 5.1: Preparative methods and surface area of metal oxides nanotubes

Starting Materials	Reaction condition	Washing solution	Products	Dia.[a] (nm)	Length (nm)	S_{BET} (m²/g)	Ref.
Nanoparticles from Ti(O-C₃H₇)₄ using sol-gel, S_{BET} = 150 m²/g	2.5-20 M NaOH; 20-110°C for 20 hr	0.1 M HCl	TiO_2	~8	~100	400	10
Nanoparticles from precipitates by dropping NH₄OH into TiOCl₂ solution, S_{BET} = 160 m²/g	5 M NaOH ; 100 - 200°C for 12 hr	Water	TiO_2[b]	15-20	200-250	180-270	44
Nanoparticles from Ti(O-C₄H₉)₄ using sol-gel, S_{BET} = m²/g	10 M NaOH ; 130°C for 24-72 hr	0.1 M HCl	TiO_2	8-10	100-200	-	45
Nanoparticles from Ti(O-C₃H₇)₄ using sol-gel, S_{BET} = not given	5-15 M NaOH; 100 - 180°C for 48 hr	0.1 M HCl	TiO_2	5-30	~1000	267-365	46
Nanoparticles from precipitates by dropping (NH₄)₂SO₄ or HCl into TiCl₄ solution and then NH₄OH neutralization, S_{BET} = 160 m²/g	10 M NaOH; 110°C for 20 hr	0.1 M HCl	TiO_2	~10	500	107-451	47
Nanoparticles of P25 of commercial Degussa AG, TiCl₄ hydrolysis in H₂/O₂ flame, S_{BET} = 50 m²/g	10 M NaOH ; 110 - 150°C for 24 hr	0.1 M HCl	TiO_2	10-30	-	200-400	48
Nanoparticles from Ti(O-C₃H₇)₄ using sol-gel, S_{BET} = 150 m²/g	10 M Na OH; 120-190°C for 22 hr	0.05 M H₂SO₄	TiO_2	~75	~10,000	400	49
0.5 M FeCl₃ solution	0.02 M (NH₄)H₂PO₄, 220°C for 48 hr	Water	Fe_2O_3	40-80	250-400	-	50
0.1 M aqueous Zn(NO₃)₂ solution	HMPT,[d] 95°C 48 hr	-	ZnO[c]	40-60	3000	-	51

[a]Dia stands for diameter of nanotubes. [b] simple reflux method. [c] reflux method, nanotubes grown on ZnO coated thin film Si-substrate in presence of nanorods. [d] HPMT stands for hexamethylene tetramine.

Acid (HCl, HNO_3 or H_2SO_4) treatment/wash removes the electrostatic repulsion and results in the formation of titania nanotubes due to the lamellar sheets rolling up. The mechanism suggests that the extent of acid treatment might be the key factor in determining how the sheets roll up thus affecting overall morphology [57,89,93]. The porosity of the titania nanotubes increases with the HCl concentration reaching its maximum at 0.2 M HCl, at which the equilibrium pH value is 1.6; with a further increase in HCl concentration the porosity is found to decrease [93]. The decrease of tube size with further increases in HCl concentration can be attributed to the rapid removal of electrostatic charges that results in folding of the sheets into granules, rather than tubes [93].

The following mechanism has been proposed for titania nanotube formation by Bavykin and co-workers [94]:

Dissolution of TiO$_2$ precursors
$$TiO_2 + 2NaOH \rightarrow 2Na^+ + TiO_3^{2-} + H_2O \qquad (5.2.1)$$

Dissolution-crystallization of nanosheets
$$2Na^+ + TiO_3^{2-} \rightarrow [Na_2TiO_3]_{nsh} \qquad (5.2.2)$$

Curving of nanosheets
$$2Na^+ + TiO_3^{2-} + [Na_2TiO_3]_{nsh} \rightarrow [Na_2TiO_3]_{ntb} \qquad (5.2.3)$$

Washing of nanotubes
$$[Na_2TiO_3]_{ntb} + H_2O \rightarrow [H_2TiO_3]_{ntb} + 2NaOH \qquad (5.2.4)$$
$$\downarrow$$
$$[TiO_2]_{ntb}$$

Reaction (5.2.4) demonstrates the exchange of the sodium ion between solid and solution during washing. Crystallization (deposition) of Ti from solution into the titanate nanosheet raises the mechanical stress, which can lead to the curving and wrapping of multilayered nanosheets. The driving force for curving of the nanosheets is believed to be the mechanical stresses arising from imbalances in the widths of the multilayers. The rate of sheet crystallization affects the rate of curving, and hence the diameter of the resulting nanotubes. The proposed model is in agreement with experimentally observed dependences of the resulting nanotubes on their synthesis conditions.

Different from the formation mechanism of titania nanotubes, Fe_2O_3 nanotubes are formed by a coordination-assisted dissolution process [95]. The presence of phosphate ions is the crucial factor that induces the formation of a tubular structure, which results from the selective adsorption of phosphate ions on the surfaces of hematite particles and their ability to coordinate with ferric ions.

$$Fe_2O_3 + 6H^+ \rightarrow 2Fe^{3+} + 3H_2O \qquad (5.2.5)$$

$$Fe^{3+} + H_2PO4^- \rightarrow [Fe(H_2PO_4^-)]^{3-x} \qquad (5.2.6)$$

Single-crystalline maghemite (γ-Fe_2O_3) nanotubes are obtained by a reduction and re-oxidation processes with hematite (α-Fe_2O_3) nanotubes as precursors. This approach provides a new strategy to synthesize single-crystalline nanotubes of nonlamellar-structured materials, which could be generally applicable to the synthesis of other inorganic tubular nanostructures.

5.3 Fabrication of Titania Nanotube Arrays by Anodization

Fabrication of titania nanotube arrays via anodic oxidation of titanium foil in fluoride based solutions was first reported in 2001 by Gong and co-workers [58]. Further studies focused on precise control and extension of the nanotube morphology [21], length and pore size [22], and wall thickness [3]. Electrolyte composition plays a critical role in determining the resultant nanotube array architecture and, potentially, its chemical composition. Electrolyte composition determines both the rate of nanotube array formation, as well as the rate at which the resultant oxide is dissolved. In most cases, a fluoride ion containing electrolyte is needed for nanotube array formation. In an effort to shift the band gap of the titania nanotube arrays so that they more fully respond to full spectrum light various doping strategies have been pursued [23,98,99] including the use of an organic anodization bath, and incorporation of anionic species during the anodization process.

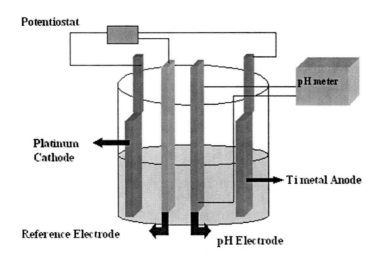

Fig. 5.4: Illustrative drawing of a three-electrode electrochemical cell in which the Ti samples are anodized. Fabrication variables include temperature, voltage, pH and electrolyte composition.

5.3.1 First Generation Nanotubes Formed in Aqueous Electrolytes *HF-based Electrolytes*

Anodization of titanium, see **Fig. 5.4**, is commonly conducted in a two-electrode electrochemical cell with a platinum foil as cathode at a constant potential. Anodization experiments commonly use magnetic agitation of the electrolyte, which reduces the thickness of the double layer at the metal/electrolyte interface, and ensure uniform local current density and temperature over the Ti electrode surface. Foils were anodized at different anodizing voltages, 3, 5, 10 and 20 V, in 0.5% wt HF aqueous solution [20]. At low anodizing voltage, the morphology of the porous film is similar to that of porous (sponge-like) alumina, with a typical pore size of 15 nm to 30 nm. As the voltage is increased, the surface becomes particulate, or nodular, in nature. As the voltage is further increased to 10V, the particulate appearance is lost, with discrete, hollow, cylindrical tube-like features appearing. Nanotube samples prepared using 10V, 14V and 20V anodization voltages have, respectively, inner diameters of 22 nm, 53 nm and 76 nm; wall thickness 13 nm, 17 nm and 27 nm; and lengths 200 nm, 260 nm and 400 nm. The titanium samples were anodized for 45 minutes, resulting in uniform nanotube arrays grown atop the supporting titanium metal foils, with an electrically insulating barrier layer separating the nanotubes from the conducting titanium foil. The nanotube structure is lost at anodizing voltages greater than 23 V, with a sponge-like randomly porous structure being realized. During anodization the color of the titanium oxide layer normally changed from purple to blue, light green, and then finally light red. The addition of acetic acid to the 0.5% HF electrolyte in a 1:7 ratio results in more mechanically robust nanotubes without changing in their shape and size [100,101].

Tapered conical shape nanotubes

In 0.5% hydrofluoric (HF) solution (pH ≈ 1.0) with an anodization voltage between 10 V to 23 V nanotube arrays of well-defined shape were obtained with length and diameter proportional to anodization voltage. Keeping the anodization voltage constant throughout the experiment results in straight nanotubes. In order to achieve a tapered conical shape, the anodization voltage was ramped up from 10 V to 23 V at rates from 0.43 V/min to 2.6 V/min to obtain continuous increase of pore size from top to bottom of nanotubes [21]. Two sets of samples were prepared: Set-1 by increasing the voltage linearly from 10 V to 23 V and then holding the voltage constant at 23 V so as to keep the total anodization time constant at 40 minutes. Set-2 by anodizing the samples initially at 10 V for 20 minutes before starting a voltage ramp of either 0.5 V/min or 1.0 V/min, and then keeping the sample at a constant 23 V for two minutes (total anodization time 35 minutes for the 1.0 V/min ramp, and 47 minutes for the 0.5 V/min ramp); **Fig. 5.5** shows the resulting FE-SEM images of the resulting tapered nanotubes. **Figure 5.5(a)** shows nanotubes resulting from a 0.43 V/min anodization voltage ramp of thirty minutes, followed by a constant 23 V anodization for 10 minutes. **Figure 5.5(b)** shows the nanotubes fabricated by anodizing the sample for 20 minutes at 10 V followed by ramping the voltage at the rate of 1.0 V/min, and finally holding the voltage at 23V for 2 minutes. For comparison the image of a straight nanotube prepared using a constant 23 V anodization is shown in **Fig. 5.5(c)**. The images are illustrative of the ability to fabricate tapered nanotubes of titania by linearly varying the anodization voltage.

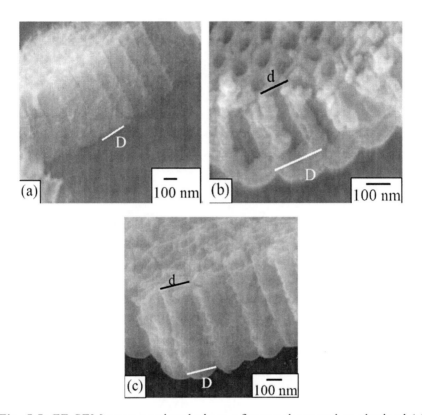

Fig. 5.5: FE-SEM cross-sectional views of tapered nanotubes obtained **(a)** using the ramp rate 0.43 V/min. to raise the voltage from 10V to 23V within 30 minutes and then holding the voltage at 23 V for 10 minutes and **(b)** by initially anodizing the sample at 10V for 20 minutes and then increased the voltage linearly at the rate of 1.0 V/min to 23V, and finally kept at 23V for 2 minutes. **(c)** shows a view of straight nanotubes obtained by applying a constant 23V for 45 minutes. Here, d denotes diameter of apex, and D diameter of cone base.

In all cases the average outer-diameter of the tube base is ≈ 166 nm, which is nearly equal to that of a tube fabricated by applying a constant 23 V anodization voltage. The average inner-diameters of the tapered end of the tubes in Set-1 are approximately 70, 80, 85, and 100 nm for, respectively, sweep rates of 0.43, 0.65, 0.87 and 2.6 V/min. The average inner-diameter of the tube from Set-2 are ≈ 36 nm and 42 nm for the 0.5 V/min and 1.0 V/min ramps, respectively. In the latter case, a sweep-rate greater than 1.0 V/min led to collapse of the nanotubes. Tapered nanotubes could not be achieved when sweeping the anodization voltage from 23 V to 10 V, followed by a constant 10 V anodization for a total anodization time of 40 minutes. Irrespective of the sweep rates, the resulting tubes were straight with a constant 22 nm inner diameter and 200 nm length, dimensions equal to those achieved for a 10 V anodization.

Wall Thickness Variation

In the growth of nanotubes via anodic oxidation of titanium, chemical dissolution and electrochemical etching are two crucial factors in nanotube growth. Varying the electrolyte bath temperature can change the rate of both etching processes [3]. Nanotube arrays were grown by potentiostatic anodization of titanium foil at 10 V in an electrolyte of acetic acid + 0.5% HF mixed in 1:7 ratio, kept at four different electrolyte bath temperatures, 5°C, 25°C, 35°C and 50°C. **Figure 5.6** shows FE-SEM images of the morphology of titania nanotubes fabricated by anodization at 10 V at (a) 5°C and (b) 50°C. The pore diameter is essentially the same (22 nm) for the 10 V anodized titania nanotube arrays fabricated at these different temperatures, whereas the wall thickness changes by a approximately a factor of four and the tube-length changes by approximately a factor of two. The wall thickness increases with decreasing anodization temperature from 9 nm at 50°C to 34 nm at 5°C. As the wall thickness increases with decreasing anodization temperature the voids in the interpore areas fill; with the tubes becoming more interconnected the discrete tube-like structure approaches a nanoporous structure in appearance. The length of the nanotubes increases with decreasing anodization bath temperature from 120 nm at 50°C to 224 nm at 5°C. **Table 5.2** shows the variation in 10 V nanotube array wall-thickness and tube-length as a function of anodization temperature. FESEM images of titania nanotubes fabricated by anodization at 20 V at (a) 5°C and (b) 25°C, with resulting inner pore diameters of 76 nm showed that the nanotube wall thickness increases from 17 nm at 25°C to 27 nm at 5°C, confirming the trend of increasing nanotube wall-thickness as a consequence of lower anodization temperature [3].

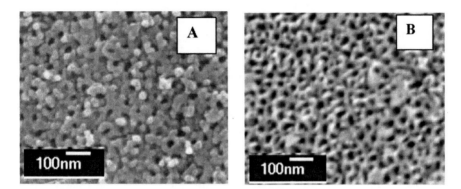

Fig. 5.6: FESEM images of 10 V titania nanotube arrays anodized at **(a)** 5°C, and **(b)** 50°C. The pore size is nearly 22 nm for all samples. In **(a)** the average wall thickness is 34 nm, and in **(b)** is 9 nm.

Table 5.2   Average wall-thickness and tube-length of 10 V titania nanotube arrays anodized at different bath temperatures.

Anodization temperature	Wall thickness (nm)	Tube-length (nm)
5°C	34	224
25°C	24	176
35°C	13.5	156
50°C	9	120

Addition of Boric Acid in HF Electrolyte

During anodization of titanium foil in a 2.5% HNO_3 plus 1% HF water solution electrolyte, with or without addition of boric acid, the applied anodic potential was initially ramped from 0 to 20 V at a rate of 6 V/min; the anodization potential was then held constant at 20 V for 4 h [102,103]. An initial ramp of the voltage was used because initial application of a 20 V anodization potential resulted in high current densities not allowing the formation of an oxide coating due to dielectric breakdown. **Figure 5.7** shows the current density, as a function of anodizing time, after the potential has reached 20 V, for both types of electrolyte. In HNO_3-HF electrolyte, the current density rapidly decreases with formation of the barrier layer, which then slightly increases with formation of the porous structure, and then stays relatively constant with time. In contrast, for boric acid (0.5M) containing HNO_3-HF electrolyte, the current densities decrease relatively slowly to minimum, and then afterwards slowly increase reaching a plateau at approximately 110 minutes.

The surface morphology of nanotube-array sample anodized in an electrolyte containing 2.5% HNO_3 and 1% HF at 20 V for 4 h showed a uniform, clean, regular structure with the nanotubes having an average pore size of about 100 nm and a wall thickness of ≈ 20 nm. The length of nanotubes is found to be ≈ 400 nm. The TiO_2 nanotube-array anodized in an electrolyte of 0.5 M H_3BO_3-2.5% HNO_3-1%HF at 20 V for 4h was found to have a precipitate layer 400 nm thick. The nanotube structures could be exposed after washing with a dilute HF solution. In these samples, there is a greater degree of pore irregularity, with sizes ranging from 10 nm to 120 nm. The average wall thickness of the nanotubes is 20 nm, and nanotube length is about 560 nm.

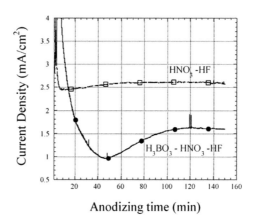

Fig. 5.7: Time variation of titanium electrode etching current density for: **(a)** Nitric acid/hydrofluoric acid electrolyte; **(b)** Boric acid/nitric acid/hydrofluoric acid electrolyte.

5.3.2   Second Generation Nanotubes Formed in Buffered Electrolytes *KF (or NaF)-based Aqueous Electrolytes*

Nanotube arrays several microns in length can be fabricated using KF (or NaF with identical results) electrolytes of variable pH [22], a summary of which is presented in **Table 5.3**. Prior to KF addition, the desired pH was obtained by adding NaOH, sulfuric acid (pH 1-2), sodium hydrogen sulfate, or citric acid (pH 2.5-7.5). The F$^-$ concentration was held fixed at 0.1 mol/L. In 0.1 mol/L F$^-$ and 1 mol/L H$_2$SO$_4$ medium, the potential window for nanotube formation is 10~25V (Samples 01 to 08). Outside of this potential range no nanotubes were formed (Samples 01 and 08). In Sample 01 (at 5V), the electrochemical etch rate was slow due to lower applied potential and only a few pits can be seen on the sample surface. In Sample 08 (at 30V), the electrochemical etch is much faster which prevents nanotube formation. Only a highly disturbed porous structure was obtained in this case [22]. The nanotube pore size was found to be proportional to the potential applied (Samples 02 to 05) and independent of the anodization time (Samples 04 and 07) and the electrolyte concentration (Samples 04 and 08). Increasing the potential from 10V to 25V increased the diameter of the resulting nanotubes from 40 nm to 110 nm. No significant difference was observed in the pore size for anodization times of 1 hr (Sample 04) and 6.5 hr (Sample 07) or for electrolyte concentrations 1 mol/L H$_2$SO$_4$ (Sample 04) and 2 mol/L H$_2$SO$_4$ (Sample 08).

Electrolyte pH affects both the behavior of the electrochemical etch, and chemical dissolution owing to the hydrolysis of titanium ions. With increasing pH the hydrolysis content increases, which slows the rate of chemical dissolution. As shown in **Fig. 5.8** (Samples 10, 13 for 10V and 12, 17 for 25V) and **Table 5.3,** longer nanotubes can be formed in higher pH solution. For a potential of 25V, with pH increasing from strong acidity (Sample 05, pH<1) to weak acidity (Sample 17, pH 4.5), nanotube length increased from 0.56 μm to 4.4 μm; for 10V, the length increased from 0.28 μm (Sample 2, pH<1) to 1.4 μm (Sample 18, pH 5.0). For a particular pH, the length increases with applied potential (Samples 10-12 and 16-17). When the potential increased from 10V to 25V, the length increased from 0.59 μm to 1.5 μm for pH=2.8 and from 1.05 μm to 4.4 μm for pH=4.5. At a particular pH, the pore size of the nanotubes was found to be increasing with anodization potential as shown in the inset of **Fig. 5.8** (Sample 10, 11, 12). However, the pore size was independent of the pH at a particular potential.

Table 5.3   Electrolyte pH and composition, anodization conditions, and size of the resulting nanotubes.

No.	Electrolyte[1]				pH[2]	V	t	D	L	Q[3]
	F⁻	SO₄²⁻	PO₄³⁻	Cit		(V)	(hr)	(nm)	(μm)	
01	0.1	1.0	-	-	<1	5	1	10±2	-	No NT
02	0.1	1.0	-	-	<1	10	1	40±5	0.28±0.02	NT
03	0.1	1.0	-	-	<1	15	1	80±9	-	NT
04	0.1	1.0	-	-	<1	20	1	100±11	0.48±0.03	NT
05	0.1	1.0	-	-	<1	25	1	110±12	0.56±0.04	NT
06	0.1	1.0	-	-	<1	30	1	-	-	No NT
07	0.1	1.0	-	-	<1	20	6.5	100±11	0.43±0.03	NT
08	0.1	2.0	-	-	<1	20	1	100±11	0.45±0.03	NT
09	0.1	1.0	-	0.2	1.3	10	20	30±5	0.32±0.03	NT
10	0.1	1.0	-	0.2	2.8	10	20	30±5	0.59±0.05	NT
11	0.1	1.0	-	0.2	2.8	15	20	50±5	1.00±0.05	NT
12	0.1	1.0	-	0.2	2.8	25	20	115±10	1.50±0.04	NT
13	0.1	1.0	-	0.2	3.8	10	20	30±5	0.80±0.06	NT
14	0.1	1.0	-	0.2	3.8	10	60	30±5	1.80±0.06	NT
15	0.1	1.0	-	0.2	3.8	10	90	30±5	2.30±0.08	NT
16	0.1	1.0	-	0.2	4.5	10	20	30±5	1.05±0.04	NT
17	0.1	1.0	-	0.2	4.5	25	20	115±5	4.40±0.10	NT
18	0.1	1.0	-	0.2	5.0	10	20	30±5	1.40±0.06	NT
19	0.1	1.0	-	0.2	5.0	25	20	115±5	6.00±0.40	NT
19	0.1	1.0	0.1	0.2	6.4	10	24	-	-	No NT
20	-	2.0	-	-	<1	10	24	-	-	No NT

(1)   Electrolyte components are in mol/L
(2)   pH < 1 represents a 1.0 or 2.0 mol/L H_2SO_4 medium.
(3)   Quality Q of resulting nanotubes. NT: nanotubes uniformly across substrate. No NT: no nanotubes or partly developed nanotube/porous structures.
(4)   Cit: citrate; t: time; D: inner diameter of nanotube; L: length of nanotube.
(5) SO_4^{2-} is from addition of H_2SO_4 or $NaHSO_4$; PO_4^{3-} is addition of potassium hydrogen phosphate $K_2HP_3O_4$; Cit denotes citric acid from its salt, $HO(CO_2Na)(CH_2CO_2Na)_2 \cdot 2H_2O$.

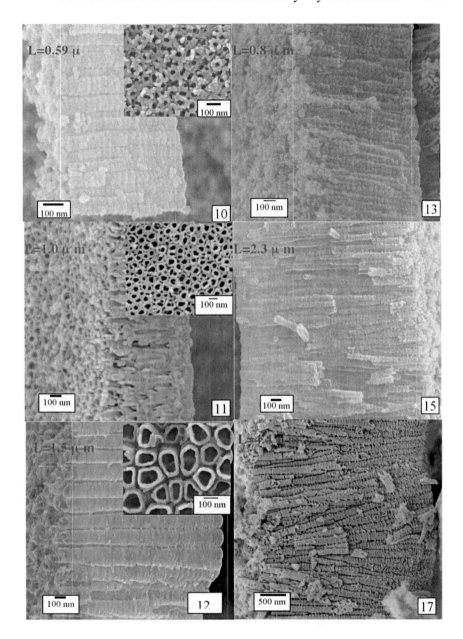

Fig. 5.8: Lateral view of the nanotubes formed in different pH solutions (pH>1). Variation of pore size with anodization potential for pH 2.8 is shown in the inset (samples 10 to12). The anodization conditions, or each sample, are listed in **Table 5.3**.

In strongly acidic solutions (pH<1), increasing the anodization time does not increase the nanotube length as shown by Samples 04 and 07. In weak acid electrolytes the nanotube length is time dependent as shown by Samples 13 to 15 (**Fig. 5.8**; 13, 15). As anodization time increases from 23 hr to 90 hr the nanotube length increase from 0.8 μm to 2.3 μm. On increasing pH values the hydrolysis content increases, resulting in a significant amount of hydrous titanic oxide precipitated on the nanotube surface. Our studies showed that the best pH range for formation of relatively longer nanotubes is between pH 3 and pH 5; lower pH forms shorter but clean nanotubes, while higher pH values result in longer tubes that suffer from unwanted precipitates. Alkaline solutions are not favorable for the self-organized nanotube formation. In the case of Sample 19 (**Table 5.3**), where the anodization was done in pH 6.4 with 0.1 mol/L PO_4^{3-}, no nanotubes but a layer of dense hydrous titania salts were found. Stronger acidity is required for nanotube formation in the presence of phosphate, owing to the formation of undissolvable titanic phosphates. No nanotube array formation has been achieved without F⁻, even in 2 mol/L H_2SO_4 solution (sample 20).

5.3.3 Third Generation Nanotubes Formed in Polar Organic Electrolytes

Formamide and Dimethyl formamide Electrolytes

It appears the key to successfully achieving very long nanotube arrays is to minimize water content in the anodization bath to less than 5%. With organic electrolytes, the donation of oxygen is more difficult in comparison to water thus reducing the tendency to form oxide. At the same time the reduction in water content reduces the chemical dissolution of the oxide in the fluorine containing electrolytes and hence aids longer-nanotube formation. Illustrative electrolyte compositions include formamide (FA; 99%) and/or N-methylformamide (NMF; 99%) solutions containing 1-5 wt % of deionized water and 0.3-0.6 wt % NH_4F (98%) [64]. **Figure 5.9** shows nanotubes nearly 70 μm long grown in a FA based electrolyte by anodization for 48 hours at a constant potential of 35 V. The average outer diameter of these nanotubes was determined to be 180 nm, wall-thickness ≈ 24 nm, resulting in an aspect ratio of ≈ 390. Lower anodization potentials result in shorter nanotubes with smaller diameters. The increase in nanotube length with larger anodization voltage is attributed to the increased driving force for ionic transport through the barrier layer at the bottom of the pore resulting in faster movement of the Ti/TiO$_2$ interface into the Ti metal. In the FA/NMF electrolytes, an increase in the outer nanotube diameter was found to increase with anodization voltage. This observation agrees with the reported behavior of nanotube arrays formed in aqueous electrolytes. Nanotubes with a smaller pore diameter and $\approx 10\%$ greater lengths were obtained using NMF rather than FA electrolytes.

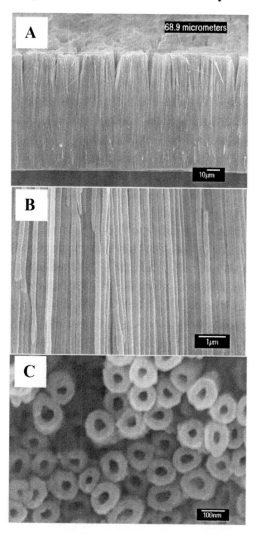

Fig. 5.9: FESEM images of TiO$_2$ nanotubes grown in formamide based electrolyte at 35 V for 48 hours showing: **(a)** cross-section at lower magnification, **(b)** cross-section at high magnification, and **(c)** top surface image.

In comparison to aqueous electrolytes, the range of applied anodization potentials over which nanotube arrays are obtained is significantly extended in the FA/NMF based electrolytes, with nanotubes formed at voltages between 10 and 50 V (compared to 10 V to 29 V for nanotube of micron length in KF or NaF electrolytes). At 60 V, the anodization was found to be unstable with sharp fluctuations in the current, anodization at voltages below 10 V were not attempted. The anodization duration is also an important variable. The length of the nanotubes increased with time up to a certain maximum length beyond which it declined. For example, increasing the anodization duration at 35 V to 164 hours resulted in a nanotube length of 10 µm compared to 30 µm obtained after 88 hours, and 70 µm after 48 hr. The time to maximum length is a function of the anodization voltage with the nanotubes reaching this maximum sooner at lower potentials.

The effect of five different cationic species on the formation of TiO_2 nanotube arrays by anodization of titanium in formamide-water mixtures containing fluoride ions has been studied [26]. The cation choice is a key parameter influencing both the nanotube growth rate and resulting nanotube length. Under similar conditions, electrolytes containing the tetrabutylammonium cation resulted in the longest nanotubes (~94 µm), while the shortest nanotubes (~3 µm) were obtained when H^+ ions were the sole cationic species in the anodization electrolyte. This difference in nanotube growth is attributed to the inhibitory effect of the quarternary ammonium ions that restrict the thickness of the interfacial (barrier) oxide layer; a thinner interfacial oxide layer facilitates ionic transport thus enhancing the nanotube growth. The aspect ratio of the resulting nanotubes is also voltage dependent, with the highest aspect ratio of ≈ 700 obtained at an anodizing voltage of 20 V in an electrolyte containing tetrabutylammonium ions. **Table 5.4** summarizes the effect of different cations and cation concentrations on the nanotube morphology for formamide based electrolytes containing 5% water at different anodization potentials. A trend of decreasing wall-thickness at higher fluoride ion concentrations is attributable to the enhanced chemical etching afforded by higher F^- concentrations. At larger anodization voltages, the driving force for ionic transport through the barrier layer at the bottom of the pore is greater and results in faster movement of the Ti/TiO$_2$ interface into the Ti metal. This enhanced pore deepening effect resulted into higher nanotube length obtained at larger anodization voltages. The real time potentiostatic anodization behavior, **see Fig. 5.10**, of Ti anodized at 20 V in 95% formamide + 5% water solutions containing 0.27 M of NH$_4$F, (C$_4$H$_7$)$_4$ NH$_4$F, HF and C$_6$H$_6$(CH$_3$)$_3$ NH$_4$F respectively. **Figure 5.11** shows the real time anodization behavior of Ti foil anodized at 20 V in electrolytes containing an identical concentration of NH$_4$F (0.27M) but with different amounts of water ranging from 100% water (no formamide) to 2.5% water + 97.5% formamide.

Table 5.4   Effect of cation type, cation concentration, and anodization duration on the morphological features of TiO$_2$ nanotube arrays for a given potential. In every case, the electrolytes contain 5% water in formamide and the anodization was performed at room temperature, 22°C. The anodization duration is the approximate time required to obtain maximum nanotube length.

Anodization voltage 35 V

Cation	Molar (M) cation concentration	Anodization Duration (hr)	Outer Diameter (nm)	Wall Thickness (nm)	Nanotube Length (μm)
H$^+$	0.14	101	256	21	5.6
H$^+$	0.27	48	214	20	7.3
NH$_4^+$	0.14	88	208	17	29.2
NH$_4^+$	0.27	30	159	15	37.4
Na$^+$	Saturated solution (~0.04)	48	48	18	9.6
Bu$_4$N$^+$	0.27	48	190	22	68.9

Anodization voltage 20 V

Cation	Molar (M) cation concentration	Anodization Duration (hr)	Outer Diameter (nm)	Wall Thickness (nm)	Nanotube Length (μm)
H$^+$	0.27	48	99	22	2.9
NH$_4^+$	0.14	55	90	19	14.4
NH$_4^+$	0.27	24	90	17	19.6
Bu$_4$N$^+$	0.27	34	90	16	35.2

Anodization Voltage 15 V

Cation	Molar (M) cation concentration	Anodization Duration (hr)	Outer Diameter (nm)	Wall Thickness (nm)	Nanotube Length (μm)
NH$_4^+$	0.14	110	81	29	8.2
Bu$_4$N$^+$	0.27	46	80	15	20.0
BnMe$_3$N$^+$	0.27	42	70	18	7.2

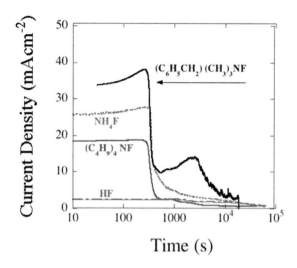

Fig. 5.10: Current-time behavior during 20 V potentiostatic anodization of a Ti foil (99.8 % pure) in a formamide solution containing 5% H_2O and identical molar concentrations (0.27M) of fluoride ion bearing compounds with four different cationic species: Hydrogen (H^+), Ammonium (NH_4^+), Tetrabutylammonium ($[C_4H_9]_4N^+$), and Benzyltrimethylammonium ($[C_6H_5CH_2][CH_3]_3N^+$).

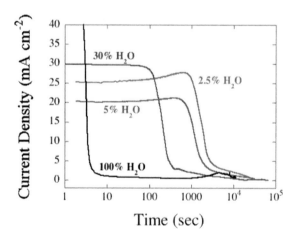

Fig. 5.11: The anodization current-time behavior of a Ti foil (99.8% pure) anodized at 20 V in an electrolyte containing 0.27 M NH_4F in formamide of variable water content.

Dimethyl Sulfoxide Electrolytes

Figure 5.12 shows FESEM images of samples prepared in dimethyl sulfoxide (DMSO; 99.6%) and 2% hydrofluoric acid (HF; 48% aqueous solution) electrolyte at 40V for 69 hours [24]. These nanotubes have a length of approximately 45 μm, pore diameter ≈ 120 nm, and wall thickness ≈ 15 nm. The roughness factor for these nanotubes is ~1800. The top-surface of the as-prepared samples is typically covered with broken tubes and other debris from the anodization bath. The pores appear clogged in such cases, although a complete clogging does not take place as this prevents the electrolyte species from reaching the bottom of the tubes thus terminating the anodization process. The chemical etching rate of the oxide by the fluoride is low in non-aqueous organic electrolytes allowing the reaction products to stay at the surface of the nanotubes. The debris could be removed using sonication. Anodization in the range 20 V to 60 V yielded a regular, well-aligned nanotube-array architecture.

For a 70 hour anodization at, respectively, anodization potentials of 20V and 60V an increase in length from about 10 μm to 93 μm was observed. The resulting pore diameters were, respectively, ~50 nm, 120 nm and 150 nm for 20, 40 and 60 V potentials. The annealed 60 V 93 μm long tubes have an outer diameter of approximately 200 nm, hence a roughness factor of ≈ 3200, with a length to outer diameter aspect ratio of ≈ 465. As the HF concentration varied from 1% to 4% the length of the nanotubes grown at 20 V increased from 4.4 μm to 29 μm. The titanium foil when subjected to anodization at 20 V in 0.5% HF in deionized water before anodizing at 40 V in DMSO containing 2% HF solution yielded 82 μm long nanotubes, almost a factor of two increase in length from that obtained without using the pre-anodization step.

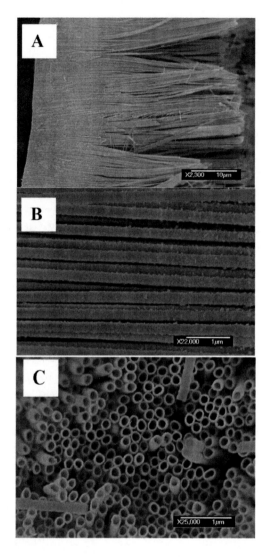

Fig. 5.12: FESEM images of nanotube array grown at 60 V in DMSO containing 2% HF.

Ethylene Glycol Electrolytes

The titania nanotube arrays were observed in the anodization voltage range 20 to 65V in the NH_4F concentration range (0.1 – 0.5 wt%) and H_2O (1% - 4%) in ethylene glycol (EG) [24,27]. **Figure 5.13** depict the self alignment exhibited by the nanotube arrays, in bottom

[**Fig. 5.13(a)**], top [**Fig. 5.13(b)**] and side [**Fig. 5.13(c)** and **Fig. 5.13(d)**] views of an illustrative sample anodized at 60V in 0.3 wt % NH$_4$F and 2% H$_2$O in ethylene glycol for 17 hrs. For 17 hour anodization of titanium foil at 60 V, it was found that (i) for a given concentration of water, the length increases with increasing NH$_4$F up to 0.3 wt % (ii) for NH$_4$F concentrations up to 0.3 % wt the nanotube, length increases with increasing H$_2$O concentration up to 2 vol % and (iii) in the regime of 0.1 – 0.3 wt % NH$_4$F and 1-2 vol % H$_2$O the used solution, in comparison to use of a fresh solution, exhibited an increase in nanotube length ranging from 15 μm – 70 μm for the same applied potential and duration. While used solutions result in sharply higher growth rates for anodization potentials 60 V and above, at lower anodization potentials (20-40 V) both the fresh and used solutions result in similar growth rates.

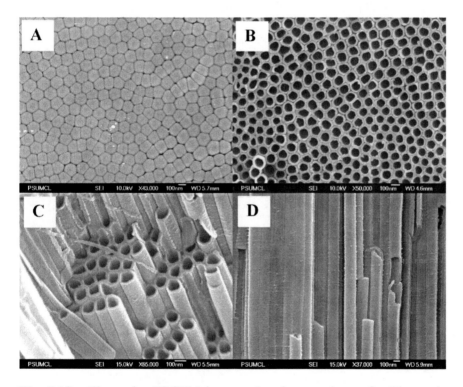

Fig. 5.13: Illustrative FESEM images showing topology of a Ti sample anodized in an electrolyte comprised of 0.3 wt % NH$_4$F and 2 vol % H$_2$O in ethylene glycol at **(a-c)** 60 V, and **(d)** 65 V.

For a 17 hr anodization in a fresh electrolyte mixture of 0.3 wt % NH_4F and 2 vol % H_2O in ethylene glycol, the length obtained at 20, 40, 50, 60 and 65 V was, respectively, 5 μm, 30 μm, 45 μm, 165 μm and 106 μm. The rate of reaction was studied by anodizing titanium foil samples at 60 V, in a fresh (un-used) solution of 0.3% wt NH_4F and 2 vol % H_2O in ethylene glycol for different durations. The samples anodized for 4 hr, 17 hrs, 21 hrs, 48 hrs and 96 hrs exhibited a length of ~ 58 μm, 160 μm, 188 μm, 289 μm and 360 micron respectively, cross sectional FESEM images of the resulting samples is shown in **Fig. 5.14**. The nanotubes of 360 μm length, 165 nm outer diameter, have an aspect ratio of approximately 2200. Extended ultrasonic cleaning, of approximately 1 minute duration, resulted in the nanotubular film separating from the underlying Ti substrate, resulting in a self-standing membrane comprised of a close-packed array of vertically oriented titania nanotubes.

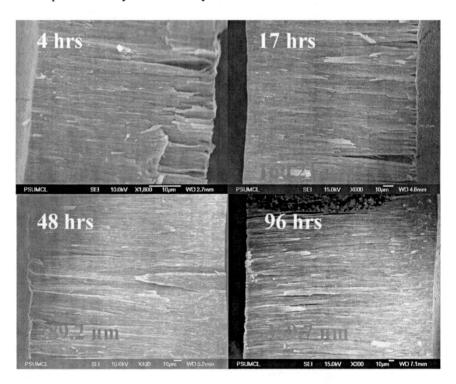

Fig. 5.14: Cross sectional FESEM images indicating length of titania nanotube arrays achieved as a function of anodization duration (60 V, ethylene glycol containing 0.3 wt % NH_4F and 2 vol % H_2O).

During a double-sided anodization process, where both sides of the starting Titanium foil are exposed to the anodizing electrolyte, by starting with 1 mm thick Ti foil have obtained 2 mm thick nanotube array membrane, comprised of two 1 mm long nanotube arrays, see **Fig. 5.15**.

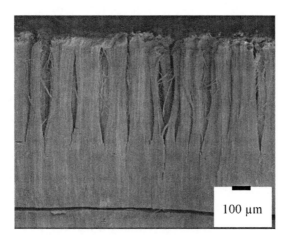

Fig. 5.15: FESEM image of (top half) completely anodized 2.0 mm thick Ti foil sample. A maximum individual nanotube array length of 1005 µm was obtained upon anodizing the Ti foil sample at 60 V for 216 hours in 0.5 wt % NH_4F and 3.0 % water in ethylene glycol. The black line seen towards the bottom marks the separation between the two ≈ 1 mm long nanotube arrays formed during the double-sided anodization of the sample.

5.3.4   Fabrication of transparent TiO_2 nanotubes arrays

Methods have been developed for fabrication of the highly-ordered titania nanotube arrays from titanium thin films atop a substrate compatible with photolithographic processing, notably silicon or FTO coated glass [104]. The resulting transparent nanotube array structure, illustrated in **Fig. 5.16**, is promising for applications such as anti-reflection coatings and dye sensitized solar cells (DSSCs). **Fig. 5.17** shows the typical anodization behavior of a 400 nm Ti thin film anodized at 10 V in an HF based electrolyte. For a fixed HF concentration, the dimensions of the tube vary with respect to voltage; for a fixed anodization potential, the dimensions vary with respect to electrolyte HF concentration. The anodization-potential range over which nanotube-arrays could be formed depended upon the concentration of HF in the electrolyte. For example, at a concentration of 0.5% HF nanotubes were formed in a potential range of 10 V-15 V. The anodization voltage window for successfully achieving the nanotube arrays is 6 V to 10 V for 0.25% HF concentration, and 10 V to 18 V for 1% HF concentration. With reference to the anodization behavior (400 nm Ti thin film anodized at 10V in an HF-based electrolyte) seen in **Fig. 5.17** within a few seconds, ≈ 25 s, after application of the voltage, the measured current density reduced from > 15 mA/cm^2 to a local-minimum of 1.25 mA/cm^2 (point P1 on the plot), with the field-assisted oxidation of the Ti metal surface reducing the current. The structure of the film at point P1 of **Fig. 5.17** is shown in **Fig. 5.18(a)**; as evident from the figure fine pits or cracks form on the oxide surface and act as pore nucleation sites. These pits and cracks arise due to the chemical and field-assisted dissolution of the oxide at local points of high energy. The reduced oxide layer thickness at these points results in a current increase; **Fig. 5.18(b)** corresponds to point P2 on the plot where the crack/pit density has reached saturation as evidenced by the current maximum. Beyond this point the current gradually drops due to a corresponding increase in porous structure depth. A porous structure is clearly seen in the **Fig. 5.18(c)**, corresponding to point P3, with pore diameters of ~20 nm. **Fig. 5.18(d)**, point P4, shows the transition between a porous structure and the nanotubular structure. **Fig. 5.18(e)**, point P5, shows the resulting nanotube-array with pore diameters of $20 - 30$ nm. Nanotube-array length increases to point P4 (≈ 360 s), as evident from the decrease in the current. Between P4 and P5 the nanotube array length remains essentially constant.

Fig. 5.16: The key stages in fabrication of a transparent TiO_2 nanotube array film: (top) Sputter deposition of a high quality Ti thin film; (middle) anodization of resulting film, and (bottom) heat treatment to oxidize remaining metallic islands.

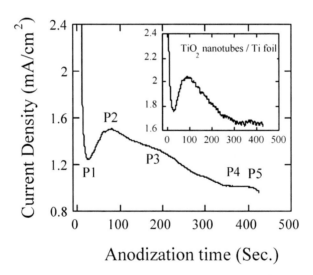

Fig. 5.17: Real time observation of anodization behavior of a 400 nm Ti thin film anodized at 10V in the HF - aqueous electrolyte (acetic acid and 0.5 vol.% HF mixed in ratio of 1:7). Inset shows a typical current density versus time response observed for a titanium foil (with one face protected with polymer coating) anodized at the same potential and electrolyte.

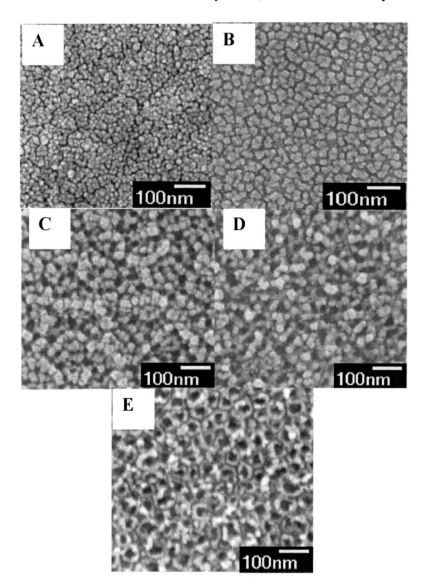

Fig. 5.18: SEM images of samples taken at points marked P1, P2, P3, P4, and P5 in **Fig. 5.17**.

As the anodization proceeds the metal below the oxide barrier layer is progressively consumed. Beyond a certain stage the metal layer becomes thin or discontinuous enough to create highly resistive electrical current pathways. Hence beyond P5 (**Fig. 5.17**) the current drastically reduces, finally dropping to zero as the metal film becomes completely discontinuous. We make note that only a small dip in current can be seen at P5 as the sample was taken out of the anodization bath at this point. Keeping the sample inside the anodization bath beyond point P5 destroys the nanotube structure due to the its chemical dissolution in HF electrolyte. Real time observation of the current enabled us to remove samples from the anodization bath as soon as the dip in the current plateau region, just beyond P5, was observed. SEM images of samples kept beyond point P5 showed that the nanotube layer was severely damaged or completely eliminated. Thus, strict process control is necessary to obtain intact nanotube-arrays without an underlying metal layer. However, if there is a need to keep a continuous metal layer underneath the nanotubes the samples should be removed from the anodization bath at a point between P4 and P5. Close monitoring of the electric current response during potentiostatic anodization helps determine the optimum anodization parameters and serves as a process control tool.

The inset of **Fig. 5.17** shows a typical current versus time plot obtained during anodization (same conditions as thin film) of a 250 μm thick Ti foil with one face protected with a polymer coating. It can be seen that the current-time behavior is no different from that of the thin films. We note that if both sides of the foil are exposed to anodization the current behavior will be significantly different from that seen in **Fig. 5.17**. In this case, the changes in the current after the initial dip cannot be discerned due to the anodization process progressing at different levels on both sides of the sample. It was noticed in both thin films and metal foils (**Fig. 5.17**) that the current shows periodical fluctuations of a small magnitude between point P2 and P5.

In strongly acidic solutions (pH<1) both the nanotube growth rate and dissolution rate are increased, therefore increasing the anodization time does not increase the nanotube length. In short nanotubes (obtained at 10V), the time required to form nanotubes is about 6 minutes. Further increasing the anodization time results in a more uniform nanotubular structure, both in shape and size, but with little change in the nanotube length. Increasing pH decreases the chemical dissolution rate, and apparently prolongs the time needed to reach equilibrium between the rate of nanotube growth and the dissolution rate; in weak acid electrolytes, therefore the nanotube length appears to be time dependent. In such solutions, even without having a protective covering on one side of foil, we still observed the same pattern in current vs time plot as shown in **Fig. 5.17** [105].

5.3.5   Mechanistic Model of Nanotube Array Formation

The key processes for anodic formation of titania [106-109] are: (1) oxide growth at the surface of the metal occurs due to interaction of the metal with O^{2-} or OH^- ions [110]. After the formation of an initial oxide layer, these anions migrate through the oxide layer reaching the metal/oxide interface where they react with the metal. (2) Metal ion (Ti^{4+}) migration from the metal at the metal/oxide interface; Ti^{4+} cations will be ejected from the metal/oxide interface under application of an electric field that move towards the oxide/electrolyte interface.   (3) Field assisted dissolution of the oxide at the oxide/electrolyte interface [110]. Due to the applied electric field the Ti-O bond undergoes polarization and is weakened promoting dissolution of the metal cations. Ti^{4+} cations dissolve into the electrolyte, and the free O^{2-} anions migrate towards the metal/oxide interface, see process (1), to interact with the metal [111]. (4) Chemical dissolution of the metal, or oxide, by the acidic electrolyte also takes place during anodization. Chemical dissolution of titania in the HF electrolyte plays a key role in the formation of a nanotubes versus nanoporous structure.

To help understand the process of nanotube formation, FESEM images of the surface of the samples anodized at 20 V for different durations were taken and analyzed.   At the start the anodization the initial oxide layer [111], formed due to interaction of the surface Ti^{4+} ions with oxygen ions (O^{2-}) in the electrolyte, can be seen uniformly spread across the surface. The overall reactions for anodic oxidation of titanium can be represented as

$$2H_2O \rightarrow O_2 + 4e + 4H^+ \tag{5.3.1}$$
$$Ti + O_2 \rightarrow TiO_2 \tag{5.3.2}$$

In the initial stages of the anodization process field-assisted dissolution dominates chemical dissolution due to the relatively large electric field across the thin oxide layer [111]. Small pits formed due to the localized dissolution of the oxide, represented by the following reaction, act as pore forming centers.

$$TiO_2 + 6F^- + 4H^+ \rightarrow TiF_6^{2-} + 2H_2O \qquad (5.3.3)$$

Then, these pits are converted into bigger pores and the pore density increases. After that, the pores spread uniformly over the surface. The pore growth occurs due to the inward movement of the oxide layer at the pore bottom (barrier layer) due to processes (5.3.1)-(5.3.3). The Ti^{4+} ions migrating from the metal to the oxide/electrolyte interface dissolve in the HF electrolyte. The rate of oxide growth at the metal/oxide interface and the rate of oxide dissolution at the pore-bottom/electrolyte interface ultimately become equal, thereafter the thickness of the barrier layer remains unchanged although it moves further into the metal making the pore deeper. FESEM images show the formation of small pits in the inter-pore regions that eventually leads to pore-separation and tube formation. The thickness of the tubular structure ceases to increase when the chemical dissolution rate of the oxide at the mouth of the tube (surface) becomes equal to the rate of inward movement of the metal/oxide boundary at the base of the tube. Higher anodization voltages increase the oxidation and field-assisted dissolution hence a greater nanotube layer thickness can be achieved before equilibrating with chemical dissolution.

With the onset of anodization, a thin layer of oxide forms on the titanium surface, **Fig. 5.19(a)**. Small pits originate in this oxide layer due to the localized dissolution of the oxide, **Fig. 5.19(b),** making the barrier layer at the bottom of the pits relatively thin which, in turn, increases the electric field intensity across the remaining barrier layer resulting in further pore growth as seen in **Fig. 5.19(c)**. The pore entrance is not affected by electric field assisted dissolution and hence remains relatively narrow, while the electric field distribution in the curved bottom surface of the pore causes pore widening, as well as deepening of the pore. The result is a pore with a scallop shape [112]. As the Ti-O bond energy is high (323 kJ/mol), in the case of titania it is reasonable to assume that only pores having thin walls can be formed due to the relatively low ion mobility and relatively high chemical solubility of the oxide in the electrolyte, hence un-anodized metallic portions can initially exist between the pores. As the pores become deeper the electric field in these protruded metallic regions increases enhancing the field assisted oxide growth and oxide dissolution, hence simultaneously with the pores well-defined inter-pore voids start forming **Fig. 5.19(d)**. Thereafter, both voids and tubes grow in equilibrium. The nanotube length increases until the electrochemical etch rate equals the chemical dissolution rate of the top surface of the nanotubes; after this point is reached the nanotube length will be independent of anodization duration.

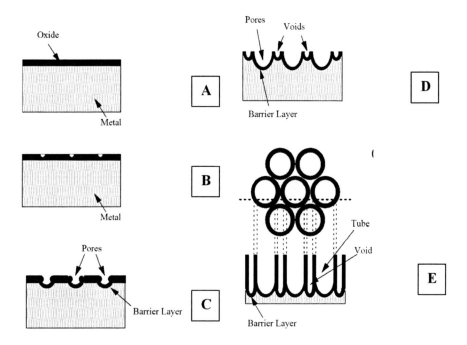

Fig. 5.19: Schematic diagram of the evolution of straight nanotubes at constant anodization voltage: **(a)** Oxide layer formation, **(b)** pit formation on the oxide layer, **(c)** growth of the pit into scallop shaped pores, **(d)** metallic part between the pores undergoes oxidation and field assisted dissolution, **(e)** fully developed nanotubes with a corresponding top view.

This chemical dissolution, the key for the self-organized formation of the nanotube arrays, reduces the thickness of the oxide layer (barrier layer) keeping the electrochemical etching (field assisted oxidation and dissolution) process active. No nanotubes can be formed if the chemical dissolution is too high or too low. The electrochemical etch rate depends on anodization potential as well as concentration of electrolytes. If the electrochemical etch proceeds faster than the chemical dissolution the thickness of the barrier layer increases, which in turn reduces the electrochemical etching process to the rate determined by chemical dissolution. The chemical dissolution rate is determined by the F^- concentration and solution pH [reaction (5.3.3)]. With increasing F^- and H^+ concentrations chemical dissolution increases. Recent investigations have shown that only in a certain F^- concentration range can nanotube arrays be achieved. The anodic potential at which nanotubes are formed is related to the F^- concentration, with higher potentials requiring electrolytes of higher F^- concentration.

In polar organic electrolytes (like FA, NMF, DMSO and EG), the water is usually the source of oxygen in anodizing solutions [113]. While the exact mechanism by which water contributes oxygen to an anodic oxide film is not well-understood, strong evidence has been found for hydroxyl ion injection [113] from the electrolyte into the anodic oxide film during anodization. When more water is present, hydroxyl ions are injected into the body of the oxide layer and affect the structure sufficiently to impede ion transport [114] through the barrier layer, which is necessary for further movement of the metal-oxide interface into the metal. When less water is present, the difficulty in extracting oxygen and/or hydroxyl ions from the solution limits the rate of growth of the overall oxide film. Also, the barrier oxide layer exhibits increased ionic conductivity caused by the non-stoichiometry induced by the reduced hydroxyl ion availability to the oxide [24-27]. The amount of hydroxyl ion injection is dependent on the solvent structure. It is no coincidence that the solvents where high growth rates of nanotubular TiO_2 films have been obtained, namely formamide (~4 μm/min), N-methylformamide(~4 μm/min) and ethylene glycol (~15 μm/min) possess a high degree of structuring. Ethylene glycol has a clearly pronounced spatial net of hydrogen bonds. In DMSO, which is not as structured, final nanotube lengths as large as 100 μm are obtained but the growth rates are much slower (<1.5 μm per minute). For similar electrolytic compositions, the growth rates are also much smaller in Dimethylformamide (DMF), which is unstructured. When the solvent structure provides an environment in which titanium ions in the surface of the oxide bond to oxygen ions produced at the oxide surface, minimal hydroxyl ion injection occurs and ionic transport through the barrier layer is relatively rapid. With the highly-polar FA and NMF electrolytes the dramatically increased nanotube-array length appears to be due to an accelerated rate of nanotube growth from the substrate. The dielectric constant of FA and NMF are, respectively, of 111 and 182.4, much greater than that of water which has a dielectric constant of 78.39 [115]. For a given potential, higher electrolyte capacitance induces more charges to be formed on the oxide layer improving extraction of the Ti^{4+} ions, while the higher electrolyte polarity allows HF to be easily dissolved facilitating its availability at the TiO_2-electrolyte interface.

In formamide electrolyte containing fluoride ion, the starting anodization current does not drop instantly as observed in aqueous bath. The gas evolution which is indicative of electronic conduction was observed at the anode. The anodization current drops steeply thereafter due to the initial formation of an insulating oxide layer, see **Fig. 5.10**. In this region, electronic conduction decreases due to the blocking action of the formed oxide, and ionic conduction increases. Once the oxide layer is completely formed over the entire exposed surface of the anode, electronic conduction becomes negligible and ionic conduction dominates the mechanistic behavior. Nanotube formation reduces the surface area available for anodization with a correlated decrease in current density, while deepening of the pore occurs.

The conductivity of the electrolytes also plays a role in controlling nanotube array growth. Ethylene glycol containing 2% water and 0.35 % NH_4F have a conductivity of 460 μS/cm which is much lower than the conductivity of the formamide based electrolytes (>2000 μS/cm) [27]. The total applied anodization voltage is the sum of the potential difference at the metal-oxide interface, the potential drop across the oxide, the potential difference at the oxide-electrolyte interface, and the potential drop across the electrolyte, which is non-negligible for the ethylene glycol based electrolytes. Therefore, maximum nanotube length obtained at lower voltages (say 20V) in formamide bath could only be achieved at higher voltages in EG baths. In contrast, with DMSO electrolytes of relatively small conductivity, the increased nanotube-array length appears to be largely due to limited chemical dissolution of titania by the hydrofluoric acid. For a given rate of pore formation the chemical dissolution of the oxide at the pore mouth by F⁻ determines the tube length. To control this dissolution reaction, the H^+ ion concentration was reduced by limiting the water content to the level of water contained in HF containing solution. This water ensured the field assisted etching of the Ti foil at the pore bottom and, additionally, protophilic DMSO accepts a proton from HF reducing its activity. This allowed the DMSO nanotubes to grow deep into the titanium foil without any significant loss from the pore mouth.

5.4 Doped Titania Nanotube Arrays

5.4.1 Flame Annealed Nanotubes

On annealing Ti metal foils in a natural gas flame at 850°C Khan and co-workers found the diffuse reflectance spectra of these samples to be significantly broadened [116]; in addition to a shift in the primary absorption threshold from 414 nm to 440 nm, a second optical absorption threshold appeared at 535 nm, which was used to extract a band gap of 2.32 eV. However, on flame annealing their samples using a propane/butane-oxygen mixture, Augustynski and coworkers described a shift of the spectral photoresponse into the visible region up to 425 nm but their samples did not exhibit a secondary band edge [117].

The effect of flame annealing on two different nanotube array geometries have been investigated [98]. The first geometry consisted of nanotubes with an average pore size of 22 nm, an average wall thickness close to 20 nm and a length of ~200 nm (10V, acetic acid + 0.5% HF solution). The as-anodized nanotubes were amorphous; to induce crystallinity they were subsequently annealed at 450°C in oxygen ambient for 6 hours. In second geometry, long nanotubes were synthesized by anodic oxidation of titanium foils in a new electrolyte containing potassium fluoride (0.1M), tetrabutylammonium hydroxide (0.05 M), trisodium citrate (0.2 M) and sodium hydrogen. All samples were prepared using a potential of 25 V, over a 17-hour anodization. The pH of the electrolyte, adjusted by adding sodium hydrogen sulfate, was 4.5 for the duration of the process. Anodization was conducted using a two-electrode electrochemical cell with a platinum cathode. The as prepared samples were amorphous; crystallinity was induced by annealing at 600°C in an oxygen atmosphere for 6 hours. The final nanotubes had a pore size of 100 nm, a wall thickness close to 20 nm and a length of 4.4 µm.

Flame annealing of the nanotube-array samples was performed in air after the crystallinity-inducing annealing step by exposing them to the reductive region of a propane burner for 3 min. The temperature of the titania surface while exposed to propane flame was found to be 1020°C ± 25°C. The duration of the flame anneal was kept brief to preserve the nanotubular structure, which is destroyed upon prolonged exposure to temperatures in excess of 650°C due to thermal oxidation of the underlying titanium foil. The sharp contrast in the morphology of the long nanotubes with and without the flame anneal is depicted in **Fig. 5.20.**

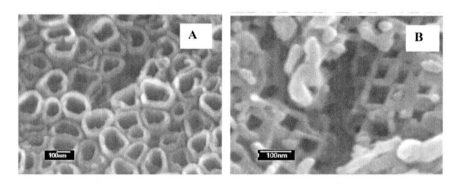

Fig. 5.20: FESEM images of nanotube array top surface, anodized at 25 V, **(a)** before and **(b)** after flame annealing.

5.4.2   Dopant Introduction via Modification of Anodization Bath Chemistry

Recently band-gap engineering of TiO_2 by anionic doping has been receiving attention. Asahi and co-workers [42] performed density of states (DOSs) calculations on the effect of substitutional doping and identified nitrogen as the most effective dopant due to its comparable ionic radius and because its p-states contribute to band gap narrowing by mixing with the p states of oxygen [43]. However, Lee et al. [118] contradicted this assessment in their report, where their density functional calculations indicate that while nitrogen doping produces isolated N 2p states above the valence band maximum of TiO_2, the mixing of N with O 2p states is too weak to result in appreciable band-narrowing. There has recently been a surge of interest in this area documenting different experimental approaches towards nitrogen doping of titania [119-121].   The possibility exists for electrochemical incorporation of anionic dopants, speci.cally nitrogen, during the anodization process [99]. As previously reported in the anodic fabrication of nanoporous alumina using a neutral organic electrolyte [122], it was hypothesized that a significant amount of organic material could be incorporated into anodic TiO_2 films by using organic electrolytes. Such in situ chemical doping of the TiO_2-nanotube arrays might prove useful in band-gap engineering of the resulting material. Fluroine doped titania nanotubes were prepared by anodizing titanium foil in 1:1 DMSO and ethanol solvent containing 4% HF at 20V (vs Pt). FE-SEM images of an anodized sample are shown in **Fig. 5.21(a)**. FE-SEM images of a sample identical to that in **Fig. 21(a)** that has been washed in dilute HF prior to imaging is shown in **Fig. 5.21(b)**.

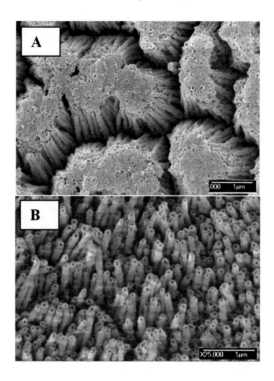

Fig. 5.21: FE-SEM images of titanium foil sample anodized in DSMO and ethanol mixture solution (1:1) containing 4% HF at +20 V (vs. Pt) for 70 h at room temperature **(a)** before and **(b)** after washing in dilute HF.

5.4.3   CdS - Coated Nanotubes

Other efforts concerned with shifting of the TiO_2 band gap have focused on the photoelectrode sensitization through combination with narrow-band gap semiconductor films [123]. Such sandwich electrodes may be advantageous as electron injection may be optimized through confinement effects, and a sensitizer, a 1.5 eV edge absorber, is well approximated by a narrow band gap semiconductor material [124]. Since the conduction band of bulk CdS is *ca.* 0.5 V more negative than that of TiO_2, this coupling of the semiconductors should have a beneficial role in improving charge separation. TiO_2 nanotube electrodes were prepared at 20V in a HF and acetic acid solution. These amorphous samples were then crystallized at 480°C for 6 hours in oxygen gas ambient. A CdS film was then deposited upon the crystallized TiO_2 nanotube-array by cathodic reduction, using a conventional three-electrode system comprising an Ag/AgCl reference electrode and Cd counter electrode [112]. A mixed solution of saturated elemental sulfur in benzene with 0.6 M CdCl2 in dimethyl sulfoxide (DMSO) was used as the electrolyte. The solution was bubbled with flowing N_2 for 30 minutes prior to electro-deposition in order to remove O_2 and any moisture within the solution. The cathodic potential was kept constant at -0.5V for different deposition times. After electrodeposition the samples were thoroughly rinsed with acetone, methanol and D.I. water. The prepared CdS- TiO_2 electrodes were annealed at 350°C and 400°C for 60 minutes in a N_2 atmosphere to investigate the influence of annealing on their photoelectrochemical response. It was suggested that when a cathodic potential is applied to the TiO_2 nanotube electrode, it will reduce sulfur to S^{2-} on the electrode surface, while the applied electric field induces Cd^{2+} to migrate towards the electrode hence under proper conditions CdS will form at the electrode surface [125]. **Figure 5.22(a)** shows an illustrative FE-SEM image, top surface view, of a TiO_2 nanotube-array upon which just a few CdS nanoparticles, ≈ 20 nm diameter, have been deposited (–0.5V for 5 minutes). **Figure 5.22(b)** shows the topology after a 30 minute (–0.5V) electrodeposition of the CdS nanoparticles.

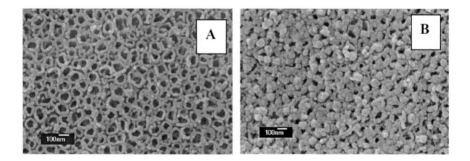

Fig. 5.22: Top surface FESEM view of the prepared CdS-TiO_2 nanotube-array electrode after CdS electrodeposition at –0.5V for 30 minutes.

5.5   Material Properties

5.5.1   Structural and Elemental Characterization

The properties of titania depend on the crystallinity and isomorph type, hence their application also varies. For example, anatase phase is preferred in charge separating devices while rutile is used predominantely in gas sensors and as dielectric layers. Rutile has minimum free energy compared to any other titania polymorphs hence given the necessary activation energy, all other polymorphs including anatase transfer into stable rutile through first order phase transformation. However the temperature at which metastable anatase to stable rutile transformation takes place depends upon several factors including impurities present in the anatase, primary particle size, texture and strain in the structure. Hence, porosity and/or surface area reduction occur due to the sintering effects associated with nucleation-growth type of phase transformations.

As-anodized titania nanotubes are amorphous and are crystallized at high temperature anneal. GAXRD patterns of the 20V HF electrolyte sample annealed at different temperatures in dry oxygen ambient are shown in **Fig. 5.23** [126]. In the diffraction patterns, the anatase phase starts appearing at a temperature of 280°C. As the 250°C annealed sample was amorphous (only reflections from titanium support can be seen), it is clear that the sample was crystallized in anatase phase at a temperature between 250 and 280°C. At a temperature near 430°C rutile phase appears in the x-ray diffraction pattern. After this temperature, rutile (110) peak grows whereas the anatase (101) peak diminishes. Complete transformation to rutile occurs in the temperature range 620°C to 680°C. It can also be seen from **Fig. 5.23** that the reflection from the titanium support is getting reduced at temperatures between 430 and 580°C and they fully vanish at around 680°C. This shows that the oxidation followed by crystallization of titanium support takes place at these temperatures. With respect to variation of the size of the anatase and rutile crystallites with temperature, it was found that the anatase grain size initially increases with temperature but between ~480°C to 580°C the grain size decreases to increase again after 580°C. At the same time the grain size of rutile progressively increases with temperature after its nucleation. At 430°C a rutile fraction of 31% compared to anatase was formed. It increased to 75% at 480°C and further to 92% at 580°C on annealing for 3 hours.

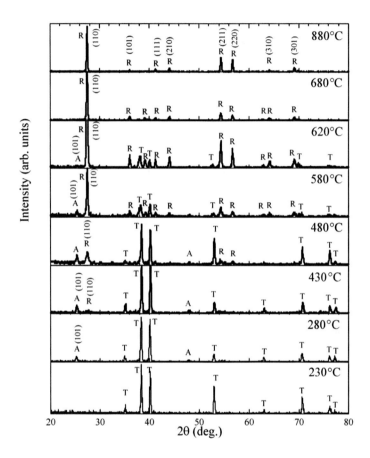

Fig. 5.23: GAXRD patterns of the nanotube samples annealed at temperatures ranging from 230 to 880°C in dry oxygen ambient for 3 h. A, R, and T represent anatase, rutile, and titanium, respectively.

The evolution of the surface morphology as a result of high temperature annealing has also been studied. For nanotube arrays atop Ti foil, the structure of the 20V sample was found to stable till around 580°C (for 10V sample, it is about 500°C). No discernible change in the pore diameter or wall thickness was observed even after annealing for 3 hours at this temperature. It was observed that at temperatures in the range 550 to 580°C depending on the sample, small protrusions come out through the porous structure. Above this temperature the tubular structure completely collapsed leaving dense rutile crystallites. To help Annealing a 20V sample to 820°C at a heating rate of 10°C/min and then suddenly cooled down using the same rate in oxygen ambient were found to disturb the tubular structure, but the sample still possesses porosity. The walls of the tubes coalesced together to form worm like appearance. HRTEM image of the wall and contact points of crystallized nanotubes indicated that the crystallite in the wall has a length of ~35nm and a width of around ~12 nm. Different regions of the walls were examined using HRTEM and all crystallites were found to be anatase from Fourier transform analysis. On comparing energy dispersive x-ray spectra (EDS) of the as deposited and the one fired at 580°C in oxygen ambient, it was shown that the relative intensity of oxygen peak with respect to titanium K_α peak increased on annealing in presence of oxygen which is an indication of the improvement in the stoichiometry of the sample.

The as-anodized titania films fabricated from Ti films deposited on glass, taken out of the anodization bath at P5 of **Fig. 5.17**, having an extremely thin discontinuous metal layer underneath the nanotubes, were annealed at 260°C, 280°C and 500°C for 6 hours in dry oxygen ambient. Their GAXRD patterns showed only one exception, the absence of rutile phase in the thin film samples annealed at 500°C [104]. This result is in striking contrast to that found with nanotube-arrays formed from Ti thick-film foils, where an earlier study noted that rutile phase appears at 430°C and both rutile and anatase co-exist till around 620°C (shown in earlier section). However we find that thin film samples with a continuous metal layer underneath the nanotubes behave in a way similar to that of the foil samples, with both rutile and anatase phases co-existing at 480°C. These results support the contention that rutile phase grows at the interface between the barrier layer and titanium metal where the metal is thermally oxidized. The constraints imposed by the nanotube walls make it difficult for the anatase crystals situated there to undergo phase transformation to rutile.

The x-ray patterns of samples obtained in H_3BO_3-HNO_3-HF and HNO_3-HF baths, annealed at 550°C for 6 h with a heating and cooling rate of 1°C/min in oxygen ambient, are similar to one observed for short nanotubes. After annealing, the phase-structure of the architecture can be viewed as an anatase nanotube-array atop a rutile barrier layer. In comparison, a TiO_2 film made by 550°C thermal annealing is primarily rutile phase with traces of anatase phase. The normalized reference intensity ratio (RIR) method was used to estimate the weight fraction of anatase, rutile, and titanium in the resulting samples. The calculated RIR result of the H_3BO_3-HNO_3-HF prepared sample is anatase 33.6%, rutile 58.7%, and titanium 7.7%. The calculated RIR result of the HNO_3-HF prepared sample is anatase 1.7%, rutile 66.5%, and titanium 31.7%. Considering a similar x-ray sampling depth for both samples, the higher weight percentage of titanium in the HNO_3-HF anodized sample indicates a thinner barrier layer, and nanotube-array length. Consequently the thinner barrier layer gives rise to a lower measured anatase weight percentage for the sample obtained in HNO_3-HF.

A XPS scan for the HNO_3-HF sample, after 550°C annealing, indicated the elements Ti (23.0%), O (64.3%), N (1.0%), F (0.5%), and C (8.0%). For the H_3BO_3-HNO_3-HF electrolyte sample, 550°C annealed, a scan point out Ti (27.4%), O (65.3%), N (0.3%), F (0.8%), and C (4.7%). Both carbon and some of the oxygen can be viewed as surface contamination, while the small amounts of N and F originate from the electrolytes used for sample preparation. XPS analysis of the H_3BO_3-HNO_3-HF sample before annealing specifies Ti (26.9%), O (60.2%), N (1.7%), F (6.6%), and C (4.5%). Chemical state analysis indicates the sample is comprised of Ti^{4+} bonded with oxygen (TiO_2), contaminated with N, F and C compounds; no boron was detected in the samples. The O1s spectra of the samples (in boric acid bath) showed a single peak at 530.8 eV. However, in the HNO_3-HF sample there is an indication of a second peak at 532 eV, revealing the presence of two forms of oxygen. The $Ti2p_{3/2}$ peak has a binding energy of 459.0 eV for both samples, indicating Ti present in the samples is in the form of TiO_2. For these measurements the sampling depth of the x-rays is 8 nm, thus the Ti substrate cannot be detected. The position of $2p_{3/2}$ peak of Ti in the form of TiO_2 was in consistent with the formation of a crystalline TiO_2 [127].

GAXRD patterns of the 6 μm nanotube arrays sample fabricated using KF (or NaF, the two acids result in equivalent architectures) based electrolytes annealed at different temperatures up to 700°C are shown in **Fig. 5.24** [5,22]. It can be seen that the nanotubes maintain the amorphous behavior on annealing at 230°C. The crystallization occurs in anatase phase at a temperature near 280°C. It may be noted that crystallization of the samples prepared using HF electrolyte without any additives also showed the same crystallization temperature [126]. Apparently, electrolyte concentration or pH has no influence on the crystallization temperature of the nanotubes. As the temperature increases the crystallinity increases as more amorphous regions became crystalline. The rutile phase started appearing at 530°C, with the rutile phase dominating in samples annealed at 700°C. Upto 580°C, rutile concentration increases significantly as evident from the higher ratio of anatase 101 peak to rutile 110 peak. No sign of nanotube disintegration was observed in the FESEM images of the 580°C annealed samples. This is in contrast with shorter nanotubes grown at pH<1 which started disintegrating at this temperature [126]. An EDX spectrum of an amorphous nanotube (10V, pH=4.5) showed no elements other than titanium and oxygen.

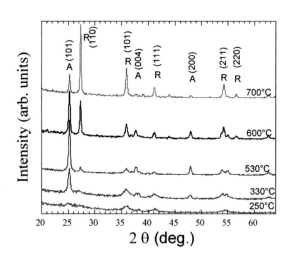

Fig. 5.24: Glancing angle X-ray diffraction patterns of a 6.0μm long nanotube array as a function of annealing temperatures (oxygen ambient).

A transmission electron microscope (TEM) image of a single nanotube grown from a sample prepared at pH 5.0 and annealed at 600°C, see **Fig. 5.25,** indicates the presence of anatase phase in the walls; rutile phase is not found in the tube walls. The TEM data in conjunction with the GAXRD patterns (discussed earlier) confirms the crystallization model proposed in our earlier work on short nanotubes prepared using HF electrolyte. According to this model, anatase crystals are formed at the nanotube-Ti substrate interface region as a result of the oxidation of the metal at elevated temperatures and in the nanotubes. The rutile crystallites originate in the oxide layer (formed by the oxidation of titanium metal) underneath the nanotubes at high temperatures through nucleation and growth as well as phase transformation of anatase crystallites existing in the region. The constraints imposed by the nanotube walls, however, make critical radii needed for rutile nucleation very large [5]. This prevents the anatase phase at the nanotube walls from undergoing transition to rutile phase. Therefore, nanotubes, annealed at temperatures between 530°C and 580°C, can be considered as anatase crystallites stacked in cylindrical shape on a rutile foundation.

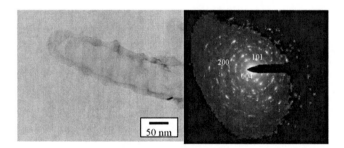

Fig. 5.25: TEM image of a single nanotube from a sample prepared at pH 5.0 and annealed at 600°C is given with inset showing diffraction pattern.

5.5.2. Characterization of Doped Titania Nanotubes
Flame Annealed Samples

An XPS scan of flame annealed nanotubes revealed the presence of carbon in all samples and a summary of the carbon content and carbon state information is provided in **Table 5.5** [98]. Fluorine was present in all samples, at surface concentration of nearly 2 atomic %, decreasing to about 0.2 atom % in the interior. The presence of both sodium and fluorine are directly related to the chemistry of the anodizing baths. Based on the analysis of the C1s peak, incorporated carbon was present in C-C (285.3 eV), CO (286.5 eV), COO (289.0 eV) and C-N bonds. The Ti-C signal at 281.9 eV was not observed. The carbon content of the short nanotubes, which is initially quite small, becomes appreciable upon flame annealing. A significant amount of carbon (~3%) is present in the long nanotube sample even prior to flame annealing, which is attributed to the presence of a large number of organic ions such as citrate and tetrabutylammonium in the anodizing bath. In long nanotubes, flame annealing introduces additional carbon into a structure where carbon preexists in appreciable quantities. Hence, flame annealed long nanotubes have the highest carbon content (>5 %) of the samples studied.

Table 5.5   Carbon content and chemical state information (from XPS).

Sample	Depth (nm)	Total C (atom %)	C-C	C-O	COO
Short NT	0	0.7	-	-	-
	100	0.3	-	-	-
Flame annealed short NT	0	3.3	2.5	0.4	0.5
	100	2.8	1.9	0.6	0.3
Flame annealed long NT	0	5.6	3.9	0.8	0.9
	100	5.2	3.8	0.9	0.6
Long NT	0	3.5	2.3	0.5	0.7
	100	3.0	1.9	0.6	0.5
Flame annealed Ti foil	0	4.0	2.7	0.6	0.7
	100	3.8	2.5	0.7	0.5

Nitrogen-doped Titania

Titanium foils were potentiostatically anodized at 25V in an electrolyte of pH 3.5 containing 0.4M ammonium nitrate NH_4NO_3 and 0.07M HF acid; with reference to **Fig. 5.26**, Sample A was removed after 17 s of anodization, while Sample B was anodized for 240 s. Sample C was anodized for 6 h at 20V in an electrolyte of pH

3.5 containing 2.5M NH_4NO_3 and 0.07M HF. Such anodization chemistry restricts the electrolytic ions to nitrogen and fluorine bearing species, allowing control of the possible elements that can be incorporated into the anodic titania films. The potential and pH regimes chosen were such as to facilitate nanotube array formation. The maximum current at the onset of the anodization was limited by the compliance of the power supply used to perform the anodization. In the first 25 s, after application of the voltage, the measured current density reduced from 4120 mA/cm^2 to a local-minimum between 15 and 25 mA/cm^2, with the field-assisted oxidation of the Ti metal surface reducing the current. In the potential range under consideration, this behavior is typical for the anodization of Ti in fluoride ion containing acidic electrolytes; however, the magnitude of the anodization currents are much greater. The larger anodization currents are attributed to the stronger oxidizing and etching action of the nitrate ion containing electrolyte. High-resolution N 1s XPS spectra of Samples A (x = 0.23), B (x = 0.09) and C (x = 0.02) is shown in **Fig. 5.26**. XPS data confirms that all the incorporated nitrogen is substitutional on the oxygen site. The nitrogen peak at 396.8 eV was observed and assigned to atomic β-N, indicating a chemically bound N$^-$ state. Fluorine was present in the amorphous as-anodized samples, with the final concentration of incorporated F$^-$ sensitive to the annealing conditions. Annealing processes (in air) lasting longer than six hours at temperatures above 600°C resulted in fluorine atoms being completely resubstituted by oxygen. The depth profile of a 250 nm thick film with a surface nitrogen concentration x = 0.05 (the sample was anodized at 20 V for 120 s in a pH 4.5 electrolyte containing 0.4 M ammonium nitrate and 0.07 M Hydrofluoric acid then annealed per the other samples) indicated that the doping of nitrogen is inhomogeneous with the maximum nitrogen being incorporated close to the surface then linearly decreasing with increasing depth inside the film.

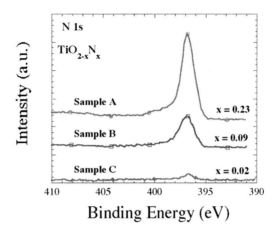

Fig. 5.26: N 1s XPS spectra for samples A-C with respective nitrogen doping levels [99].

The precise reactions involving the decomposition of ammonium ions and nitrate ions at the anodic surface to form N-doped titania are currently unclear, and the subject of ongoing studies. However, the anodization of aluminum in nitric acid has been studied previously and is known to be relatively complex [128]. A study of the interaction of aluminum with nitrate ions in thin oxide films formed in nitrate ion containing electrolytes indicated that the adsorption of nitrate ions on the oxidized surface of aluminium was followed by their reduction inside the oxide film [129]. Parhutik and co-workers [130] reported the incorporation of electrolyte anions in the anodic film formed by anodization of Al in HNO_3 solution. Furthermore, it was reported that the anion concentration, in the growing oxide, reaches a maximum value at the moment when intensive pore growth starts and the oxide is thin, i.e. when the anodizing time is very brief and the surface topology strongly dependent upon the applied forming conditions. Similar behavior was observed for Ti anodization, with maximum nitrogen incorporation occurring in a film anodized for a mere 17 s (Sample A). Thus, a trade-off exists between the morphology and the level of nitrogen doping. Shorter anodization periods result in higher concentrations of incorporated nitrogen, whereas longer anodization periods are required for evolution of the nanotube-array architecture.

CdS-Coated Nanotubes

In the GAXRD pattern of CdS coated (short and HF fabricated) TiO_2 nanotube array annealed at 350°C for 1 hour, Chen and co-workers [112] observed a prominent TiO_2 Bragg peak along with weak Bragg reflections at 2θ values of 26.55, 30.75, 44.04, 52.16, 54.67, corresponding to the (111), (200), (220), (311), and (222) Bragg reflections of cubic CdS, respectively. The general scan spectrum of XPS of CdS-TiO2 electrodes showed sharp peaks for Ti, O, Cd, S, and also C. The Cd 3d core level XPS spectrum has two peaks at 405.3 eV (3 $d_{5/2}$) and 411.9 eV (3 $d_{3/2}$), in good agreement with published values [131]. The S 2p core level spectrum indicated that there are two chemically distinct species in the spectrum. The peak at 161.9 eV is for sulfide, the structure occurs because of a split between $2p_{3/2}$ and $2p_{1/2}$; the split is near 1.18 eV and the area ratio is 2:1, in agreement with published values of the S 2p signal for CdS. Measured atomic concentrations of the as-prepared samples suggested that when the sulfate/O ratio is 1, the sulfide/Cd ratio is 0.86; this means that the CdS nanoparticles obtained are slightly Cd rich, which is expected for CdS under normal synthesis conditions.

5.6   Optical Properties of Titania Nanotubes Arrays

5.6.1. FDTD Simulation of Light Propagation in Nanotube Arrays

The titania nanotube arrays can be grown over a wide range of pore diameters, wall thicknesses, and lengths, with each topology showing different light absorption and photocatalytic properties leading to different values of photoconversion efficiency [3,5]. With such a variety of geometric variables, knowledge of the light absorbing behavior of the nanotube-array geometries prior to sample fabrication would be desirable. Therefore, the computation electromagnetic technique finite difference time domain (FDTD) was used to simulate the light-absorbing properties of the nanotube arrays as a function of feature size [132]. The simulations were performed for titania nanotube array films with no metal layer underneath the nanotubes (transparent, Type-I) and also for the nanotubes grown on titanium foil (opaque, Type-II). Note that in the former case, Type-I, the glass substrates were not included in the simulations and hence the nanotube film can be considered self-standing. **Figure 5.27(a),(b)** show, respectively, the two models used to represent self-standing titania nanotube array (transparent, Type-I) and the nanotube array on titanium foil (opaque, Type-II). The distance between two adjacent tubes was taken as 10 nm. The FDTD space was terminated with an Absorbing Boundary Condition (ABC) made of an uniaxial perfect matching layer to eliminate the reflection of any fields from these boundaries. The Type-II model contains a perfect electrical conductor layer at the bottom of the nanotube array to represent the titanium layer. Therefore, in the case of Type-I, transmittance and in the case of Type-II, reflectance are used to determine the absorbance of light by the nanotube array. In all simulations reported to date the distance between two adjacent tubes was taken as 10 nm.

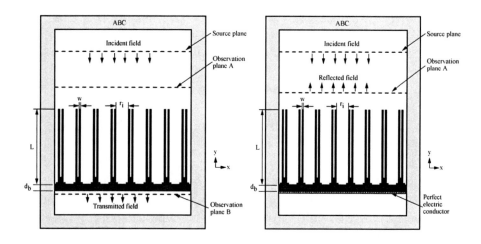

Fig. 5.27:   Geometry of two-dimensional FDTD models used for determining the propagation of a tranverse electromagnetic wave through: **(a)** A self-standing titania nanotube array film (Type-I), **(b)** titania nano-tube array film on Ti substrates (Type-II).

The validity of the FDTD simulations were established by comparison of the calculated and experimentally measured transmittance of a Type-I film of different porosity, see **Fig. 5.28.**

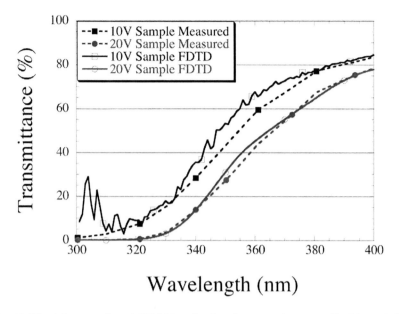

Fig. 5.28: Measured and FDTD calculated transmittance of a Type-I film made using different anodization voltages. The 10V sample is 200nm long, the length of 20V sample is 360 nm.

Figure 5.29 shows the propagation of a TEM wave through a titania nanotube array on Ti foil (Type-II). For illustration purposes, a derivative Gaussian wave (center frequency $= 8 \times 10^{14}$ Hz, bandwidth $= 2 \times 10^{14}$ Hz) is used as the excitation source. The tube length, pore diameter, wall thickness, and barrier layer thickness are, respectively, 1000 nm, 100 nm, 20 nm, and 100 nm. **Figure 5.29(a)** shows the wave originating from the source and moving towards the nanotubes. When the wave front hits the top surface of the nanotube array, **Fig. 5.29(b)**, most of the incident energy is transmitted into the nanotubes with a negligible portion reflecting back. The reflected wave can be seen in **Fig. 5.29(b-c)** as a faint horizontal line on the top of the nanotubes. The wave dissipates as it travels through the nanotube-array to reach the barrier layer, **Fig. 5.29(c)**. **Figure 5.29(d-f)** show the wave reflecting back from the conducting Ti layer at the bottom of the nanotube-array. Note that the reflected wave contains multiple wave fronts as the derivative Gaussian pulse contains radiation over a wide frequency range; the individual frequencies travel at different velocities through the barrier layer and nanotube array due to the frequency-dependent variation in the titania permittivity. The measured absorbance spectrum of a titania nanotube array (length—200 nm, pore size—22 nm, wall thickness—13 nm and barrier layer thickness—100 nm) was compared with the simulated results. Both curves were found similar except that the absorption edge of measured spectra is shifted slightly to the higher wavelength region compared to the simulated spectra. This is due to the fact that the barrier layer has rutile crystallites and the nanotube walls consist of anatase crystallites. The band gap of the rutile is lower (3.0 eV) compared to the anatase (3.2 eV). The rutile phase at the barrier layer leads to the shifting of the absorption edge to higher wavelength, a property not taken into account by the FDTD simulations.

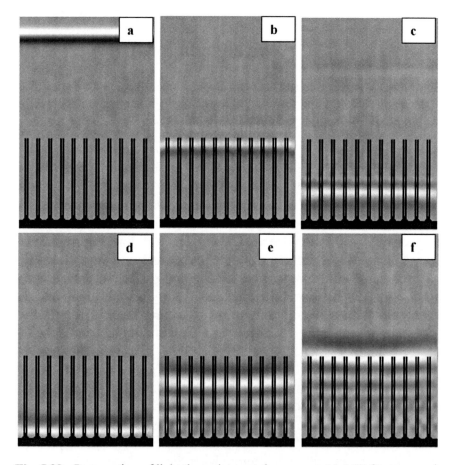

Fig. 5.29: Propagation of light through nanotube array at **(a)** 4.67 femtoseconds (fs), **(b)** 9.34 fs, **(c)** 11.68 fs, **(d)** 14.01 fs, **(e)** 16.35 fs, and **(f)** 17.51 fs. The variation in field strength is represented by the different shades of gray.

   With respect to the applied properties of Type-II samples, it should be noted that to induce crystallinity the nanotube array samples are annealed at elevated temperatures in an oxygen environment. The diffusion of oxygen into Ti foil is consistent with the Fick's second law, hence a gradient in the oxide composition exists from the top of the barrier layer to the Ti metal. Consequently there is a gradient in the complex permittivity spectrum of the oxide layer underneath the nanotubes and hence light is bent before it is reflected back from the metal. This gradient was considered during the simulation process by linearly increasing the permittivity values of the barrier layer so the permittivity at the bottom of barrier layer is 10 times larger than the top. Hence when light is reflected back from the metal it is more readily absorbed, therefore the intensity of the reflected light is low, on a unit length basis, compared to that of the transmitted wave in Type-I samples. As a result, a clear difference in the absorbance can be seen between the Type-II samples (**Fig. 5.30**) and the Type-I samples (**Fig. 5.31**). The increased light absorption in Type-II samples makes them more suitable for water photolysis experiments, while the Type-I films are better suited for application in solar cells.

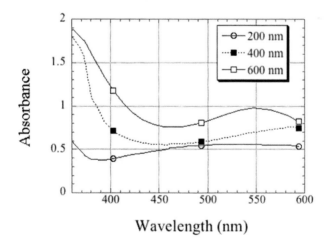

Fig. 5.30: Absorbance of titania nanotube films as a function of tube length for arrays grown on Ti foil (Type-II); the inner diameter of the tube is 20 nm, wall thickness 10 nm, and barrier layer thickness 100 nm.

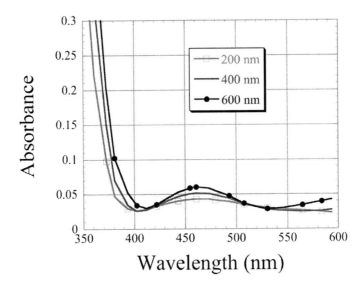

Fig. 5.31a: Absorbance of titania nanotube array films as a function of tube length (Type-I; the inner diameter of the tube was 20 nm, wall thickness 10 nm, and barrier thickness 100 nm).

The transmittance of light through self-standing titania nanotube array films (Type-I) are calculated as a function of tube length while keeping wall thickness, pore diameter, and barrier layer thickness constant. **Figure 5.32** plots the transmittance of the film as a function of excitation wavelength and tube length for nanotubes of length 200 nm, diameter 22 nm and barrier layer thickness 100 nm. The transmittance reaches a value over 95% at wavelengths greater than 380 nm. The spacing between the interference patterns, created by the interaction of the transmitted wave and the wave reflected back from the top of the nanotubes, reduces with increasing nanotube length. In the region below about 330 nm the absorption is so high that the nanotube length has little influence. Here the light is completely absorbed by the nanotubes within a path length of a few tens of nanometers. Above this wavelength region the transmitted fields depend on the nanotube length. It was found that for a given nanotube length, wall thickness and barrier layer thickness, the transmittance increases slightly with the increasing pore size. With the increase in porosity the air column volume increases and the solid material volume decreases, yielding reduced effective refractive indices.

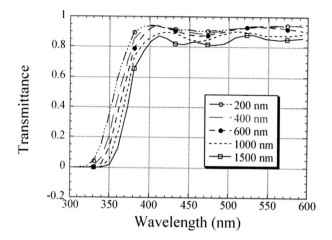

Fig. 5.32b: Transmittance of titania nanotube array films as a function of tube length (Type-I; the inner diameter of the tube was 20 nm, wall thickness 10 nm, and barrier thickness 100 nm).

5.6.2. Measured Optical Properties

The transmittance spectrum of a titania nanotube-film (transparent) on glass is shown in **Fig. 5.33**. The optical behavior of the TiO$_2$ nanotube-arrays is quite similar to that reported for mesostructured titanium dioxide [133]. The difference in the envelope-magnitude encompassing the interference fringe maxima and minima is relatively small compared to that observed in titania films deposited by rf sputtering, e-beam and sol-gel methods [134].

The absorbance (or optical density) of the films were estimated from the transmittance 'T' using the relation: $A = -log(T)$. Here we assumed that all the incident light is either transmitted or absorbed, reflection or scattering being negligible. The Napierian absorption coefficient of the sample was calculated using Lamberts law, $\alpha = 2.303 \left(A/d \right)$, where '$d$' is the thickness of film, which can be determined using the relation:

$$d = \frac{\lambda_1 \lambda_2}{2\left[\lambda_2 n(\lambda_1) - \lambda_1 n(\lambda_2)\right]} \qquad (5.6.1)$$

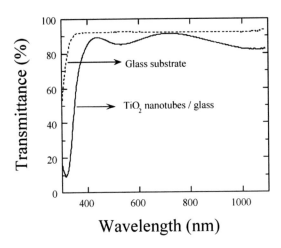

Fig. 5.33: Transmittance spectra of glass (Corning 2947) substrate, and 450°C annealed nanotubular titania film atop same glass (Corning 2947) substrate.

where λ_1 and λ_2 are the wavelengths corresponding to the two adjacent maxima or minima and $n(\lambda_1)$ and $n(\lambda_2)$ are the refractive indices at λ_1 and λ_2 respectively. The refractive indices of the titania nanotube film were calculated using the transmittance spectrum in the range 380 nm to 1100 nm employing Manifacier's envelope method [135]:

$$n(\lambda) = \sqrt{S + \sqrt{S^2 - n_0^2(\lambda)n_S^2(\lambda)}} \tag{5.6.2}$$

$$S = \frac{1}{2}\left[n_0^2(\lambda) + n_S^2(\lambda)\right] + 2n_0n_S\frac{T_{max}(\lambda) - T_{min}(\lambda)}{T_{max}(\lambda) \times T_{min}(\lambda)} \tag{5.6.3}$$

where n_0 and n_s are the refractive indices of air and film respectively, T_{max} is the maximum envelope, and T_{min} is the minimum envelope. From the transmittance spectrum, the refractive index of glass is calculated as a function of wavelength using the relation:

$$n_S(\lambda) = \frac{1}{T_S(\lambda)} + \sqrt{\frac{1}{T_S^2(\lambda)} - 1} \tag{5.6.4}$$

where T_s is the transmittance of glass.

Figure 5.34 shows the refractive index of the thin film titania nanotubes, and for comparison the glass substrate, calculated using equations (5.6.3) and (5.6.4). The optical behavior of the TiO$_2$ nanotube-arrays is quite similar to that reported for mesostructured titanium dioxide [134]. The average refractive index of the nanotube array (450°C annealed) was found to be 1.66 in the visible range, 380 to 800 nm. The thickness, as calculated by inserting the values of refractive indices and the wavelength corresponding to two consecutive maximum or minimum (**Fig. 5.34**) in equation (5.6.2), was found to be 340 nm. This agrees with the value of 300 nm for the total thickness of the nanotube array including the barrier layer, determined from SEM images.

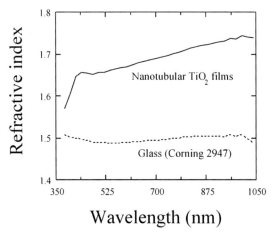

Fig. 5.34:   Refractive index variation of 450°C annealed nanotubular titania film, and for comparison a glass (Corning 2947) substrate, in the range 380 to 1050nm. The TiO$_2$ film has an average refractive index in the visible range of 1.66.

The porosity of the nanotube array architecture was determined from the relation [136]:

$$Porosity\ (\%) = \left[1 - \frac{n^2 - 1}{n_d^2 - 1}\right] \times 100 \tag{5.6.5}$$

where n (=1.66) and n_d (= 2.5) are the refractive indices of the nanotube structure (annealed at 450°C) and nonporous anatase films respectively. The porosity of the nanotube structure was calculated as 66.5%, which is close to the calculated value of 67% for nanotube arrays grown on titanium foil using a 10 V anodization potential [4]. The low refractive index is due to the high porosity of the nanotube architectures, with nanotube diameters much less than the wavelength of light in the visible range, which reduces the light reflection from the surface of the array.

The absorption coefficient α and the band gap E_g are related through the equation [137]

$$(\alpha h v)^s = h v - E_g \tag{5.6.6}$$

where v is the frequency, h is Plank's constant, and $s = 0.5$ for indirect band gap material.

The Tauc plot, $\sqrt{\alpha h v}$ vs. $h v$, obtained after substituting the value of α in this equation is shown in **Fig. 5.35**. The optical band gap, obtained by dropping a line from the maximum slope of the curve to the x-axis, is 3.34 eV. It may be noted that XRD results showed only anatase phase in the transparent titania nanotube array film. The reported band gap value of anatase phase in bulk is 3.2 eV [138]. A slight blue shift in the value might be due to a quantization effect in the nanotubular film where the wall thickness is about 12 nm. A band tail to 2.4 eV is observed. The degree of lattice distortion is likely to be relatively higher for nanotube-array films, thus causing aggregation of vacancies acting as trap states along the seams of nanotube walls leading to a lower band-to-band transition energy.

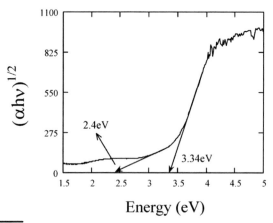

Fig. 5.35: $\sqrt{\alpha \hbar \omega}$ vs. $\hbar \omega$ plot of a 450°C annealed nanotube-array film. An indirect band gap of 3.34 eV and band tailing up to 2.4 eV is observed in the film.

As seen from the structural studies, nanotube array films that retain a metal layer underneath have anatase phase residing in the nanotube walls and rutile phase in the barrier layer. The absorbance spectra of these opaque films were compared with those of nanotubes grown on metal foils (both annealed at 480°C) in the wavelength range 320-800 nm, as shown in **Fig. 5.36.** Although there is no significant difference between the behavior of thin films and foil samples, the shift in the absorption edge towards higher wavelengths on annealing the samples is evident from this figure. The presence of rutile phase, which has a band gap of ~3.0 eV, makes the absorption edge close to 400 nm [138].

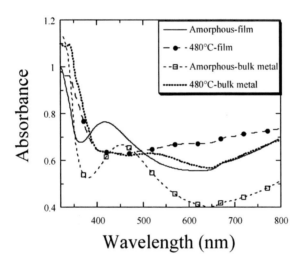

Fig. 5.36: Diffuse reflectance spectra of thin film (with a significant Ti metal layer underneath) and bulk metal samples. Both film and foil samples were prepared in the same electrolyte and annealed in the identical conditions. Only as-anodized (indicated as amorphous) and 480°C annealed samples are shown.

The UV-Vis spectra of the electrochemically doped nitrogen and fluorine doped TiO_2 thin films revealed that the presence of fluorine did not result in a discernable change in optical absorption, whereas N-doping exhibited slightly higher optical absorption in the wavelength range from 400-510 nm [99]. The optical absorption is a function of both film thickness and nitrogen concentration. The film with the highest nitrogen concentration, Sample A with $x = 0.23$, is also the thinnest film owing to the fact that it was anodized for only 17s. The improvement in optical absorption is manifested most clearly for films of similar thickness. All N-doped films exhibited a shift in the primary absorption threshold with the magnitude of this shift increasing with the concentration of incorporated nitrogen.

The normalized visible reflectance spectra of a plain TiO_2-nanotube array electrode, as well as CdS modified TiO_2 nanotube-array electrodes are shown in **Fig. 5.37**. The reflectance onset was determined by linear extrapolation from the inflection point of the curve toward the baseline. The CdS coating on nanotube-array has red-shifted the absorption edge into the visible region, with the absorption tail extending to 500 nm; the band gap calculated from this reflectance edge is about 2.53 eV. After annealing (N_2, 350°C, 1 hr) its absorption behavior has further red-shifted, with the reflectance tail extending to 515 nm, a calculated edge band gap of 2.41 eV, a typical band gap value for bulk CdS. The absorption edge corresponds to a nanoparticle size of approximately 10 nm to 20 nm [139]. With annealing the CdS particles aggregate, causing the spectrum to red-shift, a behavior previously attributed to the formation of valence-band tail states [140].

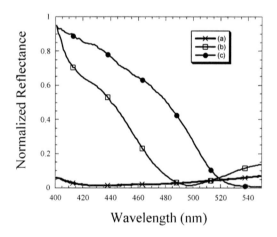

Fig. 5.37: Normalized visible reflectance spectra of CdS-TiO_2 nanotube-array electrodes. **(a)**: TiO_2 nanotube-array electrode; **(b)** 20 minute −0.5 V electro-deposited CdS modified TiO_2 nanotube-array electrode, as fabricated; **(c)** electrode of **(b)** after annealing at 350°C for 60 minutes in N_2.

5.7 Photoelectrochemical and Water Photolysis Properties

Figure 5.38 illustrates the experimental setup for water photo-electrolysis measurements with the nanotube arrays used as the photoanodes from which oxygen is evolved. The I–V characteristics of ≈ 400 nm long 'short' titania nanotube array electrodes, photocurrent density vs. potential, measured in 1M KOH electrolyte as a function of anodization bath temperature under UV (320–400 nm, 100mW/cm^2) illumination are shown in **Fig. 5.39**. The samples were fabricated using a HF electrolyte. At 1.5V the photocurrent density of the 5°C anodized sample is more than three times the value for the sample anodized at 50°C. The lower anodization temperature also increases the slope of the photocurrent—potential characteristic. On seeing the photoresponse of a 10 V 5°C anodized sample to monochromatic 337 nm 2.7 mW/cm^2 illumination, it was found that at high anodic polarization, greater than 1V, the quantum efficiency is larger than 90%.

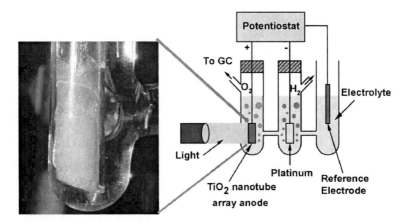

Fig. 5.38: Illustration of experimental setup for hydrogen generation by water photoelectrolysis.

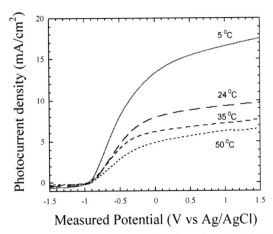

Fig. 5.39: Variation of photocurrent density (in 1M KOH solution) vs. measured potential vs. Ag/AgCl for 10V samples anodized at four anodization bath temperatures, 5, 25, 35 and 50°C. The samples were measured under 320–400nm 100mW/cm² illumination.

The photoconversion efficiency of light energy to chemical energy in presence of external applied potential was calculated using the following expression [133]:

$$\% \, \eta = [(\text{Total power output} - \text{electrical power output}) \\ / \, \text{light power}] \times 100 = j_p \, [(E^0_{\text{rev}} - |E_{\text{app}}|) / I_0] \times 100 \qquad (5.7.1)$$

$E_{\text{app}} = E_{\text{meas}} - E_{\text{aoc}}$ where the total power output is $j_p E^0_{\text{rev}}$ and the electrical power input is $j_p \, |E_{\text{app}}|$, j_p is the photocurrent density in mA/cm². E^0_{rev} is the standard reversible potential which is 1.23V NHE⁻¹. E_{meas} is the electrode potential (vs Ag/AgCl) of the working electrode at which photocurrent was measured under illumination of light and E_{aoc} is the electrode potential (vs Ag/AgCl) of the same working electrode at open circuit condition under same illumination and in the same electrolyte solution. I_0 is the intensity of incident light in mW/cm².

The titania nanotube array architecture results in a large effective surface area in close proximity with the electrolyte thus enabling diffusive transport of photogenerated holes to oxidizable species in the electrolyte. Separation of photogenerated charges is assisted by action of the depletion region electric field. Minority carriers generated within a 'retrieval' length from the material surface, that is a distance from the surface equal to the sum of the depletion layer width and the diffusion length, escape recombination and reach the electrolyte. The relevant structure sizes of the titania nanotube arrays, i.e. half the wall thickness, are all smaller than 20 nm which is less than the retrieval length of crystalline titania [141], hence bulk recombination is greatly reduced and the quantum yield enhanced. Due to light scattering within a porous structure incident photons are more effectively absorbed than on a flat electrode [142]. However while bulk recombination is reduced by the nanotube architecture, photogenerated minority carriers can be trapped by surface states.

The photoconversion efficiency as a function of potential for different 'short' photoanodes is shown in **Fig. 5.40**. A maximum conversion efficiency of 6.8% is obtained for nanotubes anodized at 5°C. For this sample, gas chromatographic analysis verified that the volume ratio of the evolved hydrogen and oxygen was 2:1, which confirmed water splitting. With the nanotube array photoanodes held at constant voltage bias, determined by the peak position in the photoconversion efficiency curve with respect to the Ag/AgCl electrode, during 1800 s of exposure, 48 mmol of hydrogen gas was generated. Normalizing this rate to time and incident power we find a hydrogen generation rate of 960 mmol/h W, or 24 mL/hW. Oxygen bubbles evolving from the nanotube array photoanode do not remain on the on the sample, hence the output remains stable with time irrespective of the duration of hydrogen production.

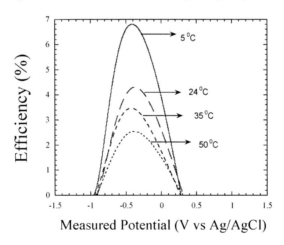

Fig. 5.40: Photoconversion efficiency under 320–400nm 100mW/cm^2 illumination as a function of measured potential [vs. Ag/AgCl] for 10V samples anodized at four temperatures; 5°C, 25°C, 35°C and 50°C.

Under UV (320–400 nm, 100 mW/cm^2) illumination a maximum photoconversion efficiency of 7.9% was obtained for 'short' nanotube arrays anodized in boric acid contained electrolyte [103], with a hydrogen generation rate of 42 mL/h W. Under full spectrum illumination (AM 1.5, 100 mW/cm^2), a photoconversion efficiency of 0.45% was obtained. The enhanced photoresponse of the boric acid anodized sample is not due solely to a modified nanotube array structure since the maximum nanotube array length achieved is about 600 nm. It is possible boron, which is difficult to identify by XPS, remains inside the titania matrix and affects its charge transfer properties.

Figure 5.41(a) shows the I-V characteristics of 6 μm nanotube arrays annealed at different temperatures under UV (320 nm to 400 nm) illumination with an intensity 100mW/cm^2 on the surface; the dark current in all cases is approximately 10^{-7} to 10^{-6} A. The photocurrent increases with increasing annealing temperature to 675°C, after which it reduces with samples annealed at 700°C showing a low photocurrent ($\sim 10^{-4}$ A). The corresponding light energy to chemical energy conversion (photoconversion) efficiencies is shown in **Fig. 5.41(b)**. The efficiency of about 12.25% was obtained for samples annealed in the range 580°C to 620°C. The increase in photocurrent and efficiency are due to the increased crystallinity of the nanotube-walls, with the reduction of the amorphous regions and grain boundaries in turn reducing the number of charge carrier recombination centers. However, at temperatures near 675°C the densification of the bottom part of the nanotubes starts isolating the undestroyed nanotubes from the metal electrode reducing the number of charge carriers reaching the electrode.

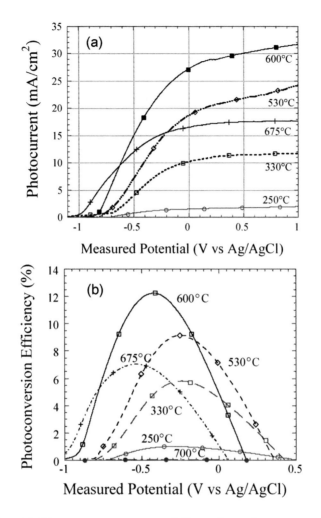

Fig. 5.41: **(a)** Photocurrent generated from 6 μm long nanotube-arrays with respect to annealing temperature, and **(b)** the corresponding photoconversion efficiencies.

The effect of nanotube-array length on the photoresponse, with all samples annealed at 530°C was also studied; both photocurrent magnitude and photoconversion efficiency are seen to increase with length [5]. On exposing 6 μm nanotube-array samples annealed at 600°C to individual wavelengths of 337 nm (3.1 mW/cm^2) and 365 nm (89 mW/cm^2), the quantum efficiency was calculated as 81% and 80% respectively. The high quantum efficiency clearly indicates that the incident light is effectively utilized by the nanotube-arrays for charge carrier generation. For a 6 μm nanotube array annealed at 600°C, the hydrogen evolution rate is about 76 mL/hr.W which, to the best of our knowledge, is higher than any reported hydrogen generation rate for any oxide material system by photoelectrolysis.

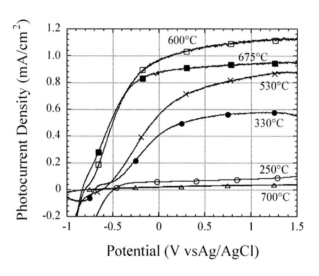

Fig. 5.42: Photocurrent generated from 6 μm long nanotube-arrays with respect to annealing temperature under AM 1.5 illumination.

Figure 5.42 shows the I-V characteristics under simulated spectrum AM 1.5 illumination of 6 μm long nanotube arrays annealed at different temperatures. The photocurrent increases with increasing annealing temperature to approximately 620°C, after which it reduces with samples annealed at 700°C showing a low photocurrent ($\sim 10^{-4}$ A). The highest visible spectrum efficiency of about 0.6% was obtained for samples annealed in the range 580°C to 620°C. For a 6 μm nanotube array annealed at 600°C, the hydrogen generation rate is 1.75 mL/W•hr.

We note the unique, highly ordered titania nanotube array structure enables the conductive electrolyte, in this work 1M KOH, to permeate the entire internal and external surfaces, hence there is a constant electrostatic potential along the length of the tubes (no RC ladder effect). Therefore long-range electron transport is dominated by diffusion rather than drift. In this case, for a nanotube array of length d, the diffusion driving force is nearly constant and approximately equal to $2k_BT/d$ [143]. The nanotube-array architecture, with a wall thickness of 20 nm, ensures that the holes are never generated far from the semiconductor-electrolyte interface. Furthermore since the wall thickness is much less than the minority carrier diffusion length $L_p \approx 100$ nm in TiO_2 [144], charge carrier separation takes place efficiently.

The increased crystallinity of the samples annealed at elevated temperatures reduces the number of grain boundaries, improves connectivity between grains and eliminates any amorphous regions that provide defects acting as carrier recombination centers. The width of the anatase crystallites in the walls is restricted by the wall thickness, approximately 20 nm. The potential drop within the wall can be represented as:

$$\Delta\phi_0 = kTr_0^2/6eL_D^2 \tag{5.7.2}$$

where r_0 is half the width of the wall, T is the temperature, and L_D is the Debye length given by

$$L_D = \left[\varepsilon_0\varepsilon kT/2e^2N_D\right]^{1/2} \tag{5.7.3}$$

where N_D is the number of ionized donors per cm^3 [145]. This potential drop across the wall thickness may not be enough to separate the photogenerated electrons and holes, however due to the nanoscale dimensions of the walls the holes can reach the surface through diffusion, which takes place on a scale of picoseconds. Minority carriers generated within a 'retrieval' length from the material surface, that is a distance from the surface equal to the sum of the depletion layer width and the diffusion length, escape recombination and reach the electrolyte. The relevant dimensional features of the titania nanotube arrays, i.e. half the wall thickness, are all smaller than 10 nm which is less than the retrieval length of crystalline titania [141], hence bulk recombination is greatly reduced and the quantum yield enhanced. Furthermore charge carriers near the electrolyte-nanotube interface region are readily accessible to the electrolyte species due to overlapping wave functions [145].

The short circuit photocurrent density (320–400 nm illumination, 100mW/cm^2) of the sample anodized in 1:1 DMSO and ethanol containing 4% HF solution, **Fig. 5.43** curve **(a)**, is more than six times the value for the sample obtained in a 1% hydrofluoric acid aqueous solution, **Fig. 5.43** curve **(b)**. Furthermore the slope of the photocurrent-potential curve is significantly enhanced in the organic electrolyte sample. The enhanced photoresponse of the sample anodized in DMSO and ethanol may be due to the distinct tube structure. Nanotube-arrays obtained in a 1% hydrofluoric acid aqueous solution are approximately 500 nm in length [24], which is much shorter than the nanotubes obtained in the organic electrolyte. A maximum photoconversion efficiency of 10.7% is obtained for nanotubes anodized in the DMSO containing organic electrolyte. Under visible spectrum AM 1.5 illumination, the photoresponse of TiO$_2$ nanotube-array photoanodes is shown in **Fig. 5.44**. The short circuit photocurrent density (i.e. 1.08mA/cm^2) of the sample anodized in organic electrolyte is found to be about four times the value for the sample prepared in aqueous solution [23].

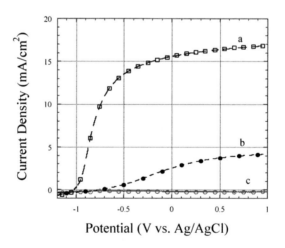

Potential (V vs. Ag/AgCl)

Fig. 5.43: Photocurrent density versus applied potential in 1 M KOH solution under UV (320 nm to 400 nm) illumination (96 mW/cm^2). Anodic samples prepared as: **(a)** Titanium foil anodized at 20 V for 70 h in DSMO and ethanol mixture solution (1:1) containing 4% HF. **(b)** H$_2$O-HF electrolyte at 20 V for 1 h. Both samples were annealed at 550°C 6 h in oxygen atmosphere prior to testing. Dark current for each sample is shown in **(c)**.

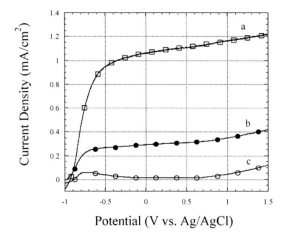

Fig. 5.44: Variation of full spectrum AM 1.5 illumination (100 mW/cm²) photocurrent density versus potential in 1 M KOH solution. Anodic samples prepared as: **(a)** Titanium foil anodized at 20 V for 70 h in DSMO and ethanol mixture solution (1:1) containing 4% HF.   **(b)** H_2O-HF electrolyte at 20 V for 1 h. Both samples were annealed at 550°C 6 h in oxygen atmosphere prior to testing. Dark current for each sample is shown in **(c)**.

Figure 5.45(a) shows the I-V characteristics under UV (320 nm to 400 nm) spectrum illumination with an intensity 100 mW/cm² of nanotube arrays fabricated using a formamide electrolyte, 0.56 grams of NH_4F in a solution mixture of 5 ml de-ionized water + 95 ml formamide, anodized at 15 V, 20 V, and 25 V anodization potentials. Prior to photocurrent measurement all samples were annealed at 525°C for 1 hr in oxygen. **Figure 5.45(b)** shows the corresponding photoconversion efficiency of the nanotube array samples. The use of different anodization voltages resulted in variation of nanotube array length and tube outer diameter (hence packing fraction): 15 V resulted in a nanotube array 8.2 μm long, 80 nm outer diameter; 20 V resulted in a nanotube array 14.4 μm long, 94 nm outer diameter; 25 V resulted in a nanotube array 16 μm long, 140 nm outer diameter. The 20 V sample, 14.4 μm long 94 nm outer diameter, achieves a photoconversion efficiency of 14.42%. There is of course an optimal length and geometric area, which we have yet to determine, where the absorption of the incident light is balanced by recombination of the photogenerated electron-hole pairs. Too short a nanotube array and the light is not fully absorbed [25,146]; too long a nanotube array and the efficiency suffers from recombination of the photogenerated electron-hole pairs.

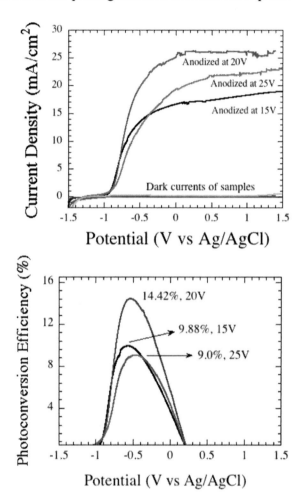

Fig. 5.45: **(a)** Photocurrent density and **(b)** corresponding photoconversion efficiency, of nanotube array samples fabricated using a formamide electrolyte, 0.56 grams of Ammonium Fluoride in a solution mixture of 5 ml de-ionized water + 95 ml formamide, at indicated anodization voltages. Prior to photocurrent measurement all samples were annealed at 525°C for 1 hour in oxygen.

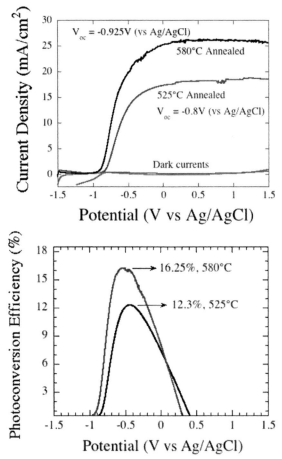

Fig. 5.46: **(a)** Photocurrent density and **(b)** corresponding photoconversion efficiency of nanotube array samples fabricated in an electrolyte of 1.2 g of NH_4F in a solution of 5 ml de-ionized water + 95 ml formamide at 35 V. The samples were annealed at indicated temperatures for 1 hour in oxygen prior to measurement. The resulting nanotube array samples were 30 μm in length, with an outer diameter of 205 nm.

The photocurrent density of nanotube array samples fabricated in an electrolyte of 1.2 g of NH_4F in a solution of 5 ml de-ionized water + 95 ml formamide at 35 V is shown in **Fig. 5.46(a)**. The resulting nanotube array samples were 30 μm in length, with an outer diameter of 205 nm. The samples were annealed at 525°C and 580°C for 1 hour in oxygen prior to measurement. The 580°C annealed sample had an open circuit voltage V_{OC} of -0.925 V (vs. Ag/AgCl); the 525°C annealed sample had an open circuit voltage

V_{OC} of -0.80 V (vs. Ag/AgCl). **Figure 5.46(b)** shows the corresponding photoconversion efficiency of the two samples.

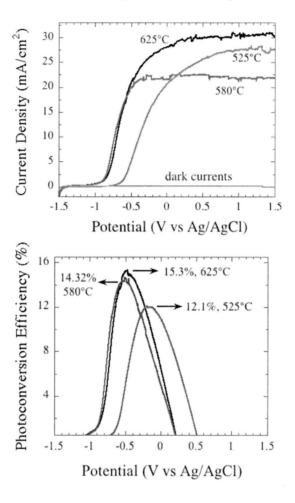

Fig. 5.47: (a) Photocurrent density and **(b)** corresponding photoconversion efficiency of nanotube array samples fabricated in an ethylene glycol electrolyte, 0.25 wt % NH_4F and 1% H_2O at 60 V for 6 hours. The samples were annealed at indicated temperatures for 1 hour in oxygen prior to measurement. The resulting nanotube array samples were approximately 24 µm in length, with an inner pore diameter of 110 nm and outer diameter of 160 nm.

The photocurrent density of nanotube array samples fabricated in an ethylene glycol electrolyte, 0.25 wt % NH_4F and 1% H_2O at 60 V for 6 hours is shown in **Fig. 5.47(a)**. The resulting nanotube array samples were approximately 24 μm in length, with an inner pore diameter of 110 nm and outer diameter of 165 nm. The samples were annealed at 625°C, 580°C, and 525°C for 1 hour in oxygen prior to measurement. **Figure 5.47(b)** shows the corresponding photoconversion efficiency of the three samples.

The roughness factor, i.e. the physical surface area of the film per unit of projected area measures the internal surface area of the electrode and is of crucial significance in light harvesting. Assuming an idealized nanotubular structure with inner diameter D, wall-thickness W and tube-length L, the purely geometric roughness factor is calculated as

$$G = [4\pi L\{D + W\} / \{\sqrt{3}\,(D+2W)^2\}] + 1 \qquad (5.7.4)$$

This calculation assumes all surfaces of the nanotubes to be perfectly smooth. In reality, the surfaces are not smooth and the actual roughness factor, as determined by dye desorption was found to be 1.8-3.0 times the geometric roughness. **Figure 5.48(a)** is a plot of the calculated geometric roughness factor for nanotubes with three different geometries. The plot shows that higher surface area is more easily obtained with nanotubes of smaller pore size. This is because even though each single nanotube of a smaller pore diameter has a smaller surface area than a nanotube of the same length with a large pore diameter, more of the smaller nanotubes may be packed in an equivalent area. **Figure 5.48(b)** is a plot of the photoconversion efficiency and maximum photocurrent density obtained with nanotube array photoelectrolytic cells as a function of the geometric surface area. A larger geometric surface area entails the availability of a larger number of active reaction sites for chemical reactions to occur and allows photogenerated holes to access a large number of solution ions. Greater nanotube length enables higher absorption of incident photons and results in a larger surface area for fixed nanotube pore size and outer diameter. When a sufficient bias is applied, nearly all photogenerated carriers are collected. The maximum photocurrent is a measure of the generation and collection of charge carriers and thus correlates well with the surface area of the nanotubes. In contrast the photoconversion efficiency also depends on the separation processes

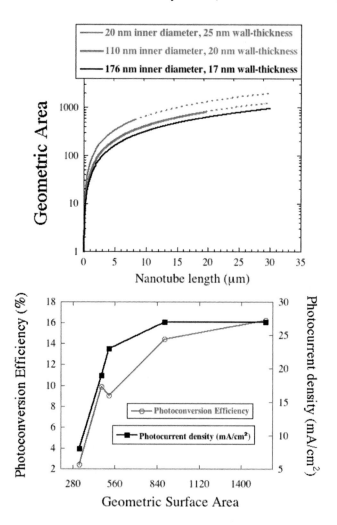

Fig. 5.48: **(a)** The calculated geometric roughness factor as a function of nanotube length. The solid, versus dashed lines, denote nanotube array lengths considered in this work. **(b)** Photoconversion efficiency during water photolysis and maximum photocurrent obtained using TiO$_2$ nanotube array photoanode as a function of geometric roughness factor of the nanotube arrays.

of the photogenerated electron-hole pair, i.e. its recombination characteristics.   At   low   values   of   applied   bias   a   higher photoconversion efficiency is obtained when the electron-hole separation process is superior, a property manifest in the slope of the potential-current plot. The recombination characteristic is better for thin, defect free, fully crystallized structures and is a complex function of the temperature and duration of crystallization, the wall-thickness of the nanotubes, barrier-layer thickness and the incorporation (if any) of dopants from the electrolyte during anodization.

Comparing **Figs. 5.45, 5.46,** and **5.47** it can be seen that the maximum photocurrent obtained with the best ethylene glycol nanotubes (~ 30 mA/cm^2 at 100 mW/cm^2 UV illumination) is greater than that obtained with nanotubes formed in formamide. However nanotubes anodically formed in formamide have a higher potential-current slope and thus exhibit the best photoconversion efficiency among the samples investigated.   Higher annealing temperatures increase the thickness of the barrier layer and the rutile content in the barrier layer, properties that interfere with charge transfer to the back contact, but also improve the crystallinity of the nanotube walls. When the nanotube length exceeds ten micrometers the great majority of the charge carriers are photogenerated in the nanotube walls [146] hence the photoelectrochemical properties of the nanotubes are significantly improved by the increased crystallinity of the nanotube walls. In the temperature regime 500-620°C, a trend is observed whereby higher photoconversion efficiencies are obtained for nanotubes annealed at higher temperatures.   However, the temperature of annealing cannot be increased indefinitely since oxide growth below the nanotube layer during a high temperature anneals eventually distorts and finally destroys the nanotube array layer [126].

The *I-V* characteristics of CdS-sensitized TiO2 nanotube electrodes are presented in **Fig. 5.49** for an illumination intensity of 1 sun (AM 1.5, 100 mW/cm^2, light of wavelength below 400 nm was filtered). The measurements were done in 1.0 M Na$_2$S electrolyte solution, an efficient hole scavenger for CdS in which the electrodes are stable. For the as-prepared CdS-TiO2 electrode, photocurrent onset occurs at -1.30V vs. Ag/AgCl, a -0.60 V negative shift compared to the plain TiO2 nanotube-array electrode. In comparison to the plain TiO$_2$ nanotube-array electrode, addition of the CdS film increased the photocurrent from 0.16 mA/cm^2 to 0.55 mA/cm^2. Electrochemically synthesized CdS thin film comprised of nanoparticles can be used to sensitize the TiO2 nanotube-array making it more responsive to the visible spectrum, with an obvious application to solar cells [124,139]. The photocurrent response is sensitive to the annealing temperature, as discernable in **Fig. 5.49**. The sample annealed at 350°C reaches ≈ 1.42 mA/cm^2, and the 400°C annealed sample reaches ≈ 2.51 mA/cm^2, respectively 9 and 16 times higher than that of bare TiO$_2$ nanotube array electrode. The *I-V* curves of these samples gradually lose their well-defined photocurrent saturation as the annealing temperature increases. Higher temperature annealing may result in pore blockage due to sintering of the CdS nanoparticles, reducing the area accessible to the hole scavenging electrolyte solution.

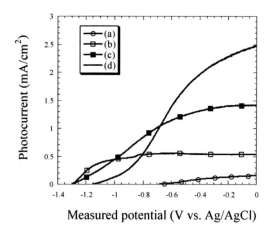

Fig. 5.49: Photocurrent versus voltage in 1 M Na$_2$S under AM 1.5 (1 sun), 100 mW/cm^2 illumination: **(a)**: bare TiO$_2$ nanotube electrode. **(b)** As-prepared electrodeposited CdS film (-0.5 V, 30 min.) upon TiO$_2$ nanotube-array electrode. **(c)** CdS (-0.5 V, 30 min.)-TiO$_2$ electrode after annealing at 350°C in N$_2$ for 60 minutes. **(d)** CdS (-0.5 V, 30 min.)-TiO$_2$ electrode after annealing at 400°C in N$_2$ for 60 minutes.

5.8 Ti-Fe-O Nanotube Array Films for Solar Spectrum Water Photoelectrolysis

As we have previously discussed, for sustained water splitting to occur several semiconductor properties such as bandgap, flat band potential, and corrosion stability must be simultaneously optimized

[147,148]. Hematite, α-Fe$_2$O$_3$, is a n-type semiconductor with a ~2.2 eV bandgap that is sufficiently large enough for water splitting but small enough to collect a significant fraction, $\approx 40\%$, of solar spectrum energy. However iron oxide suffers from low electron mobility, generally in the range of 0.01 cm^2/V•s [149] to 0.1 cm^2/V•s [150] resulting in rapid electron-hole recombination, and an insufficiently negative flat band potential as needed for spontaneous water splitting [151]. Therefore α-Fe$_2$O$_3$ solar spectrum photoconversion efficiencies reported to date are far from the theoretical maximum value of 12.9% for a material of this band-gap [152].

As seen previously n-type TiO$_2$ exhibits excellent charge transfer properties and photochemical stability, furthermore its bandgap includes the redox potential for the H$_2$O/•OH reaction allowing photogeneration of oxygen through water splitting [153]. However the bandgap of TiO$_2$, ~ 3.0 eV for rutile and 3.2 eV for anatase, limits its activation to UV radiation which accounts for only $\approx 5\%$ of solar spectrum energy. Efforts to shift the bandgap of TiO$_2$ while maintaining its excellent charge transfer properties and photocorrosion stability have primarily focused on metal doping with the resulting materials demonstrating, at best, mixed photocatalytic activities.

In addition to the bandgap the material architecture of a semiconductor photoanode plays a critical role in determining the resultant photoconversion efficiencies. Recombination in low mobility materials can be minimized by decreasing the relevant architectural feature size(s) to less than the minority carrier diffusion length; for α-Fe$_2$O$_3$ the hole diffusion length is 2-4 nm [154]. This is of particular interest considering the ability to fabricate vertically oriented, highly ordered TiO$_2$ nanotube arrays having wall thicknesses of ≈ 4 nm [26]. The precisely oriented nature of the crystalline (after annealing) nanotube arrays makes them excellent electron percolation pathways for vectorial charge transfer between interfaces. Where the use of vertically oriented titania nanotube arrays have been studied an extraordinary enhancement of the extant TiO$_2$ properties has been found [155-161]. Such a material architecture in α-Fe$_2$O$_3$ offers the intriguing prospect of efficient charge transfer in a material well suited for capturing solar spectrum energy. Ideally we would like to combine the charge transport and photocorrosion properties of the TiO_2 nanotubes with the bandgap of α-Fe_2O_3. Since both Ti and Fe can be anodized in a fluoride ion containing ethylene glycol solution, the possibility exists for obtaining highly-ordered, vertically oriented thin-wall Ti-Fe oxide nanotube arrays, the material architecture we desire for application to water photoelectrolysis, by anodization of TiFe metal films.

Experimental

Ti-Fe metal films were deposited on fluorine doped tin oxide (FTO) coated glass substrates by simultaneous co-sputtering from titanium and iron targets. The substrate temperature was maintained at 350°C – 400°C. Details of sputtering conditions, the resulting thickness of the Ti-Fe metal films and their crystallinity are given in **Table 5.6**. The metal films of high titanium or high iron concentration are crystalline, with the former a hexagonal titanium lattice and the latter a cubic iron lattice. Films with moderate levels of iron were amorphous. Irrespective of the crystallinity of the starting film the anodized films were amorphous with a very thin residual metal layer underneath. Distinct nanotube arrays were formed from samples having iron concentrations less than about 70%. Films of higher Fe content, up to 100% Fe, were fabricated and characterized, however they showed minimal photo-electrochemical properties. We consider Ti:Fe metal films in the ratio 31:69, 56:44, 63:37, 74:26, 80:20, 93.4:6.6, and 96.5:3.5 as determined from the relative sputtering rates, with identifying sample numbers given by the Fe content.

Table 5.6: Assigned sample name, iron and titanium target powers and corresponding sputter rates, substrate temperature, thickness of resulting Ti-Fe film, Ti-Fe film composition.

Sample	Fe target power (W) / sputter rate (nm/s)	Ti target power (W) / sputter rate (nm/s)	Substrate Temp. °C	Film Thickness (nm)	Film Composition	Crystallinity of Metal Starting Film
69	300 / 5.5	100 / 2.5	350	400	69 % Fe – 31 % Ti	Crystalline
44	300 / 5.5	300 / 7.0	350	1000	44 % Fe - 56 % Ti	Crystalline
37	250 / 4.1	300 / 7.0	350	500	37 % Fe – 63 % Ti	Amorphous
26	200 / 2.5	300 / 7.0	350	400	26 % Fe – 74 % Ti	Amorphous
20	150 / 1.3	300 / 7.0	400	750	20 % Fe – 80 % Ti	Amorphous
6.6a	100 / 0.5	300 / 7.0	400	750	6.6 % Fe – 93.4 % Ti	Crystalline
6.6b	100 / 0.5	300 / 7.0	400	1500	6.6 % Fe – 93.4 % Ti	Crystalline
3.5	50 / 0.25	300 / 7.0	400	750	3.5 % Fe - 96.5 % Ti	Crystalline

Anodization of the Ti-Fe films was performed at a constant voltage of 30 V in ethylene glycol containing 0.3 wt % NH_4F and 2.0 vol % de-ionized water. The amorphous as-anodized samples were crystallized by oxygen annealing at 500°C for two hours. The photoelectrochemical properties were investigated using a three-electrode configuration with a Ti-Fe-O photoanode as the working electrode, saturated Ag/AgCl as a reference, and platinum foil as a counter electrode. A 1.0 M NaOH solution was used as the electrolyte. Sun light was simulated with a 300W xenon lamp (Spectra physics) and AM 1.5 filter (oriel). The light intensity was set using a NREL calibrated crystalline silicon solar cell, equivalent to AM 1.5 light at 100 mW/cm^2. Incident photon to current conversion efficiency (IPCE) measurements were performed with the spectral irradiance of the light from the 300W xenon lamp, integrated with a parabolic reflector, passing through an AM 1.5 filter and monochromator determined using an Oriel calibrated silicon photodiode.

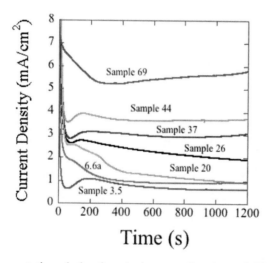

Fig. 5.50: Current-time behavior during anodization of Ti-Fe films, see **Table 5.6**, at 30 V in an ethylene glycol electrolyte containing 0.3 wt % NH_4F and 2.0 vol % de-ionized water.

Figure 5.50 compares the real time constant voltage anodization behavior of the Ti-Fe films, with the samples identified by their iron content, anodized at 30 V in EG + 0.3 wt % NH_4F + 2.0 % de-ionized water; a systematic variation in anodization behavior is seen with decreasing Fe content. The sharp drop in the anodization current in the first 100 s is due to the formation of an initial electrically insulating oxide layer, followed by an increase in the current due to oxide pitting by the fluoride ions. The current then gradually decreases to plateau at a steady state value corresponding to maximum nanotube length. Illustrative top, cross-sectional and bottom-side FESEM images of Samples 44, 20 and 3.5 are shown in **Fig. 5.51**.

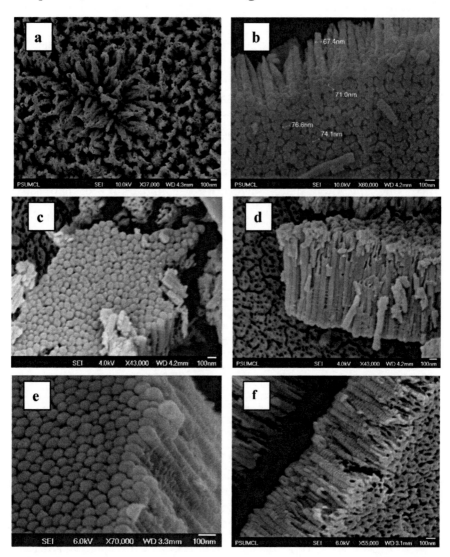

Fig. 5.51: Illustrative FESEM images of Ti-Fe-O nanotube array samples. Sample 44, top surface and lateral view are seen in **(a)** and **(b)**. Sample 20, image of tube bottom **(c)** and lateral view **(d)**. Sample 3.5, image of tube bottom **(e)** and lateral view **(f)**.

TiO$_2$ nanotube array formation by anodization in aqueous electrolytes is purely a subtractive process; one starts with a titanium film of thickness X and achieves a nanotube array of thickness less than X. In contrast, using an ethylene glycol electrolyte one can begin with a Ti film of thickness X and achieve a nanotube array of thickness greater than X, with material removed from the pores going into wall formation. Nanotube arrays obtained from constant voltage anodization of Ti foil in ethylene glycol containing 0.3 wt % NH$_4$F and 2.0 vol % de-ionized water, have shown lengths up to 3.5x greater than the starting Ti foil thickness. A similar effect is seen for anodization of the Ti-Fe films of higher Ti content. The nanotube lengths obtained from Sample 44 films are close to the starting film thickness, whereas for Sample 3.5 the resulting nanotubes are nearly twice the length of the starting metal film thickness. We note the addition of water to the ethylene glycol electrolyte as a critical ingredient. While the exact mechanism by which water contributes oxygen to an anodic oxide film is not well understood, evidence suggests hydroxyl ion injection from the electrolyte into the anodic oxide film [152,153].

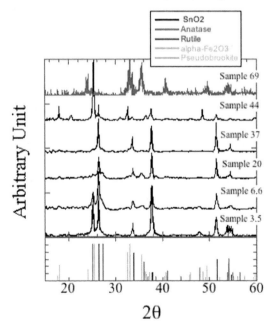

Fig. 5.52: GAXRD patterns of Ti-Fe-O nanotube array samples annealed at 500°C in dry oxygen for 2 hr, see **Table 5.6** for fabrication details. Standard patterns for anatase, rutile, pseudobrookite, α-Fe$_2$O$_3$ (hematite), tin oxide (substrate) are shown for phase identification.

Film Structure

As-anodized Ti-Fe-O nanotube films are amorphous, partially crystallized by annealing in oxygen atmosphere at 500°C for two hours with a ramp up and down rate of 1°C /min. GAXRD patterns of the annealed films are seen in **Fig. 5.52**. All studies on TiO_2 nanotube array films made by anodization have indicated the tubes to be anatase, fixed atop a rutile base [154]. The Ti-Fe-O samples with low iron concentrations show the presence of anatase phase; Sample 3.5 is largely anatase with a dominant (110) plane, and a small amount of rutile. The anatase phase disappears at higher iron concentrations (see the patterns of Samples 20 and 37). The presence of the rutile peak is evident in Samples 6.6 and 20, which is consistent with the fact that low amounts of iron in titania aid rutile formation [155,156]. The predominant (104) plane of α-Fe_2O_3 coincides with the second intense plane of FTO, used as the conductive layer on the glass substrate, and its subsequent two less intense planes, (110) and (116), appear at $2\theta = 35.6$ and $2\theta = 54.1$ respectively. The XRD pattern of Sample 26 is omitted from **Fig. 5.52** because of its relatively weak peaks. In Sample 37 the proportion of rutile increases slightly, and anatase is almost negligible. In general, Samples 20 and 37 showed poor crystallinity, Sample 44 showed pseudobrookite phase, while Sample 69 showed hematite phase. In general, it appears that the presence of iron inside the TiO_2 matrix degrades the crystallization process since the proportion of rutile does not increase with decreasing amounts of anatase, an outcome possibly due to the fixed 500°C annealing temperature. However higher annealing temperatures spoil the FTO - TiFeO interface, resulting in samples of high series resistance demonstrating essentially nil photoelectrochemical properties.

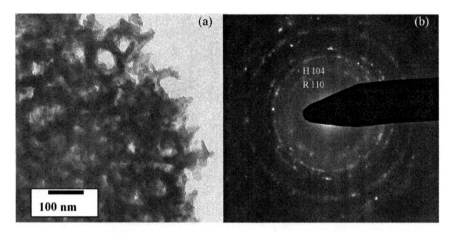

Fig. 5.53: **(a)** TEM image of Sample 20 annealed at 600°C, and **(b)** selected area diffraction pattern (H –hematite and R-rutile).

The crystalline nature of the nanotube walls is critical to applications involving light absorption, electrical carrier generation, and carrier transport therefore the crystalline nature of the nanotube walls were studied using HRTEM. Sample 20, with moderate iron doping was used for the study. TEM studies confirmed the relatively poor crystallinity of samples annealed at 500°C, hence further studies were conducted on a Sample 20 annealed at 600°C for 2 hrs in oxygen. As seen in **Fig. 5.51**, and as described in our earlier articles on TiO_2 nanotube arrays formed in HF aqueous electrolytes, the nanotube morphology appears as a column of stacked torus (doughnut) shaped rings as seen in the TEM image of **Fig. 5.53(a)**. A selected area electron diffraction (SAED) pattern from the corresponding region is shown in **Fig. 5.53(b)**. Although the pattern shows mainly rutile phase, reflections from hematite phase also can be seen. **Figure 5.54(a)** shows a HRTEM image of a nanotube wall, with the polycrystalline nature evident from the figure. The crystallites have a wide size distribution. The crystallite marked R has a rutile structure the lattice image of which is given in the upper left inset. A Fast Fourier Transform (FFT) of regions R and H are given in **Fig. 5.54(b)** and **5.54(c)** indicating the presence of rutile and hematite phases in the respective regions.

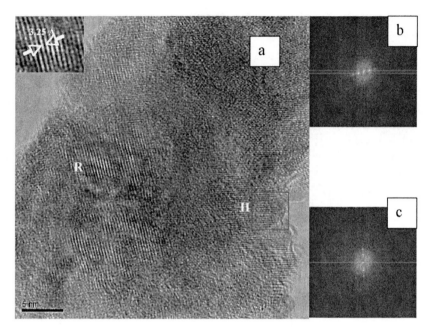

Fig. 5.54: **(a)** HRTEM image of a nanotube wall (Sample 20 annealed at 600°C) showing rutile (region R) and hematite (region H) crystallites with the lattice image of region R given in the upper-left inset; **(b)** FFT of region R showing a predominate 110 plane, and **(c)** FFT of region H showing a predominate 104 plane.

To find the distribution of iron within the nanotube walls an energy dispersive x-ray spectroscopy (EDS) line scan was performed via scanning transmission electron microscopy (STEM), see **Fig. 5.55**. The intensity of both the TiK_α and FeK_α lines are maximum at the center of the wall due to its torus shape. Despite the presence of isolated hematite crystallites, a more or less uniform distribution of iron relative to the titanium can be seen across the wall. STEM line scans were performed across a number of walls, and while the average relative intensity of the TiK_α and FeK_α lines varied from wall to wall the relative distribution across a single wall remained uniform. It appears that some of the iron goes into the titanium lattice substituting titanium ions, and the rest either forms hematite crystallites or remains in the amorphous state.

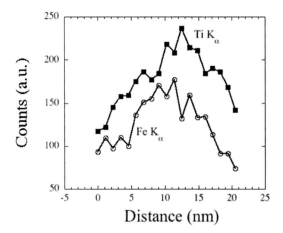

Fig. 5.55: EDS line scan profile across nanotube wall [from region similar to that shown in **Fig. 5.54(a)**].

Crystallization of the as-anodized amorphous nanotubes takes place through nucleation and growth [154]. Hence, a sufficiently high temperature and appropriate duration are needed for the complete transformation of the amorphous phase into crystalline. The removal of the amorphous phase is critical for effective device application, however even after annealing at 600°C Sample 20 showed the presence of amorphous phase. HRTEM studies indicate that rutile and hematite phases are formed in separate events depending upon the temperature and nature of the nucleation sites. In general, the nanotubes of moderate iron concentration consist mainly of rutile phase with hematite crystallites randomly distributed.

XPS survey spectra (not shown) reveal the Ti – Fe oxide films to contain Ti, O, Fe and C. The photoelectron peaks for Ti $2p$ appear clearly at a binding energy (E_b) of 458 eV, O $1s$ at E_b = 530 eV, and Fe $2p$ at E_b = 711 eV. The XPS peak for C $1s$ at E_b = 285eV was observed due to adventitious carbon from sample fabrication and/or the XPS instruments itself. **Figure 5.56(a)** is high-resolution Ti $2p$ region spectra taken on the surface of Samples 44, 26 and 3.5 showing the presence of the main doublet composed of two symmetrical peaks at E_b(Ti $2p_{3/2}$) = 458.8 eV and E_b(Ti $2p_{1/2}$) = 464.5 eV, assigned to Ti^{4+} in the spectrum of Ti – Fe oxide film [156]. A high-resolution spectrum of the Fe $2p$ region for Samples 44, 26, and 3.5 are shown in **Fig. 5.56(b)**. The spectrum indicates the existence of doublet Fe $2p_{3/2}$ and Fe $2p_{1/2}$ and their corresponding satellites that are characteristic of Fe_2O_3 [157]. The intensity of these satellite features diminishes significantly in Sample 3.5. No component related to zero valent Fe and Ti can be extracted. It is not appropriate to determine the Ti/Fe atomic ratio using XPS since the surface morphology of the film is nanotubular, or porous, and iron-based compounds are difficult to accurately predict using XPS as sophisticated curve fitting is required with extensive analysis.

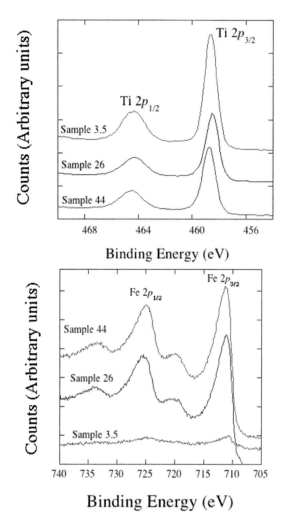

Fig. 5.56: High resolution XPS spectra from the surface of annealed Sample 44, Sample 26, and Sample 3.5: **(a)** Ti $2p$ and **(b)** Fe $2p$.

Figure 5.57 is a photograph showing the color variation in the samples as a function of iron content. **Figure 5.58** shows the absorbance of the annealed samples; as expected the absorbance edge shows a red shift with increasing Fe content. According to the bandgap structure of TiO_2 the π bonding orbit and π^* antibonding orbit formed by the $t2g$ orbit in Ti^{4+} ions, and the $p\pi$ orbit in the O^{2-} of TiO_2 crystal lattice are, respectively, the valence band and conduction band. A new energy band is formed in the TiO_2 bandgap when Fe^{3+} ions are partly substituted for the Ti^{4+} ions. The mixing of the $Ti(d)$ orbit of Ti oxide and Fe $3d$ orbit is essential to decreasing the energy gap between $Ti(d)$ and $O(p)$ orbitals of Ti oxide.

Fig. 5.57: From left, a 1.5 μm α-Fe_2O_3 film, Sample 44, Sample 20, Sample 6.6a, Sample 3.5, see **Table 5.6**.

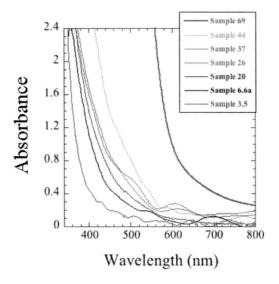

Fig. 5.58: Absorbance spectra of 500°C annealed Ti-Fe-O nanotube array films, see **Table 5.6**.

Photoelectrochemical Properties

Figure 5.59 shows current versus voltage characteristics for the nanotubular Ti-Fe-O electrodes under dark and simulated sunlight in 1.0 M NaOH. The electrodes show *n*-type behavior, i.e. positive photocurrents at anodic potentials. The dark current in each case is negligible up to 0.65 V (vs Ag/AgCl) beyond which the dark currents for water oxidation dominate, therefore no photocurrent saturation is observed. A metal-coated glass substrate was placed adjacent to translucent Sample 6.6b, reflecting the transmitted light back into the sample. This method increased the photocurrent from 1.50 mA/cm^2 to 2.0 mA/cm^2 at 0.65 V (vs Ag/AgCl). For comparison the photocurrent of a pure nanoporous α-Fe$_2$O$_3$ film 1.5 μm thick on FTO glass is also shown in **Fig. 5.59**, prepared by anodic oxidation of an iron film at 30V in ethylene glycol containing 0.3 wt% NH$_4$F and 0.5% de-ionized water. Even though this film strongly absorbs visible light the poor electron mobility results in a maximum photocurrent of 20-25 μA/cm^2 at 0.4V (vs Ag/AgCl). The highest photocurrent is exhibited by Sample 6.6, while the lowest photocurrent is demonstrated by Sample 3.5. It appears that minimal levels of Fe^{3+} ions act as trap sites, while significant amounts of Fe^{3+} ions increase the recombination rate of the photogenerated electron-hole pairs. We note Sample 44 does not comply with this trend, possibly due to the presence of the pseudobrookite phase, FeTiO$_5$.

    Two electrode measurements were performed to determine the photoconversion, light energy to chemical energy conversion, efficiency. Photoconversion efficiency η is calculated as:

$$\eta(\%) \;\; = j_p \, [(1.23 - V) / I_0] \times 100 \qquad (5.8.1)$$

where j_p is the photocurrent density (mA/cm^2), I_0 is the intensity of incident light (mW/cm^2) and V is the potential applied between the anode (Ti-Fe-O sample) and cathode (platinum). The efficiency of mirror-backed Sample 6.6b was determined to be 1.2%.

    Hydrogen generation experiments were done at constant voltage bias (1 M KOH solution), at an applied bias of 0.7 V. Under AM 1.5 100 mW/cm^2 illumination, Sample 6.6b demonstrated a sustained, time-energy normalized hydrogen evolution rate by water splitting of 7.1 mL/W•hr.   The water splitting reaction was confirmed by the 2:1 ratio of evolved hydrogen to oxygen as confirmed by a gas chromatograph (SRI, model 8610C).   No degradation in sample performance was observed under illumination over a course of several days.

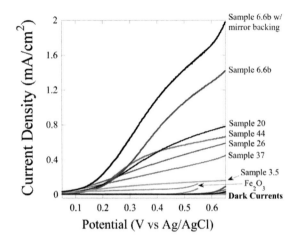

Fig. 5.59 Photocurrent density versus potential in 1 M NaOH solution for annealed Ti-Fe-O nanotube array samples, and α-Fe₂O₃ nanoporous film, under AM 1.5 (100 mW/cm²) illumination.

Steady-state wavelength-specific photocurrents were measured for the Ti-Fe-O films in a two-electrode arrangement at different applied voltages. Incident photon-to-current efficiencies (*IPCE*) are calculated using the following equation:

$$IPCE = \frac{(1240 \ eV \bullet nm)(photocurrent \ density \ \mu A / cm^2)}{(\lambda \ nm)(irradiance \ \mu W / cm^2)} \quad (5.8.2)$$

Dividing the *IPCE* by the fraction of incident photons absorbed at each wavelength gives the absorbed photon-to-current efficiency (*APCE*) [158].   Various factors including nanotube length and Fe composition factor into *APCE* calculation through the absorbance *A*, as shown below:

$$APCE = \frac{IPCE}{1-10^{-A}} \qquad (5.8.3)$$

Figure 5.60(a,b) shows APCE values corresponding to two applied voltages, 0.5 V and 0.7 V; APCE values are appreciable only in the visible range, from 400 nm to 600 nm.

Energy level band diagrams indicate electrons photogenerated in α-Fe_2O_3 cannot transfer to the titania conduction band without an applied bias. Since the redox potential for the $H_2O/\bullet OH$ reaction for oxygen evolution is above the valence band of α-Fe_2O_3 in 1M NaOH electrolyte (pH = 14), and the thickness of the α-Fe_2O_3 crystallites is comparable to the hole diffusion length, application of a small positive bias can be used to separate the photogenerated electrons and holes. As seen in **Fig. 5.60**, *APCE* values of the films decrease as we move below 400 nm towards the UV region, possibly due to the Fe^{3+} ions of α-Fe_2O_3 and iron incorporated TiO_2 crystallites becoming Fe^{2+} as associated with oxygen vacancy defects. Because of this the position of the valence band at solid-electrolyte shifts upward, crossing the redox potential of the $H_2O/\bullet OH$ reaction inhibiting hole transfer and leading to increased recombination.

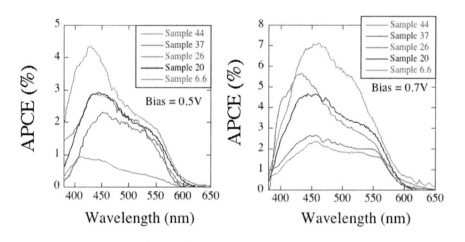

Fig. 5.60: Absorbed-photon-to-current-efficiency (*APCE*) of Ti-Fe-O nanotube array samples at: **(a)** 0.5 V bias, and **(b)** 0.7 V bias.

The photoconversion efficiency of TiO_2 nanotube arrays under UV illumination are notable, 16.5% under 320 nm – 400 nm band illumination (100 mW/cm^2). To facilitate the capture of solar spectrum energy motivates fabrication of Ti-Fe-O nanotube arrays. A photocurrent of 2 mA/cm^2 under AM 1.5 is achieved, but perhaps more importantly if the nanotube wall thickness could be reduced to less than the minority carrier diffusion length, $\approx 2\text{-}4$ nm, dramatically improved photocurrents could be expected. However while the wall thicknesses of TiO_2 nanotubes formed in aqueous based electrolytes containing fluoride ions can be controlled through the anodization bath temperature [169, 170], no such temperature dependent effect has been found. Over the various sample compositions the wall thickness of the Ti-Fe-O nanotubes held constant at ≈ 22 nm independently of anodization bath temperature (0°C to 50°C). Future efforts will focus on adjustment of the electrolyte bath chemistry to see if thin-wall Ti-Fe-O nanotubes can be achieved, that is to say having a wall thickness comparable to the hole diffusion length in $\alpha\text{-}Fe_2O_3$. The Ti-Fe-O material architecture demonstrates properties useful for hydrogen generation by water photoelectrolysis and, more importantly, it appears the general nanotube-array synthesis technique can be extended to other binary or ternary oxide compositions of interest for water photoelectrolysis.

References

1. Adachi M, Murata Y, Harada M, Yoshikawa S (2000) Formation of titania nanotubes with high photocatalytic activity. Chem Lett 29:942-943
2. Chu SZ, Inoue S, Wada K, Li D, Haneda H, Awatsu S (2003) Highly porous $(TiO_2\text{-}SiO_2\text{-}TeO_2)/Al_2O_3/TiO_2)$ composite nanostructures on glass with enhanced photocatalysis fabricated by anodization and sol-gel process. J Phys Chem B 107: 6586-6589
3. Mor GK, Shankar K, Paulose M, Varghese OK, Grimes CA (2005) Enhanced photocleavage of water using titania nanotube arrays. Nano Lett 5:191-195
4. de Taconni NR, Chenthamarakshan CR, Yogeeswaran G, Watcharenwong A, de Zoysa RS, Basit NA, Rajeshwar K

(2006) Nanoporous TiO_2 and WO_3 films by anodization of titanium and tungsten substrates: Influence of process variables on morphology and photoelectrochemical response. J Phys Chem B 110: 25347-25355

5.   Varghese OK, Paulose M, Shankar K, Mor GK, Grimes CA (2005) Water-photolysis properties of micron-length highly-ordered titania nanotube-arrays. J Nanosci Nanotechnol 5:1158-1165

6.   Uchida S, Chiba R, Tomiha M, Masaki N, Shirai M (2002) Application of titania nanotubes to a dye-sensitized solar cell. Electrochem 70:418-420

7.   Adachi M, Murata Y, Okada I,Yoshikawa Y (2003) Formation of titania nanotubes and applications for dye-sensitized solar cells. J Electrochem Soc 150:G488 –G493

8.   Mor GK, Shankar K, Paulose M, Varghese OK, Grimes CA (2006) Use of highly-ordered TiO_2 nanotube arrays in dye-sensitized solar cells. Nano Lett 6:215-218

9.   Paulose M, Shankar K, Varghese OK, Mor GK, Hardin B, Grimes CA (2006) Backside illuminated dye-sensitized solar cells based on titania nanotube array electrodes. Nanotechnol 17:1446-1448

10.   Hoyer P (1996) Formation of a titanium dioxide nanotube array. Langmuir 12:1411-1413

11.   Lakshmi BB, Dorhout PK, Martin CR (1997) Sol-gel template synthesis of semiconductor nanostructures. Chem Mater 9:857-862

12.   Imai H, Takei Y, Shimizu K, Matsuda M, Hirashima H (1999) Direct preparation of anatase TiO_2 nanotubes in porous alumina membranes. J Mater Chem 9:2971-2972

13.   Michailowski A, Al Mawlawi D, Cheng GS, Moskovits M (2001) Highly regular anatase nanotubule arrays fabricated in porous anodic templates. Chem Phys Lett 349:1-5

14.   Jung JH, Kobayashi H, van Bommel KJC, Shinkai S, Shimizu T (2002) Creation of novel helical ribbon and double-layered nanotube TiO_2 structures using an organogel template. Chem Mater 14:1445-1447

15.   Kobayashi S, Hamasaki N, Suzuki M, Kimura M, Shirai H, Hanabusa K (2002) Preparation of helical transition-metal

oxide tubes using organogelators as structure-directing agents. J Am Chem Soc 124:6550-6551

16. Tian ZR, Voigt JA, Liu J, McKenzie B, Xu H (2003) Large oriented arrays and continuous films of TiO$_2$-based nanotubes. J Am Chem Soc 125:12384-12385

17. Kasuga T, Hiramatsu M, Hoson A, Sekino T, Niihara K (1998) Formation of titanium oxide nanotube. Langmuir 14:3160-3163

18. Chen Q, Zhou WZ, Du GH, Peng LM (2002) Trititanate nanotubes made via a single alkali treatment. Adv Mater 14:1208-1211

19. Yao BD, Chan YF, Zhang XY, Zhang WF, Yang ZY, Wang N (2003) Formation mechanism of TiO$_2$ nanotubes. Appl Phys Lett 82:281-283

20. Gong D, Grimes CA, Varghese OK, Hu W, Singh RS, Chen Z, Dickey EC (2001) Titanium oxide nanotubes array prepared by anodic oxidation. J Mater Res 16:3331–3334

21. Mor GK, Varghese OK, Paulose M, Mukherjee N, Grimes CA (2003) Fabrication of tapered, conical-shaped titania nanotubes. J Mater Res 18:2588-2593

22. Cai Q, Paulose M, Varghese OK, Grimes CA (2005) The effect of electrolyte composition on the fabrication of self-organized titanium oxide nanotube arrays by anodic oxidation. J Mater Res 20:230-236

23. Ruan CM, Paulose M, Varghese OK, Mor GK, Grimes CA (2005) Fabrication of highly ordered TiO$_2$ nanotube arrays using an organic electrolyte. J Phys Chem B 109:15754-15759

24. Paulose M, Shankar K, Yoriya S, Prakasam HE, Varghese OK, Mor GK, Latempa TJ, Fitzgerald A, Grimes CA (2006) Anodic growth of highly ordered TiO$_2$ nanotube arrays to 134 μm in Length. J Phys Chem B 110:16179-16184

25. Yoriya S, Prakasam HE, Varghese OK, Shankar K, Paulose M, Mor GK, Latempa TJ, Grimes CA (2006) Initial studies on the hydrogen gas sensing properties of highly-ordered high aspect ratio TiO$_2$ nanotube-arrays 20 to 222 μm in length. Sens Lett 4:334-339

26.   Shankar K, Mor GK, Fitzgerald A, Grimes CA (2007) Cation effect on the electrochemical formation of very high aspect ratio TiO_2 nanotube arrays in formamide-water mixtures. J Phys Chem C 111:21-26

27.   Prakasam HE, Shankar K, Paulose M, Grimes CA (2007) A new benchmark for TiO_2 nanotube array growth by anodization. J Phys Chem B (in press)

28.   Serpone N, Pelizzetti E (1989) Photocatalysis: Fundamentals and Applications, Wiley, New York

29.   Schiavello M, Dordrecht H (1985) Photoelectrochemistry, Photocatalysis, and Photoreactors: Fundamentals and Developments Kluwer Academic, Boston, MA

30.   Linsebigler AL, Lu G, Yates JT (1995) Photocatalysis on TiO_2 surfaces: principles, mechanisms, and selected results. Chem Rev 95:735-758

31.   Fujishima A, Honda K (1972) Electrochemical photolysis of water at a semiconductor electrode. Nature 238:37-38

32.   Yamashita H, Harada M, Misaka J, Takeuchi M, Neppolian B, Anpo M (2003) Photocatalytic degradation of organic compounds diluted in water using visible light-responsive metal ion-implanted TiO_2 catalysts: Fe ion-implanted TiO_2. Catal Today 84:191-196

33.   Wang C, Bahnemann DW, Dohrmann JK (2000) A novel preparation of iron-doped TiO_2 nanoparticles with enhanced photocatalytic activity. Chem Commun 16:1539-1540

34.   Wang Y, Hao Y, Cheng H, Ma H, Xu B, Li W, Cai S (1999) The photoelectrochemistry of transition metal-ion-doped TiO_2 nanocrystalline electrodes and higher solar cell conversion efficiency based on Zn^{2+}-doped TiO_2 electrode. J Mater Sci 34:2773-2779

35.   Coloma F, Marquez F, Rochester CH, Anderson JA (2000) Determination of the nature and reactivity of copper sites in $Cu–TiO_2$ catalysts. Phys Chem Chem Phys 2:5320-5327

36.   Umebayashi T, Yamaki T, Itoh H, Asai K (2002) Analysis of electronic structures of 3d transition metal-doped TiO_2 based on band calculations. J Phys Chem Solids 63:1909-1920

37.   Yamashita H, Ichihashi Y, Takeuchi M, Kishiguchi S, Anpo M (1999) Characterization of metal ion-implanted titanium

oxide photocatalysts operating under visible light irradiation. J Synchrotron Radiat 6:451-452

38. Karakitsou KE, Verykios XE (1993) Effects of altervalent cation doping of TiO_2 on its performance as a photocatalyst for water. J Phys Chem 97:1184-1189

39. Choi W, Termin A, Hoffmann MR (1994) The role of metal ion dopants in quantum-sized TiO_2: Correlation between photoreactivity and charge carrier recombination dynamics. J Phys Chem 98:13669-13679

40. Lee DH, Cho YS, Yi WI, Kim TS, Lee JK, Jung HJ (1995) Metalorganic chemical vapor deposition of TiO_2:N anatase thin film on Si substrate. Appl Phys Lett 66:815-816

41. Saha NC, Tompkins HG (1992) Titanium nitride oxidation chemistry: an X-rayphotoelectron spectroscopy study. J Appl Phys 72:3072-3079

42. Asahi R, Morikawa T, Ohwaki T, Aoki K, Taga Y (2001) Visible-light photocatalysis in nitrogen-doped titanium oxides. Science 293:269-271

43. Morikawa T, Asahi R, Ohwaki T, Aoki K, Taga Y (2001) Band-gap narrowing of titanium dioxide by nitrogen doping. Jpn J Appl Phys 40:L561-L563

44. Irie H, Watanabe Y, Hashimoto K (2003) Nitrogen-concentration dependence on photocatalytic activity of $TiO_{2-x}N_x$ powders. J Phys Chem B 107:5483-5486

45. Subbarao SN, Yun YH, Kershaw R, Dwight K, Wold A (1979) Electrical and optical-properties of the system $TiO_{2-x}F_x$. Inorg Chem 18:488-492

46. Hattori A, Yamamoto M, Tada H, Ito S (1998) A promoting effect of NH_4F addition on the photocatalytic activity of sol-gel TiO_2 films. Chem Lett 27:707-708

47. Yamaki T, Sumita T, Yamamoto S (2002) Formation of $TiO_{2-x}F_x$ compounds in fluorine-implanted TiO_2. J Mater Sci Lett 21:33-35

48. Hoyer P (1996) Formation of a titanium dioxide nanotube array. Langmuir 12:1411–1413

49.   Lakshmi BB, Dorhout PK, Martin CR (1997) Sol-gel template synthesis of semiconductor nanostructures. Chem Mater 9:857–862

50.   Imai H, Takei Y, Shimizu K, Matsuda M, Hirashima H (1999) Direct preparation of anatase TiO_2 nanotubes in porous alumina membranes. J Mater Chem 9:2971-2975

51.   Michailowski A, Al-Mawlwai D, Cheng GS, Moskovits M (2001). Highly regular anatase nanotubule arrays fabricated in porous anodic templates. Chem Phys Lett 349:1–5

52.   Jung JH, Kobayashi H, van Bommel KJC, Shinkai S, Shimizu T (2002) A novel method for preparation of nanocrystalline rutile TiO_2 powders by liquid hydrolysis of $TiCl_4$. Chem Mater 14:1445–1447

53.   Kobayashi S, Hamasaki N, Suzuki M, Kimura N, Shirai H, Hanabusa K (2002) Preparation of helical transition-metal Oxide tubes using organogelators as structure-directing agents. J Am Chem Soc 124:6550–6551

54.   Tian ZR, Voigt JA, Liu J, McKenzie B, Xu HF (2003) Large oriented arrays and continuous films of TiO_2-based nanotubes. J Am Chem Soc 125:12384–12385

55.   Kasuga T, Hiramatsu M, Hoson A, Sekino T Niihara K (1998) Formation of titanium oxide nanotubes. Langmuir 14:3160–3163

56.   Chen Q, Zhou WZ, Du GH, Peng LH (2002) Trititanate nanotubes made via a single alkali treatment Adv Mater 14:1208–1211

57.   Yao BD, Chan YF, Zhang XY, Zhang WF, Yang ZY, Wang N (2003) Formation mechanism of TiO_2 nanotubes. Appl Phys Lett 82:281–283

58.   Gong D, Grimes CA, Varghese OK, Hu W, Singh RS, Chen Z, Dickey EC (2001) Titanium oxide nanotubes array prepared by anodic oxidation. J Mater Res 16:3331–3334

59.   Mor GK, Varghese OK, Paulose M, Mukherjee N, Grimes CA (2003) Fabrication of tapered, conical-shaped titania nanotube. J Mater Res 18:2588–2593

60.   Cai QY, Paulose M, Varghese OK, Grimes CA (2005) The effect of electrolyte composition on the fabrication of self-

organized titanium oxide nanotubes array by anodic oxidation. J Mater Res 20:230–236

61. Ruan CM, Paulose M, Varghese OK, Mor GK, Grimes CA (2005) Fabrication of highly ordered TiO_2 nanotube array using an organic electrolyte. J Phys Chem B 109:15754–15759

62. Macak JM, Tsuchiya H, Schmuki P (2005) High-aspect-ratio TiO_2 nanotubes. Angew. Chem. Int. Ed. 44:2100–2102

63. Macak JM, Tsuchiya H, Taveira L, Aldabergerova S, Schmuki P (2005) Smooth anodic TiO_2 nanotubes. Angew Chem Int Ed 44:7463–7466

64. Paulose M, Shankar K, Yoriya S, Prakasam HE, Varghese OK, Mor GK, Latempa TA, Fitzgerald A, Grimes CA (2006) Anodic growth of highly ordered TiO_2 nanotube arrays to 134 µm in length. J Phys Chem 110:16179–16184

65. Yoriya S, Prakasam HE, Varghese OK, Shankar K, Paulose M, Mor GK, Latempa TA, Grimes CA (2006) Initial studies on the hydrogen gas sensing properties of highly ordered high aspect ratio TiO_2 nanotube-arrays 20 µm to 222 µm in length. Sensor Lett 4:334–339

66. Liu SM, Gan LM, Liu LH, Zhang WD, Zeng HC (2002) Synthesis of single-crystalline TiO_2 nanotubes. Chem Mater 14:1391–1397

67. Lee S, Jeon C, Park Y (2004) Fabrication of TiO_2 tubules by template synthesis and hydrolysis with water vapor. Chem Mater 16:4292–4295

68. Cheng B, Samulski ET (2001) Fabrication and characterization of nanotubular semiconductor oxides In_2O_3 and Ga_2O_3. J Mater Chem 11:2901–2902

69. Wang Y, Lee JY, Zeng HC (2005) Polycrystalline SnO_2 Nanotubes Prepared via Infiltration Casting of Nanocrystallites and Their Electrochemical Application. Chem Mater 17:3899–3903

70. Zhu W, Wang W, Xu H, ShiJ (2006) Fabrication of ordered SnO_2 nanotube arrays via a template route. Mater Chem Phys 99:127–130

71.   Nakamura H, Matsui Y (1995) The preparation of novel silica gel hollow tubes. Adv Mater 7:871–872

72.   Ono Y, kanekiyo Y, Inoue K, Hojo J, Shinkai S (1999) Evidence for the Importance of a cationic charge in the formation of hollow fiber silica from an organic gel system. Chem Lett 28:23–24

73.   Jung JH, Ono Y, Hanabusa K, Shinkai S (2000) Creation of both right-handed and left-handed silica structures by sol-gel transcription of organogel fibers comprised of chiral diaminocyclohexane derivative. J Am Chem Soc 122:5008–5009

74.   Tamaru S, Takeuchi M, Sano M, Shinkai S (2002) Sol-gel transcription of sugar-appended porphyrin assemblies into fibrous silica: unimolecular stacks versus helical bundles as templates. Angew Chem Int Ed 41:853–856

75.   Kobayashi S, Hanabusa K, Hamasaki N, Kimura M, Shirai H (2000) Preparation of TiO_2 hollow-fibers using supramolecular assemblies. Chem Mater 12:1523–1525

76.   Hanabusa K, Numazawa T, Kobayashi S, Suzuki M, Shirai H (2006) Preparation of metal oxide nanotubes using gelators as structure-directing agents. Macromol Symp 235:52–56

77.   Gundiah G, Mukhopadhyay S, Tumkurkar UG, Govindaraj A, Maitra U, Rao CNR (2003) Hydrogel route to nanotubes of metal oxides and sufates. J Mater Chem 13:2118–2122

78.   Ogihara H, Sadakane M, Nodasaka Y, Ueda W (2006) Shape-controlled synthesis of ZrO_2, Al_2O_3 and SiO_2 nanotubes using carbon nanofibers as templates. Chem Mater 21:4981–4983

79.   Adachi M, Harada T, Harada M (1999) Formation of huge length silica nanotubes by a templating mechanism in the laurylamine/tetraethoxysilane System. Langamuir 15:7097–7100

80.   Wang L, Tomura S, Ohashi F, Maeda M, Suzuki, Inukai K (2001) Synthesis of single silica nanotubes in the presence of citric acid J Mater Chem 11:465–468

81.   Ajayan PM, Stephane O, Redlich P, Colliex C (1995) Carbon nanotubes as removable templates for metal oxide nanocomposites and nanostructures. Nature 375:564–566

82.   Satishkumar BC, Govindaraj AG, Vogl EM, Basumallick L, Rao CNR (1997) Oxide nanotubes prepared using carbon nanotubes as templates. J Mater Res 12:604–606

83.   Rao CNR, Nath M (2003) Inorganic nanotubes. Dalton Trans 1–24

84.   Archibald DD, Mann S (1993) Template mineralization of self-assembled anisotropic lipid microstructures. Nature 364:430–432

85.   Chen J, Xu L, Li W, Gou X (2005) α-Fe$_2$O$_3$ nanotubes in gas sensor and lithium ion battery applications. Adv Mater 17:582–586

86.   Sun Z, Yuan H, Liu Z, Han B, Zhang X (2005) A highly efficient chemical sensor material for H$_2$S: α-Fe$_2$O$_3$ nanotubes fabricated using carbon nanotube templates. Adv Mater 17:2993–2997

87.   Liu L, Kou HZ, Mo W, Liu H, Wang Y (2006) Surfactant assisted synthesis of α-Fe$_2$O$_3$ nanotubes and nanorods with shape dependent magnetic properties. J Phys Chem B 110:15218–15223

88.   Liu Z, Zhang D, Han S, Li C, Lei B, Lu W, Fang J, Zhou C (2005) Single crystalline magnetite nanotube. J Am Chem Soc 127:6–7

89.   Seo DS, Lee JK, Kimb H (2001) Preparation of nanotube-shaped TiO$_2$ powder. J Crys Gro 229:428–432

90.   Du GH, Chen Q, Che RC, Yuan ZY, Peng LM (2001) Preparation and structural analysis of titanium oxide nanotubes. Apl Phys Lett 79:3702–3704

91.   Yuan ZY, Zhou W, Su BL (2002) Hierarchical interlinked structure of titanium oxide nanofibers. Chem Commun 1202–1203

92.   Zhang Q, Gao L, Sun J, Zheng S (2002) Preparation of long TiO$_2$ nanotubes from ultrafine rutile crystals. Chem Lett 31:226–227

93.   Tsai CC, Teng H (2004) Regulation of the Physical Characteristics of Titania Nanotube Aggregates Synthesized from Hydrothermal Treatment. Chem Mater 16:4352–4358

94.   Bavykin DV, Parmon VN, Lapkin AA, Walsh FC (2004) The effect of hydrothermal conditions on the mesoporous structure of TiO_2 nanotubes. J Mater Chem 14:3370–3377

95.   Jia CJ, Sun LD, Yan ZG, You LP, Luo F, Han XD, PangYC, Zhang Z, Yan CH (2005) Single-crystalline iron oxide nanotubes. Angew Chem Int ed 44:4328–4333

96.   Li Q, Kumar V, Li Y, Zhang H, Marks TJ, Chang RPH (2005) Fabrication of ZnO nanorods and nanotubes in aqueous solutions. Chem Mater 17:1001–1006

98.   Shankar K, Paulose M, Mor GK, Varghese OK, Grimes CA (2005) A study on the spectral photoresponse and photoelectrochemical properties of flame-annealed titania nanotube-arrays. J Phys D 38:3543-3549

99.   Shankar K, Tep KC, Mor GK, Grimes CA (2006) An electrochemical strategy to incorporate nitrogen in nanostructured TiO_2 thin films: Modification of bandgap and photoelectrochemical properties. J Phys D 39:2361-2366

100.   Mor GK, Carvalho MA, Varghese OK, Pishko MV, Grimes CA (2004) A room-temperature TiO_2-nanotube hydrogen sensor able to self-clean photoactively from environmental contamination. J Mater Res 19:628-634

101.   Mor GK, Varghese OK, Paulose M, Grimes CA (2003) A Self-cleaning room temperature titania-nanotube hydrogen gas sensor. Sens Lett 1:42-46

102.   Mor GK, Varghese OK, Paulose M, Shankar K, Grimes CA (2006) A review on highly ordered, vertically oriented TiO_2 nanotube arrays: fabrication, material properties, and solar energy applications. Solar Energy Materials & Solar Cells 90:2011-2075

103.   Ruan C, Paulose M, Varghese OK, Grimes CA (2006) Enhanced photoelectrochemical response in highly ordered TiO_2 nanotube arrays anodized in boric acid containing electrolyte. Solar Energy Materials & Solar Cells 90:1283-1295

104. Mor GK, Varghese OK, Paulose M, Grimes CA (2005) Transparent highly-ordered TiO_2 nanotube-arrays via anodization of titanium thin films. Adv Funct Mater 15:1291-1296

105. Paulose M, Mor GK, Varghese OK, Shankar K, Grimes CA (2006) Visible light photoelectrochemical and water-photoelectrolysis properties of titania nanotube arrays. J Photochem Photobiol A 178:8-15

106. Zwilling V, Darque-Ceretti E, Boutry-Forveille A, David D, Perrin MY, Aucouturier M (1991) Structure and physicochemistry of anodic oxide filmes on titanium and TA6V alloy. Surf Interface Anal 27:629-637

107. Delplancke JL, Winand R (1998) Galvanostatic anodization of titanium. II. Reactions efficiencies and electrochemical behaviour model. Electrochim Acta 33:1551-1559

108. Sul YT, Johansson CB, Jeong Y, Albrektsson T (2001) The electrochemical oxide growth behaviour on titanium in acid and alkaline electrolytes. Med Eng Phys 23:329-346

109. Hwang BJ, Hwang JR (1993) Kinetic model of anodic oxidation of titanium in sulphuric acid. J Appl Electrochem 23:1056-1062

110. Parkhutik VP, Shershulsky VI (1992) Theoretical modelling of porous oxide growth on aluminium. J Phys D 25: 1258-1263

111. Thompson GE (1997) Porous anodic alumina: fabrication, characterization and applications. Thin Solid Films 297: 192-201

112. Chen S, Paulose M, Ruan C, Mor GK, Varghese OK, Kouzoudis D, Grimes CA (2006) Electrochemically synthesized CdS nanoparticle-modified TiO_2 nanotube-array photoelectrodes: Preparation, characterization, and application to photoelectrochemical cells. J Photochem Photobiol 177:177-184

113. Melody B, Kinard T, Lessner P (1998) The non-thickness-limited growth of anodic oxide films on valve metals. Electrochem Solid-State Lett 1:126-129

114. Li YM, Young L (2001) Non-thickness-limited growth of anodic oxide films on tantalum. J Electrochem Soc 148:B337-B342

115. Izutsu K (2002) Electrochemistry in nonaqueous solutions, Wiley-VCH.

116. Khan SUM, Al-Shahry M, Ingler WB (2002) Efficient photochemical water splitting by a chemically modified n-TiO_2. Science 297:2243-2245

117. Noworyta K, Augustynski J (2004) Spectral photoresponses of carbon-doped TiO_2 film electrodes. Electrochem Solid-State Lett 7:E31-E33

118. Lee JY, Park J, Cho JH (2005) Electronic properties of N- and C-doped TiO_2. Appl Phys Lett 87:011904-3

119. Chen XB, Lou YB, Samia ACS, Burda C, Gole JL (2005) Formation of oxynitride as the photocatalytic enhancing site in nitrogen-doped titania nanocatalysts: Comparison to a commercial nanopowder. Adv Funct Mater 15:41-49

120. Wu PG, Ma CH, Shang JK (2005) Effects of nitrogen doping on optical properties of TiO_2 thin films. Appl Phys A 81:1411-1417

121. Suda Y, Kawasaki H, Ueda T, Ohshima T (2005) Preparation of nitrogen-doped titanium oxide thin film using a PLD method as parameters of target material and nitrogen concentration ratio in nitrogen/oxygen gas mixture. Thin Solid Films 475: 337-341

122. Liu Y, Alwitt RS, Shimizu K (2000) Cellular porous anodic alumina grown in neutral organic electrolyte-I. Structure, composition, and properties of the films. J Electrochem Soc 147:1382-1387

123. Gerischer H, Lubke M (1986) A particle-size effect in the sensitization of TiO_2 electrodes by a CdS deposit. J Electroanal Chem 204:225-227

124. Vogel R, Hoyer P, Weller H (1994) Quantum-sized PbS, CdS, AgzS, Sb&, and Bi& particles as sensitizers for various nanoporous wide- bandgap semiconductors. J Phys Chem 98:3183-3188

125. Pandey RK, Sahu SN, Chandra S (1996) Handbook of Semiconductor Electrodeposition, Marcel Decker, New York

126. Varghese OK, Gong DW, Paulose M, Grimes CA, Dickey EC (2003) Crystallization and high-temperature structural stability of titanium oxide nanotube arrays. J Mater Res 18: 156-165

127. Marino CEB, Nascente PAP, Biaggio SR, Rocha-Filho RC, Bocchi N (2004) XPS characterization of anodic titanium oxide films grown in phosphate buffer solutions. Thin Solid Films 468:109-112

128. Yakovleva NM, Anicai L, Yakovlev AN, Dima L, Khanina EY, Buda M, Chupakhina EA (2002) Structural study of anodic films formed on aluminum in nitric acid electrolyte. Thin Solid Films 416:16-23

129. Augustynski J, Berthou H, Painot J (1976) XPS study of interactions between aluminum metal and nitrate ions. Chem Phys Lett 44:221-224

130. Parhutik VP, Makushok IE, Kudriavtsev E, Sokol VA, Khodan AN (1987) An X-ray electronic study of the formation of anodic oxide films on aluminium in nitric acid. Electrochemistry (Elektrokhymia) 23:1538-1544

131. Kundu M, Khosravi AA, Kulkarni SK (1997) Synthesis and study of organically capped ultra small clusters of cadmium sulphide. J Mater Sci 32:245-258

132. Ong KG, Varghese OK, Mor GK, Grimes CA (2005) Numerical simulation of light propagation through highly-ordered titania nanotube arrays: Dimension optimization for improved photoabsorption. J Nanosci Nanotechnol 5: 1801-1808

133. Mor GK, Shankar K, Varghese OK, Grimes CA (2004) Photoelectrochemical properties of titania nanotubes. J Mater Res 19:2989-2996

134. Asanuma T, Matsutani T, Liu C, Mihara T, Kiuchi M (2004) Structural and optical properties of titanium dioxide films deposited by reactive magnetron sputtering in pure oxygen plasma. J Appl Phys 95:6011-6016

135.   Manifacier JC, Gasiot J, Fillard JP (1976) A simple method for the determination of the optical constants n, k and the thickness of a weakly absorbing thin film. J Phys E 9: 1002-1004

136.   Vogel R, Meredith P, Kartini I, Harvey M, Riches JD, Bishop A, Heckenberg N, Trau M, Dunlop HR (2003) Mesostructured dye-doped titanium dioxide for micro-optoelectronic applications. Chem Phys Chem 4:595-603

137.   Yoldas BE, Partlow PW (1985) Formation of broad band antireflective coatings on fused silica for high power laser applications. Thin Solid Films 129:1-14

138.   Tauc J (1970) Absorption edge and internal electric fields in amorphous semiconductors. Mater Res Bull 5:721-729

139.   Sant PA, Kamat PV (2002) Interparticle electron transfer between size-quantized CdS and TiO_2 semiconductor nanoclusters. Phys Chem Chem Phys 4:198-203

140.   Kokai J, Rakhshani AE (2004) Photocurrent spectroscopy of solution-grown CdS films annealed in $CdCl_2$ vapour. J Phys D 37:1970-1975

141.   Lubberhuizen WH, Vanmaekelbergh D, Van Faassen E (2000) Recombination of photogenerated charge carriers in nanoporous gallium phosphide. J Porous Mater 7:147-152

142.   Marin FI, Hamstra MA, Vanmaekelbergh D (1996) Greatly enhanced sub-bandgap photocurrent in porous GaP photoanodes. J Electrochem Soc 143:1137-1142

143.   Vanmaekelbergh D, de Jongh PE (1999) Driving force for electron transport in porous nanostructured photoelectrodes. J Phys Chem B 103:747-750

144.   Hamnett A (1980) General discussions. Faraday Discuss Chem Soc 70:124-127

145.   Hagfeldt A, Gratzel M (1995) Light-induced redox reactions in nanocrystalline systems. Chem Rev 95:49-68

146.   Ong KG, Varghese OK, Mor GK, Grimes CA (2007) Application of finite-difference time domain to dye-sensitized solar cells: The effect of nanotube-array negative electrode dimensions on light absorption. Solar Energy Materials & Solar Cells 91:250-257

147. Aroutiounian, V.M.; Arakelyan, V.M.; Shannazaryan, G.E.; Stepanyan, G.M.; Turner, J.A.; Khaselev, O. (2002) Investigation of ceramic Fe_2O_3<Ta> photoelectrodes for solar energy photoelectrochemical converters. Int J Hydrogen Energy 27:33-38

148. Beermann, N.; Vayssieres, L.; Lindquist, S.-Eric; Hagfieldt, A. (2000) Photoelectrochemical studies of oriented nanorod thin films of hematite. J Electrochem Soc 147:2456-2461

149. Morin, F.J. (1954) Electrical properties of α- Fe_2O_3. Phys Rev 93:1195-1199

150. Gardner, R.F.G.; Sweett, F.; Tanner, D.W. (1963) The electrical properties of alpha ferric oxide—II. Ferric oxide of high purity. J Phys Chem Solids 24:1183-1186

151. Sato, N. (1998) Electrochemistry at Metal and Semiconductor Electrodes; Elsevier; Amsterdam, pg 34

152. Murphy, A.B.; Barnes, P.R.F.; Randeniya, L.K.; Plumb, I.C.; Grey, I.E.; Horne, M.D.; Glasscock, J.A. (2006) Efficiency of solar water splitting using semiconductor electrodes. Int J Hydrogen Energy 31:1999-2017

153. Grätzel, M. (2001) Photoelectrochemical cells. Nature 414:338-344

154. Kennedy, J. H.; Frese, J. K. W. (1978) Photooxidation of water at α- Fe_2O_3 electrodes. J Electrochem Soc 125: 709-714

155. Paulose, M.; Varghese, O. K.; Mor, G. K.; Grimes, C. A.; Ong, K. G. (2006) Unprecedented ultra-high hydrogen gas sensitivity in undoped titania nanotubes. Nanotechnology 17:398-402. Varghese, O. K.; Yang, X.; Kendig, J.; Paulose, M.; Zeng, K.; Palmer, C.; Ong, K. G.; Grimes, C. A. (2006) A transcutaneous hydrogen sensor: From design to application. Sensor Letters 4:120-128

156. Varghese, O. K.; Gong, D.; Paulose, M.; Ong, K. G.; Grimes, C. A. (2003) Hydrogen sensing using titania nanotubes. Sensors Actuators B, 93:338-344.

157. Mor, G. K.; Varghese, O. K.; Paulose, M.; Grimes, C. A. (2003) A self-cleaning, room-temperature titania nanotube hydrogen gas sensor. Sensor Letters 1:42-46.

158. Mor, G. K.; Carvalho, M. A.; Varghese, O. K.; Paulose, M.; Pishko, M. V.; Grimes, C. A. (2004) A room-temperature TiO_2 nanotube hydrogen sensor able to self-clean photoactively from environmental contamination. J Materials Research 19:628-634

159. Varghese, O. K.; Mor, G. K.; Grimes, C.A.; Paulose, M.; Mukherjee, N. (2004) A titania nanotube array room-temperature sensor for selective detection of hydrogen at low concentrations. J Nanosci Nanotechn 4:733-737

160. Mor, G. K.; Shankar, K.; Paulose, M.; Varghese, O. K.; Grimes, C. A. (2006) Use of highly ordered TiO_2 nanotube arrays in dye-sensitized solar cells. Nano Letters 6:215-218

161. Zhu, K.; Neale, N. R.; Miedaner, A.; Frank, A. J. (2007) Enhanced charge-collection efficiencies and light scattering in dye-sensitized solar cells using oriented TiO_2 nanotubes arrays. Nano Letters 7:69-74

162. Melody, B.; Kinard, T.; Lessner, P. (1998) The non-thickness-limited growth of anodic oxide films on valve metals. Electrochem Solid-State Lett 1:126-129

163. Krembs, G.M. (1963) Residual tritiated water in anodized tantalum films. J Electrochem Soc 110:938-940

164. Varghese, O. K.; Paulose, M.; Gong, D.; Grimes, C. A.; Dickey, E. C. (2003) Crystallization and high temperature structural stability of titanium oxide nanotube arrays. J Materials Research 18:156-165

165. Gennari, F.C.; Pasquevich, D.M. (1998) Kinetics of the anatase rutile transformation in TiO_2 in the presence of Fe_2O_3. J Mater Sci 33:1571-1578

166. Wang, R.; Sakai, R.; Fujishima, A.; Watanabe, T.; Hashimoto, K. (1999) Studies of surface wettability conversion on TiO2 single-crystal surfaces. J Phys Chem B 103:2188-2194

167. Dghoughi, L.; Elidrissi, B.; Berne`de, C.; Addou, M.; Lamrani, M.A.; Regragui, M.; Erguig H. (2006) Physicochemical, optical and electrochemical properties of iron oxide thin films prepared by spray pyrolysis. Appl Surf Sci 253:1823-1829

168.   Heimer, T.A.; Heilweil, E.J.; Bignozzi, C.A.; Meyer, G.J. (2000) Electron injection, recombination, and halide oxidation dynamics at dye-sensitized metal oxide interfaces. J Phys Chem A 104:4256-4262

169.   Grimes, C. A. (2007) Synthesis and application of highly ordered arrays of TiO_2 nanotubes. J Mater Chemistry 17:1451-1457

170.   Mor, G. K.; Prakasam, H. E.; Varghese, O. K.; Shankar, K.; Grimes, C. A. (2007) Vertically Oriented Ti-Fe-O Nanotube Array Films: Towards a Useful Material Architecture for Solar Spectrum Water Photolysis. Nano Letters 7:2356-2364

Chapter 6

OXIDE SEMICONDUCTORS: SUSPENDED NANOPARTICLE SYSTEMS

6.1 Introduction

Our discussions to this point on the use of semiconductor materials for photo-assisted water splitting have been predicated on the use on a clearly defined anode and cathode geometry as initially described by Fujishima and Honda [1] where the hydrogen and oxygen are evolved separately. While this approach offers the fundamental advantage that no gas separation step is required, the photoconversion efficiency suffers from losses inherent in the transport of the photogenerated electrons and H^+ ions from the anode to the cathode. In variations on a water-splitting theme, in 1977 Schrauzer and Guth used a TiO_2-based photocatalyst to carry out the photocatalytic decomposition of water vapor under UV light [2], resulting in the simultaneous co-generation of oxygen and hydrogen. Subsequent efforts investigated the use of various doped TiO_2 surfaces for direct photocatalytic splitting of water vapor [2-5] with the mixed evolution of hydrogen and oxygen. Effectively, rather than an electrode-based electrochemical cell each semiconductor particle can behave as its own electrochemical cell, see **Fig. 6.1**, being considered as a micro-homogenous or micro-heterogeneous system. The logical extension of this idea, bulk to planar to particulate, is a dispersion of semiconductor nano 'particles' of various geometry, e.g. sphere, tube, ribbon, or wire, see **Fig. 6.2**, throughout either liquid or vapor phase water resulting in the photocatalytic splitting thereof. The concept offers an appealing simplicity that has gained considerable academic interest.

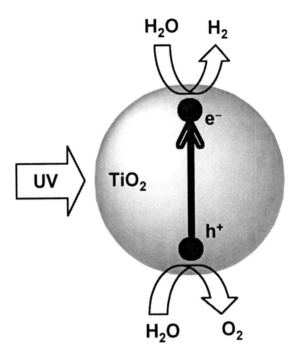

Fig. 6.1: A schematic drawing of a semiconductor particle behaving as its own electrochemical cell for water splitting.

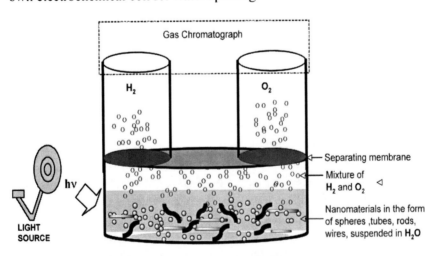

Fig. 6.2: A schematic diagram showing suspended nanomaterials in the form of spheres, rods, tubes or wires dispersed throughout some water. Under solar light illumination, hydrogen and oxygen evolved from these small particles. A membrane is placed atop the reaction vessel to separate H_2 and O_2.

Several modified TiO_2 catalysts in powder form, including $Pt-TiO_2$, $Pt-RuO_2-TiO_2$, $Pt-SrTiO_3$, $NiO-SrTiO_3$, suspended or dispersed in liquid water for solar hydrogen production have been studied [6-18]. To date no homogeneous or heterogeneous photocatalysts have been found suitable for stoichiometric water splitting, primarily due to the difficulty in oxygen formation and the rapid reverse reaction between products. Of the reported studies [6-18], efforts have focused on approaches such as metal loading, ion doping, composite formation, and the addition of electron donors or hole scavengers with an aim towards preventing the rapid recombination of photogenerated electron/hole pairs, and improving the visible light response of the semiconducting particles. As we will discuss further, these photocatalysts involve a two-step water splitting process for H_2 production and are comprised of a combination of either metal and/or two or more different oxide semiconductors. Modification of these nanoparticle photocatalysts has resulted in comparatively effective catalysts for H_2 evolution [19-26]. The high surface area of these nanoparticles, at least in principle, allow solution species to capture photogenerated electrons quickly and thereby significantly reduce the occurrence of charge recombination [25-27].

Over the past several years considerable effort has been made to design, synthesize and fabricate metal oxide nanoparticle photocatalysts capable of using visible light energy. Cu_2O, which behaves as a p-type semiconductor with a band gap energy of 2.0-2.2 eV, was the first such oxide reported for decomposing water under visible light irradiation [28]. The release of excess O_2 from Cu_2O by visible light irradiation disturbs the water splitting stoichiometry due to photoadsorption of oxygen in the bulk, and O^- and O_2^- at the surface of Cu_2O. Since the photoadsorption largely depends on the O_2 pressure as well as on the wavelength and intensity of light, at modest pressures (above $\approx 500\,Pa$) photoadsorption inhibits water splitting on Cu_2O [28]. Anion doping of wide band gap semiconductors has been an area of particular recent interest [29,30], but a robust design has yet to be realized. Several review articles [24-26,31-33] consider the importance of semiconductor particulate systems for solar hydrogen production.

Nanoparticles are usually defined as particles having one or more dimensions on the order of 100 nm or less. Nanoparticles of approximately 10 nm diameter are variously labeled as quantum dots, demonstrating size-dependent quantization of the electronic energy levels as dependent upon particle size and shape, particle-particle spacing, and nature of the outer shell. The interesting and sometimes unexpected nanoparticle properties are partly due to aspects of the material surface dominating those of the bulk. The large fraction of atoms at the surface of a material becomes significant as the size of that material approaches the nanoscale. For example, the interaction between the nanoparticle surface and a surrounding liquid is commonly strong enough to overcome differences in density, allowing it to form a suspension rather than sinking or floating. Nanoparticle synthesis methods play an important role in determining the specificity of the various properties, including physical properties and those of the electronic structure, e.g. quantum confinements and electron tunneling.

6.2 Nanoparticle-Based Photocatalytic Water Splitting

The following reactions are involved for the production of hydrogen and oxygen from water (pH = 7):

$$2H_2O + 2e^- \rightarrow H_2 + 2OH^-, \quad E°(H_2/H_2O) = -0.41 \text{ V (vs NHE)} \quad (6.2.1)$$
$$2H_2O \rightarrow O_2 + 4H^+ + 4e^-, \quad E°(H_2O/O_2) = +0.82 \text{ V (vs NHE)} \quad (6.2.2)$$

$$E°_{overall} = E°(H_2O/O_2) - E°(H_2/H_2O) = 1.23 \text{ V}$$

The reaction

$$H_2O \rightarrow H_2 + 1/2O_2$$

is endothermic and involves a change in the Gibbs free energy ($\Delta G° = 2.46$ eV or 237.2 KJ mol^{-1}). The overall reaction is a 2-electron redox process. The energy of 1.23 eV per electron transferred corresponds to a wavelength of $\lambda = 1008$ nm. Water absorbs solar radiation in the infrared region, with photon energies too low to drive photochemical water splitting. Thus any photochemical process to drive the water splitting reaction must involve a sensitizer or semiconductor that can absorb light. Semiconductors, in the presence of energy provided by light, are capable of decomposing water into hydrogen and oxygen depending upon their conduction band (CB) and valance band (VB) energy levels. The conduction band level should be more negative than the hydrogen evolution level, $E°(H_2/H_2O)$, while the valance band level should be more positive than the water oxidation level, $E°(H_2O/O_2)$.

When a semiconductor is irradiated by photons of energy equal to or greater than that of its bandgap, which it absorbs, excitation occurs and an electron moves to the conduction band leaving a hole behind in the valance band. For TiO_2 this process is expressed as:

Semiconductor: $TiO_2 + h\nu \rightarrow e^-(TiO_2) + h^+(TiO_2)$ (6.2.3)

Water: $H_2O \rightarrow H^+ + OH^-$ (6.2.4)
               $2e^-(TiO_2) + 2H^+ \rightarrow H_2$ (6.2.5)
               $2h^+(TiO_2) + 2OH^- \rightarrow 2H^+ + 1/2O_2$ (6.2.6)

The photogenerated electrons and holes can recombine in bulk or on the semiconductor surface releasing energy in the form of heat or a photon. The electrons and holes that migrate to the semiconductor surface without recombination can, respectively, reduce and oxidize water (or the reactant) and are the basic mechanism of photocatalytic hydrogen production, see **Fig. 6.3.**

The efficiency of converting light energy into hydrogen energy using suspended nanoparticle catalysts is, to date, low. The primary reasons for the low efficiencies are: {1} Rapid recombination of photo-generated electron-hole pairs. {2} Photochemical water-splitting involves at least one exthothermic reaction therefore it is relatively easy for molecular hydrogen and oxygen to recombine (backward reaction).

$$H_2(g) + 1/2O_2(g) \rightarrow H_2O(g) + 57 \text{ Kcal} \qquad (6.2.7)$$

{3} The poor visible spectrum response of corrosion resistant catalysts. For example TiO_2, which is one of the most suitable photocatalysts being chemically stable with strong catalytic activity, does not respond to visible light.

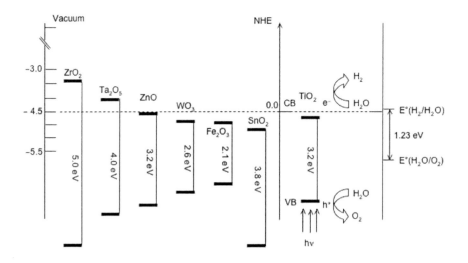

Fig. 6.3: Band position of several semiconductors in presence of aqueous electrolyte at pH = 1. Energy scale is indicated in electron volts (eV) using either normal hydrogen electrode or vacuum level as a reference. On right energy levels for TiO$_2$ photocatalytic water splitting are presented.

A significant enhancement in photocatalytic activities have been found in particles less than 10-15 nm in diameter, a behavior attributed to quantum size effects. Photochemical reactions at the semiconductor surface occur if the light-generated charge carriers reach the surface finding suitable reaction partners, protons for electrons and donor molecules for holes. For small particles, the charge carrier diffusion length is large compared to the particle size and no internal electric field (depletion layer) is necessary to separate the photogenerated electron-hole pairs. Therefore, the probability that the charge carriers will reach the surface of the particle increases as the particle size decreases.

6.3 Nanoparticle Synthesis Techniques

There are various methods available for the synthesis of 1-100 nm diameter nanoparticles. Whatever the synthesis method, it is important to consider their stability in terms of composition and size.

The surface atoms of a nanoparticle may be quite active and tend to react in the ambient environment. Smaller particles may tend to coagulate to form larger particles losing their interesting size dependent properties. Therefore synthesis methods that passivate the particles to ensure their long-term stability have gained significant interest. Liquid-phase nanoparticle synthesis routes typically yield non-aggregating nanoparticles with uniform size and shape and controlled surface properties; such desirable properties are difficult to achieve using gas-phase synthesis routes. However in contrast to liquid phase methods, gas-phase methods have been scaled for large-scale production of specific nanoparticle compositions. A few commonly used methods for metal oxide nanoparticle synthesis are shown in **Fig. 6.4**. We now briefly consider some nanoparticle synthesis routes; more detailed synthesis techniques are suggested to the reader for liquid phase [34-35] and gas phase [36-37].

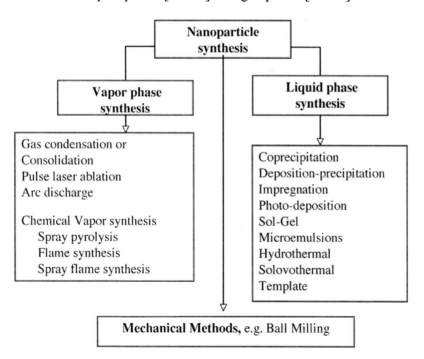

Fig. 6.4: Flow chart illustrating various nanoparticle synthesis routes.

6.3.1 Liquid-Phase Nanoparticle Synthesis
Coprecipitation [38-42]

The simultaneous occurrence of nucleation, growth, coarsening (Ostwald ripening), and agglomeration are key features of coprecipitation reactions. This method involves the simultaneous precipitation of insoluble compounds in a solution coming out from mixed, and well-defined, quantities of different ionic solutions under controlled reaction conditions. A variety of metal oxide nanoparticle compositions, and metal nanoparticles supported on metal oxides, can be prepared by the coprecipitation of sparingly soluble compounds followed by their calcination to oxide form [38-40]. The mechanistic pathways of coprecipitation processes are still unclear because of the difficulties in isolating and then studying each process separately. It is understood that a state of high supersaturation of sparingly soluble compounds in an aqueous solution dictates the key nucleation step in the precipitation process, thereby forming a large number of small particles. Coarsening and aggregation processes usually determine the morphology and size distribution of resulting particles. Agglomeration of small metal particles is inevitable in the absence of any stabilizers. Steric stabilization and colloid stabilization are the two appropriate choices concerning the chemical stability of monodispersed nanoparticles at very high or very low pH. Some agglomeration appears unavoidable in the precipitation reaction of oxides where annealing or calcination is usually required.

For mixed metal oxides obtained from their hydroxide or carbonate precursors after calcination, it is generally difficult to determine whether the as-prepared precursor is a single-phase or multiphase solid solution [35]. Non-aqueous solvents appear superior for achieving two dissimilar metal oxides such as $MM'O_2$ or $MM'O_4$ precipitates; such reactions cannot be carried out simultaneously in aqueous solution due to the large variations in pH necessary to induce precipitations [41,42]. **Table 6.1** summarizes some of the nanoparticulate semiconducting metal oxides and mixed metal oxides prepared via co-precipitation techniques. The general procedure of achieving metal loaded nanoparticles on an oxide support is shown in **Fig.6.5.**

Table 6.1 Summary of the reactions for the synthesis of oxides nanoparticle by different methods

Compounds	Starting Materials	Precipitation agent	Stabilizer	Chelating agent	Polymerizing agaent	Conditions (°C)	Product size (nm)	Ref.
Coprecipitation								
NiO	NiCl$_2$	NH$_4$OH	CTAB			500	22-28	38
ZnO	ZnC$_2$	NH$_4$OH	CTAB			500	40-60	38
SnO$_2$	SnCl$_4$	NH$_4$OH	CTAB			500	11-18	38
Sb$_2$O$_3$	SbCl$_3$	NaOH	PVA			350	10-80	39
Au/Fe$_2$O$_3$	HAuCl$_4$ Fe(NO$_3$)$_3$	Na$_2$CO$_3$	-			400	3.6	40
Fe$_3$O$_4$	FeCl$_2$, FeCl$_3$	NH$_4$OH	H$^+$			N$_2$ atm	8-50	41
MnFe$_2$O$_4$	MnCl$_2$, FeCl$_3$	N$_2$H$_4$.H$_2$O	None			500	5-25	42
Sol-gel								
ZnTiO$_3$	Zn(OAc)$_2$, Ti(^{t}OBu)$_4$			CA	EG	500-1000	18-31	53
SrTiO$_3$	Sr(NO$_3$)$_2$, Ti(^{t}OBu)$_4$			GAA	PT	750	12-25	54
BaTi$_4$O$_9$	Ba(OAc)$_2$, Ti(^{t}Opr)$_4$					600-1000	Not stated	55
La$_2$Ti$_2$O$_7$	Ti(^{t}OBu)$_4$, La(NO$_3$)$_2$			CA	EG	650-1150	55	56
Sr$_2$Nb$_x$Ta$_2$-$_x$O$_7$	NbCl$_5$, TaCl$_5$, SrCO$_3$			CA	EG	700-1000	Not stated	57

CTAB: Cetyltrimethylammonium bromide; PVA: Polyvinyl alcohol; CA: Citric Acid; EG: Ethylene Glycol; PAA: Polyacrylic acid; GAA: Glacial Acetic Acid; PT; Propanetriol

Deposition-Precipitation [40,43-45]

A modified version of a simultaneous precipitation reaction, this is a highly selective approach for generating noble metal nanoparticles on metal oxide supports, see **Fig. 6.5**. The nanoparticle size can be controlled by adjusting the reaction pH. Pt, Au, Ag nanoparticles can be deposited on a variety of metal oxide supports such as TiO$_2$, Fe$_2$O$_3$, and ZnO.

Photo-deposition [17,45-48]

This similar methods involves UV light irradiation, with light intensity, stirring speed, suspension volume and temperature influencing particle growth, **Fig. 6.5**. Shorter irradiation times can be used to limit the amount of particle deposition. Photo-deposited samples do not require calcination hence there is little chance of particle agglomeration.

Impregnation [13,45,49,50]

This method involves the repeated dipping of porous support materials into a solution containing the desired catalytic agent. It is then dried and calcined to transform the metal into insoluble form. The agent must be applied uniformly in a predetermined quantity to a preset depth of penetration. The metal loading in the finished catalyst is typically 1-5%, **Fig. 6.5.**

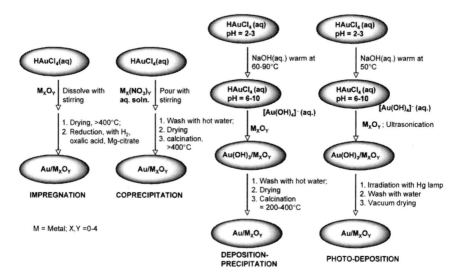

Fig. 6.5: Syntheses of metal loaded nanoparticles (Au) on metal oxide supports using impregnation, coprecipitation, deposition-precipitation, and photo-deposition methods. For Pt loaded nanoparticles H_2PtCl_6 (aq) is used.

Sol-gel Processes [51-57]

The sol is made of a stable suspended solution of metal salts or solvated metal precursors containing solid particles of nanometer diameter. Polycondensation or polyesterification results in the appearance of particles in a new phase called "the gel". Aging, drying and dehydration are steps required to achieve solid-form ultra-fine particles. Coarsening and phase transformation occur simultaneously with aging. Gel drying is associated with the removal of water or other volatile liquids from the gel network, either by thermal evaporation or near supercritical condition resulting in a monolithic structure called, respectively, a xerogel or an aerogel. Dehydration, typically by calcination, removes surface bound hydroxy functional groups. This method is ideally suited for generating metal oxide nanoparticles.

Calcining aerogels or xerogels improves sample crystallinity but normally leads to partial agglomeration. For example, condensation and supercritical drying of $Zr(NO_3)_2 \cdot 5H_2O$ or $Zr(NO_3)_2 \cdot 2H_2O$ precursors in alcohol yield amorphous ZrO_2 nanoparticles ($\sim$ 1 nm), which upon calcination at 400°C form 2-3 nm-agglomerated nanoparticles in crystalline form [51]. Similarly, TiO_2 nanoparticles can be prepared by controlled hydrolysis and condensation of $Ti(OPr)_4$ in an alcohol + HCl solution followed by supercritical drying [52].

A modified sol-gel process developed by Pechini [53] provides easy synthesis of mixed metal oxides of considerable complexity. In this method, chelating agents tend to form stable complexes with a variety of metals over fairly wide range of pH. Selective synthesis techniques are presented in **Table 6.1**. However because of the formation of gel network by the polyesterification of the chelating and polymerization agents, in which metal ions remain trapped in their matrices, this method provides limited control over particle size, shape and morphology. Synthesis of $ZnTiO_3$ is one such example where citric acid and ethylene glycol act, respectively, as chelating and polymerization agents, thus forming ≈ 30 nm particles from $Zn(OAc)_2$ and $Ti(OBu)_4$ as starting materials [53].

Microemulsion Methods [58-59]

This method involves formation of reverse micelles in the presence of surfactants at a water-oil interface. A clear homogeneous solution obtained by the addition of another amine or alcohol-based co-surfactant is termed a Microemulsion. To a reverse micelle solution containing a dissolved metal salt, a second reverse micelle solution containing a suitable reducing agent is added reducing the metal cations to metals. The synthesis of oxides from reverse micelles depends on the coprecipitation of one or more metal ions from aqueous solution. This method is similar to coprecipitation, and is useful in preparing 20-80 nm diameter ferrite nanoparticles, MFe_2O_4 [M = Fe, Mn, Co] [58]. For insoluble or unstable metal ions, nanoparticles can be prepared by the hydrolysis of a suitable precursor in common with sol-gel. Synthesis of mesoporous mixed zirconium-titanium phosphates (~50 nm particles) using cationic surfactants, octadecyltrimethylammonium chloride, has successfully demonstrated water decomposition [59].

Solvothermal/Hydrothermal Methods [60-65]

Within a sealed vessel such as an autoclave the reaction temperature of a medium can be brought to well above its boiling point by an increase in the self-generating pressure with heating. The solvothermal method is similar to the hydrothermal method except that organic solvents are used instead of water. This method can effectively prevent the product from oxidizing and has been used to synthesize a variety of non-oxide compositions. Some of these reactions involve supercritical solvents, and thus take advantage of the increased solubility and reactivity of inorganic salts at elevated temperatures and pressures. Due to the high viscosities of supercritical solvents various chemical compounds are readily dissolved that would, under ambient conditions, remain insoluble. Both these processes have advantages over coprecipitation and sol-gel, because the reaction products are usually crystalline. For example, hydrothermal synthesis of monodispersed, anatase/rutile nanocrystalline TiO_2 particles has been successfully carried out [60], as well as preparation of metal-doped titania particles [61,62]. Codoping with bromine and chlorine under hydrothermal conditions yields titania of mixed anatase/rutile phase which has shown water splitting in Na_2CO_3 under UV light irradiation [63]. Nanocrystalline TiO_2 photocatalysts of controllable anatase/rutile ratio and high surface area (113-169 m^2/g) have been prepared at low temperature by a microemulsion-mediated hydrothermal method [64]. Water splitting by NiO doped, hydrothermally synthesized pyrochlores of tantalates ($Ca_2Ta_2O_7$, $Na_2Ta_2O_6$) is reported by Ikeda and co-workers [65].

Template Methods [66-69]

Various metal and metal oxide nanoparticles have been prepared on polymer (sacrificial) templates, with the polymers subsequently removed. Synthesis of nanoparticles inside mesoporus materials such as MCM-41 is an illustrative template synthesis route. In this method, ions adsorbed into the pores can subsequently be oxidized or reduced to nanoparticulate materials (oxides or metals). Such composite materials are particularly attractive as supported catalysts. A classical example of the technique is deposition of 10 nm particles of NiO inside the pore structure of MCM-41 by impregnating the mesoporus material with an aqueous solution of nickel citrate followed by calicination of the composite at 450°C in air [68]. Successful synthesis of nanosized perovskites (ABO_3) and spinels (AB_2O_4), such as $LaMnO_3$ and $CuMn_2O_4$, of high surface area have been demonstrated using a porous silica template [69].

6.3.2 Vapor Phase Nanoparticle Synthesis

Within this method nanoparticle production can be divided into gas-to-particle and liquid/solid-to-solid routes. In the liquid/solid-to-solid route the product particles are formed from droplets or solid reactant particles via intraparticle reactions. Using this method it is possible to produce single- and multi-component materials with controlled levels of doping, however multi-component materials may be difficult to produce. In the gas-to-particle route, the most essential requirement is to create a high degree of vapor supersaturation of the source materials. Supersaturation can be achieved via physical processes such as cooling of a hot vapor or via chemical reactions of gaseous precursors, which results in the formation of condensable species. Under favorable conditions nucleation and particle growth occur that are controllable either by changing the reaction parameters or by removing the source materials. Once the particles form in the gas phase, coagulation and coalescence assume significant importance. Solid, spherical or nearly spherical particles with narrow size distributions can be produced of high purity. However in contrast to liquid phase where capping with appropriate ligand can be used to stabilize a nanoparticle dispersion, special efforts are needed to control agglomeration.

Chemical Vapor Synthesis [70-74]

Similar to chemical vapor deposition, reactants or precursors for chemical vapor synthesis are volatile metal-organics, carbonyls, hydrides, chlorides, etc. delivered to the hot-wall reactor as a vapor. A typical laboratory reactor consists of a precursor delivery system, a reaction zone, a particle collector, and a pumping system. Modification of the precursor delivery system and the reaction zone allows synthesis of pure oxide, doped oxide, or multi-component nanoparticles. For example, copper nanoparticles can be prepared from copper acetylacetone complexes [70], while europium doped yttria can be obtained from their organometallic precursors [71].

Instead of delivering the reactant into a hot wall furnace it can be injected using a nebulizer, a technique referred to as spray pyrolysis [72]. Flame synthesis is widely used for nanoparticle production due to its cost-effectiveness and versatile approach for controlled synthesis. In a flame reactor, the energy from the flame drives the chemical reactions producing clusters, which subsequently grow into nanoparticles by surface growth and /or coagulation and coalescence at elevated temperature [73]. Flame spray pyrolysis uses low vapor-pressure liquid precursors, with evaporation and ignition of the spray (liquid removal) initiated by a flame ring about the nozzle. Subsequent gas phase reaction followed by vapor condensation leads to the growth of the nanoscale particles [74].

Gas condensation [75-79]

This method involves a means of achieving supersaturation by heating solid precursors directly into vapor. The gas so produced, when passed through a cold stream of inert gas, condenses in the form of small nanocrystalline metal particles. Metal-oxides can be obtained by including O_2 in the inert gas stream. Composite materials can be achieved by mixing two or more evaporation sources. The particle size distribution is controlled through the evaporation and condensation rates [75,76]. One variation of this method is arc discharge [37,77], where metal electrodes are vaporized under high voltage acceleration. Alternatively laser ablation [78,79] can be used for generating nanoparticles of high melting point compounds.

6.3.3 Mechanical Methods of Nanoparticle Synthesis [80-81]

Grinding and mixing of powder samples with or without liquid in a rotating mill partly filled with grinding media such as balls or pebbles can be used to reduce particle sizes to the nanometer range. Magnetic and catalytic nanoparticles can usually be produced by this method. Although a simple technique of relatively low cost it produces particles of broad size distribution, while contamination from the milling equipment is often a problem.

6.4 Synthesis of Complex Oxide Semiconductors

Most of the oxide semiconductors synthesized for H_2 production from aqueous suspension are composed of two or more components, hence the design and preparation of efficient photocatalyst materials are of paramount importance. For example, in the presence of NiO $SrTiO_3$ (perovskite type) is a highly active catalyst for water decomposition. Barium tetratitanate $BaTi_4O_9$, acts as an excellent catalyst when combined with RuO_2. Ruddlesden-Popper-type hydrous layered perovskites, $H_2La_{2/3}Ta_2O_7$, have shown higher activity than anhydrous perovskites ($KTaO_3$, $La_{1/3}TaO_3$) for photocatalytic water splitting under UV irradiation [82], while intercalation of Ni cocatalyst to $H_2La_{2/3}Ta_2O_7$ via an ion-exchange reaction further increases the activity. The photocatalytic activity of Dion-Jacobson-type layered perovskite, $K[Ca_2Nb_3O_{10}]$, is significantly enhanced by Pt loading [83]. The layered oxide was exfoliated into unilamellar nanosheets, which were subsequently loaded with Pt nanoparticles via photodeposition. The Pt-loaded nanosheets were flocculated with alkali-metal ions to restore a layered aggregate. Through this route Pt nanoparticles can be incorporated in a highly dispersed fashion into the host lattice of the layered perovskite [83]. Overall water splitting is observed with RuO_x-loaded $A[Ca_2Nb_3O_{10}]$ ($A = LI^+$, Na^+, K^+) restacked perovskite nanosheets under UV light irradiation [84]. It is clear from these examples that there are advantages in using a heterogeneous "oxide semiconductor system" consisting of two or more active components. To fabricate such a system requires, in general, multiple synthesis steps with impregnation and photodeposition the two most widely used methods for loading of the co-catalysts.

The majority of perovskite or layered perovskite type materials based on Nb(V), Ta(V) and Ti(IV) involve conventional solid-state reaction preparative methods, in which appropriate amounts of precursor oxides or carbonates are ground together and then calcined at high temperatures (typically 1000-1300°C) for times sufficient to allow cation interdiffusion. The materials produced by this method have low surface area, uncontrolled grain growth, localized segregation of the components, and possible loss of stoichiometry due to volatilization of the constituent components. To a great extent these underlying issues decrease the photocatalytic activity of a given catalyst. The Pechini-type sol-gel process (polymerizable complex) based on polyesterification between citric acid and ethylene glycol provides a better alternate strategy to synthesize mixed metal oxides, such as $M_xM'_yM''_zO_7$ (x, y, z = 0-4; M, M', and M'' represent three different metals), which yield ultra-fine nanoscale particles [56, 57]. **Figure 6.6** shows a flow chart of the synthesis procedure for $La_2Ti_2O_7$ crystals by the polymerizable complex method.

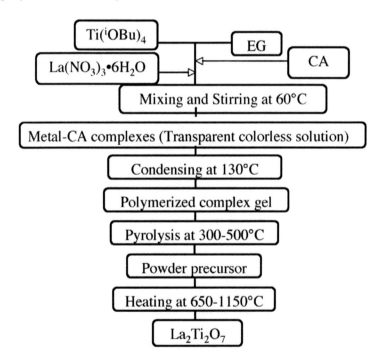

Fig. 6.6: A flow-chart for preparing $La_2Ti_2O_7$ by the Pechini-type sol-gel (polymerizable complex) method [56].

Metal, metal oxide, and mixed-metal oxide nanoparticles are strongly affected by heat treatments that lead to particle agglomeration, with a corresponding reduction in photocatalytic activities. Visible-light active zinc ferrite doped titania, TiO_2-$ZnFe_2O_4$ photocatalysts prepared by sol-gel demonstrated a decrease in photocatalytic activity with an increase in calcination temperatures over 400-600°C [85]. The sol-gel process in association with surfactant-assisted templating has been used to synthesize several Ta_2O_5-based photocatalysts loaded with NiO cocatalyst [86]. Pt- or NiO loaded solid state reaction synthesized Cr-doped $Ba_2In_2O_5/In_2O_3$ composite semiconductor has been found to be a novel photocatalyst system with enhanced activity for water splitting under UV and visible light irradiation [87].

MCM-41 is mesoporus silicate, and generally considered an excellent catalyst support; its pore size can be systematically varied and photocatalytic substances readily incorporated. Liu and Wang prepared Zr-doped MCM-41 using a hydrothermal route achieving photocatalytic water decomposition about 2.5 times higher than conventional ZrO_2 catalysts [88]. Hydrothermally prepared Cr-doped MCM-41, sol gel loaded with 25 wt% TiO_2 has been found active under visible light irradiation [89]. Doping of metal cations and/or non-metal anions into oxide semiconductors can improve their visible light photocatalytic response with, of course, phase composition and specific surface areas affecting photocatalytic activity [29,30,49,61-63,85-92]. Recently hydrothermal approaches, good for preparing uniform nanosized particles, followed by ammonia nitridation yielded (S,N) codoped-titania as a visible light active photocatalyst [90]. Aita and co-workers used a solvothermal process to prepare N-doped titania in a hexamethylenetetramine-alcohol solution [91]. Yuan and co-workers used urea and titanium chloride as starting materials in preparing N-doped TiO_2 [92]. Recently deposition of Cu_2O quantum dots on TiO_2 nanoparticles was reported for water photoreduction under sacrificial conditions [93].

6.5 Design of Oxide Semiconductors

In the past several years noble metal loading, ion doping, composite metal-oxide semiconductors, and multi-component semiconductors have been meticulously designed, fabricated, and then investigated for application to water splitting [45,46,49-59,63,65,82-89,93]. Liquid phase and gas phase syntheses routes, as described above, appear the most common synthesis approaches for production of these modified oxide semiconductor nanoparticle photocatalysts, some of which we now consider.

6.5.1 Metal Loaded Oxide Semiconductors
[6,7,9,10,12,15,17,45,94-107]

Various metals such as Cu, Ag, Au, Ni, Pd and Pt have been loaded on a variety of metal oxide semiconductors, but most of the work has focused on TiO_2. The Fermi level of metals is lower than that of TiO_2 so there is rapid transfer of photo-generated electrons from TiO_2 to the metal particles while the photo-generated holes remain in the TiO_2 valence band, resulting in efficient charge separation and consequently reduced possibility of electron-hole recombination. Although efficient charge separation by metal loading is partly able to reduce charge recombination, the occurrence of the thermodynamically favorable backward reaction inhibits generation of H_2 from pure water splitting.

Bamwenda et al. [45] employed Au and Pt loaded TiO_2 prepared by deposition-precipitation (DP), impregnation (IMP), photodeposition (PD), and colloidal mixing (CM) and compared their photocalytic activity for hydrogen production. Factors that affected H_2 production properties include synthesis methods, pretreatment conditions, metal loadings, and the oxide metal contact. The photocatlytic properties of nanoparticulate gold are found to be sensitive to the method of synthesis, with activity levels found in the following order: Au-TiO_2(FD) $\geq$ Au-TiO_2(DP) > Au-TiO_2(IMP) > Au-TiO_2(CM) and Pt-TiO_2(FD) > Au-TiO_2(DP) $\approx$ Au-TiO_2(IMP), respectively. Gold and platinum precursors calcined in air at 300°C were found to have the highest activity towards H_2 generation, followed by a decline in activity with increasing calcination temperature. The maximum H_2 yield observed for Pt-TiO_2 and Au-TiO_2 corresponded to metal loadings of, respectively, 0.3-1 wt.% and 1-2 wt.%. The roles of Au and Pt on TiO_2 seem to involve the attraction and trapping of photogenerated electrons, the reduction of protons, and the formation and desorption of hydrogen. The higher overall activity of Pt samples is probably a result of the more effective trapping and pooling of photogenerated electrons by Pt and/or because platinum sites have a higher capability for the reduction reaction.

Hydrogen production from water by irradiating suspensions of Pt, Pd and Rh-loaded WO_3 in methylviologen (MV^{2+}) solution with visible light demonstrated efficiencies, in rank order, of Rh > Pt > Pd [95]. The conduction band level of WO_3 is not negative enough to reduce H^+ to H_2, but it readily reduces MV^{2+} which in turn helps reduce water. Simultaneous evolution of O_2 was observed since the valence band level of WO_3 is below the water oxidation level. The metal dopants act as sensitizers to increase the visible light absorption efficiency of the host semiconductors, i.e. photosensitization, and are also involved in the electron-transfer reactions such as capturing and transferring of conduction band electrons and/or valance band holes, thereby preventing electron-hole recombination and behaving as active sites for H^+ reduction.

Role of Carbonate Salts [97-100]

Despite a few successful experimental results Pt-loaded TiO_2 catalysts do not prevent the occurrence of backward reactions, with either reduced or no O_2 evolution from illuminated aqueous suspensions. Mill and Porter [96] identified photoadsorbed O_2 and $O_2^{\cdot-}$ over Pt-TiO_2 as agents interrupting the water splitting process. According to Arakawa and his coworkers [97], addition of sodium carbonate to the Pt-TiO_2 suspension can significantly improve stiochiometric decomposition of liquid water. Such systems [17, 98-100] introduce catalysts that facilitate the reaction as shown:

$$Na_2CO_3 \rightarrow 2Na^+ + CO_3^{2-} \tag{6.5.1}$$
$$CO_3^{2-} + H^+ \rightarrow HCO_3^- \tag{6.5.2}$$
$$HCO_3^- + h^+ \rightarrow HCO_3^{\cdot} \tag{6.5.3}$$
$$HCO_3^{\cdot} \rightarrow H^+ + CO_3^{\cdot-} \tag{6.5.4}$$
$$2CO_3^{\cdot-} \rightarrow C_2O_6^{2-} \tag{6.5.5}$$
$$C_2O_6^{2-} + 2h^+ \rightarrow O_2 + 2CO_2 \tag{6.5.6}$$

As seen in reaction (6.5.3) photogenerated holes are consumed, making electron-hole separation more effective as needed for efficient water splitting. The evolution of CO_2 and O_2 from reaction (6.5.6) can promote desorption of oxygen from the photocatalyst surface, inhibiting the formation of H_2O through the backward reaction of H_2 and O_2. The desorbed CO_2 dissolves in aqueous suspension, and is then converted to HCO_3^- to complete a cycle. The mechanism is still not fully understood, with the addition of the same amount of different carbonates, see **Table 6.2**, showing very different results [99]. Moreover, the amount of metal deposited in the host semiconductor is also a critical factor that determines the catalytic efficiency, see **Fig. 6.7**.

Table 6.2: Rate of photocatalytic production of H_2 and O_2 over a Pt/TiO$_2$ suspension with salts additives [99].

Additives	Amount (mol)	Rate of evolution (μmol/h)[a]			
		H_2	O_2	H_2/O_2	pH
None	-	1	0	-	7.9
Na$_2$CO$_3$	0.38	316	158	2.0	11.0
	0.76	568	287	2.0	11.0
	1.14	39	13	3.0	11.0
NaHCO$_3$	0.36	4	2	2.0	8.4
Li$_2$CO$_3$	0.06	5	1	5.0	10.8
K$_2$CO$_3$	0.38	5	1	5.0	11.5
	1.45	20	10	2.0	11.8
NaOH	0.76	5	0	-	13.3
NaCl	0.76	5	0	-	6.8
Na$_2$SO$_4$	0.38	4	0	-	5.5
Na$_3$PO$_4$	0.25	5	0	-	12.0

Catalyst: (0.3 wt% Pt/TiO$_2$); 0.3 g of catalyst suspended in 350 ml of water placed in an inner irradiator quartz cell. Irradiated with high-pressure mercury lamp (400 W).
[a] Rate at steady state.

With reference to **Fig. 6.7**, note there is no photocatalytic activity without Pt loading [99]; the rate of gas evolution increases with increasing Pt content reaching a maximum at 0.3 wt%. Other metals have shown similar trends in photocatalytic activity. Therefore, it is suggested that metal loading is one of the essential requirements for photocatalytic decomposition of liquid water. However the back reaction of evolved gas on the Pt particles increases with Pt loading. To suppress the backward reaction of H_2

and O_2 recombination the addition of an iodide anion (I^-) has shown significant effect [101]. It can preferentially adsorb on the Pt surface of an aqueous suspension of Pt-TiO$_2$, thereby forming an I_2 layer. The iodine layer in turn reduces the chances of a backward reaction occurring. However adding too much carbonate salt or iodide anion beyond an optimum level reduces the beneficial effects since the adsorbed species on the catalyst surface commonly decrease the light harvesting.

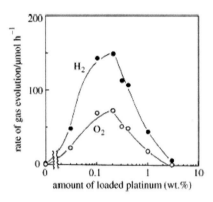

Fig. 6.7: Dependence of H$_2$ and O$_2$ evolution upon platinum loading. Water = 350 mL; Pt/TiO$_2$ = 0.3 g; Na$_2$CO$_3$ = 80 g.   Reproduced with permission from Ref. [99].

Role of Mediators [45,102-107]

Rapid recombination of photogenerated electrons and holes can be controlled by electron donors (or hole scavengers) that irreversibly react with photo-generated holes leading to enhanced hole-electron separation. Since electrons are consumed during the photocatalyic process, for sustainable hydrogen production continuous electron-donating ability is required. Methanol, ethanol, lactic acid, formaldehyde, EDTA, CN$^-$ have been widely used and proven to be effective electron donors for photocatalytic hydrogen production [45, 102-104]. For example, Wu and co-workers [102] reported the use of Cu-TiO$_2$ in aqueous methanol solutions. Nada et al. [104] employed different electron donors studying their effect on hydrogen production; the rate of hydrogen production was found to decrease in the following order: EDTA > Methanol > ethanol > lactic acid. The hydrogen evolution rate over other metal oxide semiconductors are summarized in **Table 6.3.**

Table 6.3: Rate of H_2 evolution from aqueous methanol solution [17].

Catalyst	Rate of H_2 evolution (μmol/h)	Pt-catalyst	Rate of H_2 evolution (μmol/h)	Light abs. (< nm)
TiO_2	29	Pt/TiO_2	1166	413
Ta_2O_5	381	Pt/Ta_2O_5	2583	310
ZrO_2	971	Pt/ZrO_2	1546	248

Catalyst, 1.0 g; Water 350 mL; Pt 0.1 wt%, Light source: 400 W high-pressure Hg lamp; inner irradiation type quatz cell

Various pairs of inorganic ions such as IO_3^-/I^-, Fe^{3+}/Fe^{2+}, and Ce^{4+}/Ce^{3+} have been used as redox mediators to facilitate electron-hole separation in metal loaded oxide semiconductor photocatalysts [105-107]. Two different photocatalysts, Pt-TiO_2 (anatase) and TiO_2 (rutile), suspended in an aqueous solution of NaI were employed to produce H_2 and O_2 under, respectively, the mediation of I^- (electron donor) and IO_3^- (electron acceptor) [105]. The following steps are involved in a one-cell reaction in the presence of UV light.

At the Pt-TiO_2 (anatase) conduction band
$$6H^+ + 6e^- \rightarrow 3H_2 \qquad (6.5.7)$$

At Pt-TiO_2 (anatase) valence band
$$I^- + 6OH^- + 6h^+ \rightarrow IO_3^- + 3H_2O \qquad (6.5.8)$$

At the TiO_2 (rutile) conduction band
$$IO_3^- + 3H_2O + 6e^- \rightarrow I^- + 6OH^- \qquad (6.5.9)$$

At the TiO_2 (rutile) valance band
$$6OH^- + 6h^+ \rightarrow 3/2O_2 + 3H_2O \qquad (6.5.10)$$

For hydrogen production on the photocatalyst with the more negative conduction band, I^- can scavage holes [reaction (6.5.8)] thus conduction band electrons are available to reduce protons to H_2

molecules [reaction (6.5.7)]. For oxygen production on the photocatalyst with the more positive valence level, IO_3^- can react with electrons to form I^- [reaction (6.5.9)], hence the valance band holes are available to oxidize water forming oxygen [reaction (6.5.10)].

6.5.2 Mixed Metal-Oxide Semiconductors
[3,5,10,11,13-16,108,111-133]

Domen and co-workers, who observed the effect of NiO over $SrTiO_3$ powder for water splitting into H_2 and O_2 [5], later studied the mechanism and found that NiO acts as hydrogen evolution site while oxygen evolution occurs at $SrTiO_3$ [13]. A remarkable catalytic effect of RuO_2 on powdered TiO_2 for hydrogen evolution from an ethanol-water mixture has been observed by Sakata et al. [108]. To investigate the supporting effect of RuO_2 on TiO_2, experiments varying the amount of RuO_2 on a fixed amount of TiO_2 were carried out [108]; the rate for RuO_2/TiO_2, under illumination by a 500 W Xe lamp, was found to be 30 times higher than that for TiO_2 alone and sensitive to increasing amounts of RuO_2 reaching a maximum at 0.4×10^{-3} molar ratio of RuO_2/TiO_2, beyond which a decrease in performance was observed as the concentration of RuO_2 further increased. If present in small quantities these metal oxides act as electron pools (reducing sites) where H^+ is reduced to hydrogen. However when the concentration exceeds a critical limit they act as electron-hole recombination centers. One should, of course, use caution in deciding the amount of metal oxide co-catalyst to be loaded onto semiconductor particles, also chemical additives such as carbonate salt can significantly influence the water splitting reaction over mixed oxides.

Table 6.4: Photocatalytic decomposition of water over various semiconductor catalysts in Na_2CO_3 solution or pure water.

Semiconductor (Band gap, eV)	Loading (metal or metal oxide)	Rate of gas evolution (µmol/h)[a]				Ref.
		Na_2CO_3 solution		Pure Water		
		H_2	O_2	H_2	O_2	
TiO_2 (3.0)	None	Tr.	0	Tr.	0	17
	Pt (0.3 wt%)	78	38	2	0	
	RuO_2 (1.0 wt%)	34	17	Tr	0	
	NiO (1.0 wt%)	64	32	1	0	
Ta_2O_5 (4.0)	None	Tr.	0	Tr.	0	17
	Pt (0.3 wt%)	1	0	1	0	
	RuO_2 (1.0 wt%)	68	34	32	17	
	NiO (1.0 wt%)	153	79	190	99	
ZrO_2 (5.0)	None	142	75	72	36	17
	Pt (0.3 wt%)	53	23	Tr.	0	
	RuO_2 (1.0 wt%)	12	6	11	5	
	NiO (1.0 wt%)	43	22	129	70	
$SrTiO_3$ (3.0)	None	Tr.	0	Tr.	0	17
	Pt (0.1 wt%)	10	4	9	2	
	Rh (0.1 wt%)	48	14	20	4	
	NiO (1.0 wt%)	41	20	9	4	
$BaTi_4O_9$ (3.7)	Pt (0.1 wt%)	2	Tr.	Tr.	0	17
	RuO_2 (1.0 wt%)	36	18	30	13	
$K_2Ti_6O_{13}$	Pt (0.1 wt%)	63	17	Tr.	0	17
	RuO_2 (1.0 wt%)	49	24	11	1	
$K_4Nb_6O_7$	Pt (0.3 wt%)	451	217	Tr.	0	17
	RuO_2 (1.0 wt%)	60	28	403	197	
	NiO (1.0 wt%)	41	20	211	100	
$NaTaO_3$ (4.0)	None	-	-	160	86	111
	NiO (0.05 wt%)	-	-	2180	1100	111
	Au (0.3 wt%)	-	-	642	224	124
$LiTaO_3$ (4.7)	None	-	-	430	220	112
	NiO (0.1 wt%)	-	-	98	52	112
$KTaO_3$ (3.6)	None	-	-	29	13	112
	NiO (0.1 wt%)	-	-	7	3	112
	Au (0.3 wt%)	-	-	58	25	124
$Ba_2Ta_2O_6$ (4.1)	None	-	-	33	15	113
	NiO (0.5 wt%)	-	-	629	303	
$Sr_2Ta_2O_6$ (4.4)	None	-	-	140	66	114
	NiO (0.1 wt%)	-	-	960	490	
$Sr_2Ta_2O_7$ (4.6)	None	-	-	53	18	115
	NiO (0.15 wt%)	-	-	1000	480	115
$Sr_2Nb_2O_7$ (4.6)	None	-	-	5.9	0	115
	NiO (0.15 wt%)	-	-	10	3.2	115

Table 6.4 summarizes the water splitting properties of a number of mixed oxide semiconductors. In the presence of a co-catalyst such as NiO several highly active titanates, tantalates and niobates photocatalysts are reported for water splitting under UV light irradiation; $NiO/NaTaO_3$ is one of the most active catalysts, with a crystal structure of corner-shared MO_6 octahedra. In such compounds, e.g. $NaTaO_3$, the top of the valence band is primarily oxygen 2p non-bonding in character, while the conduction band arises from the π^* interaction between the transition metal t_{2g} orbitals (Ta_{5d}) and oxygen (O_{2p}). The band position changes with distortion and connection of the TaO_6 units. A study on luminescent properties has concluded that the closer the M-O-M bond angle is to 180° the more the excitation energy is localized [109]. Eng and co-workers [110] have carried out computational and experimental studies using orbital methods and UV-visible diffuse reflectance spectroscopy to quantitatively probe the relationships between composition, crystal structure and the electronic structure of oxides containing octahedrally coordinated d^0 transition metal ions such as Ti^{4+}, Nb^{5+}, and Ta^{5+}, primarily in perovskite and perovskite-related structures. For isostructural niobate, titanate and tantalate compounds the band gap increases as the effective electronegativity of the transition metal ion decreases. The effective electronegativity decreases in the following order: $Nb^{5+} \sim Ti^{4+} > Ta^{5+}$. Eng [110] also observed that the band gap is sensitive to changes in the conduction bandwidth, which can be maximized for structures possessing linear M–O–M bonds, such as the cubic perovskite structure. As this bond angle decreases the conduction band narrows and the band gap increases. This tendency indicates that photo-generated electron hole pairs can migrate relatively easily in the corner-shared framework of TaO_6 units. Niobate photocatalysts show differences in photocatalytic properties mainly due to conduction band energy levels, which are lower than in their Ta counterparts, and the degree of excitation energy delocalization.

Table 6.5: Effect of co-catalyst loaded $NaTaO_3$ photocatalysts on water splitting [112].

Cocatalyst	Loading method	Amount (wt%)	Rate of gas evolution (μmol/h)	
			H_2	O_2
None			166	93
CoO	IMP	0.1	42	21
NiO	IMP	0.1	1700	822
Ni	PD	0.16	219	110
CuO	IMP	0.14	72	38
RuO_2	IMP	0.2	147	78
Ag	IMP	0.2	38	12
Pt	PD	0.1	2.6	0

Catalyst: 1.0 g; Pure water: 350 mL; Light source: 400-W high-pressure mercury lamp; IMP stands for impregnation and PD stands for photodeposition.

The photocatalytic activities of the alkali tantalates, prepared by solid state reaction in the presence of excess alkali, lie in the following order: $KTaO_3$ > $NaTaO_3$ > $LiTaO_3$. As given in **Table 6.5**, NiO is found to be the most effective cocatalyst when loaded onto $NaTiO_3$. The conduction band levels of $NaTaO_3$ and NiO are, respectively, −1.06 eV and −0.96 eV. Therefore, the photogenerated electrons in the $NaTaO_3$ conduction band are able to transfer to the conduction band of NiO, which is an active site for hydrogen evolution and enhances charge separation. Pt is well known as an excellent co-catalyst for hydrogen evolution when combined with metal oxides. However Pt in combination with $NaTaO_3$ resulted in minimal hydrogen evolution, and no oxygen evolution due to the occurrence of a rapid reverse reaction on the Pt. Calcination temperatures have a strong influence on the $NiO/NaTiO_3$ activity due to the balance between activation by decomposition of $Ni(NO_3)_2$ and deactivation due to the destruction of the H^+ exchanged $NaTaO_3$ surface [111,112]. Similar behavior has been observed in other NiO loaded tantalates such as $BaTa_2O_6$, $SrTa_2O_6$, $Sr_2Ta_2O_7$ and $K_3Ta_3Si_2O_{13}$ with water splitting data summarized in **Table 6.4** [113-119].

$K_3M_3Si_2O_{13}$ (M =Ta and Nb) [119] possesses a unique pillared structure, in which three corner sharing MO_6 chains are bridged by ditetrahedral Si_2O_7 units. No photocatalytic activity for water splitting was observed from $K_3Nb_3Si_2O_{13}$, 4.1 eV bandgap, with or without cocatalyst addition. The M-O-M bond angles in

MO_6 chains of $K_3Ta_3Si_2O_{13}$ (bandgap = 3.9 eV) are 178° and 163° while those of $K_3Nb_3Si_2O_{13}$ are 174° and 167°. As described above, the distortion of the MO_6 octahedra strongly affect the energy structure of the photocatalysts. The 0.2 eV difference in bandgap is enough to dramatically alter the photocatalytic activities. $K_3Nb_3Si_2O_{13}$ showed no photoluminescence, ascribed to recombination between photo-generated electron - hole pairs at luminescence centers, with measurements at temperatures as low as 77K indicating that non-radiative transitions readily occur in $K_3Nb_3Si_2O_{13}$ leading to no photocatalytic activity.

Table 6.6 : Photocatalytic activities of mixed oxides prepared by SSR and PC methods.

Semiconductor	Preparation method (Cal. Temp.)[a]	Cocatalyst (wt%)	Surface area ($m^2\,g^{-1}$)	Rate of gas evolution (μmol/h)		Ref.
				H_2	O_2	
$Li_2Ti_2O_7$	SSR (1150°C)	Ni (1.0)	1.0	440	217	56
	PC (1050°C)	Ni (1.0)	1.0	960	478	
$Sr_2Nb_2O_7$ (3.9)	SSR (1100°C)	None	0.7	5.9	0	57
	SSR (1100°C)	NiO (0.15)	0.7	10	3.2	
	PC (900°C)	None	4.3	7.5	3.5	
	PC (900°C)	NiO (0.15)	4.3	94	46	
$Sr_2Ta_2O_7$ (4.5)	SSR (1100°C)	None	0.7	52	10	57
	SSR (1100°C)	NiO (0.15)	0.7	1000	480	
	PC (900°C)	None	4.3	2787	1347	
	PC (900°C)	NiO (0.15)	4.3	94	46	
	PC (800°C)	NiO(0.15)	10.4	3517	1733	
$K_2La_2Ti_3O_{10}$	SSR(900°C)	Ni (3 at.%)	4	1255	632	120
	PC (900°C)	Ni (3 at.%)	5	2186	1131	
$Sr_5Ta_4O_{15}$ (4.75)	PC (1000°C)	NiO (0.15)	2.99	1194	722	121
$Sr_4Ta_2O_9$ (4.81)	PC (1000°C)	NiO (0.15)	2.61	32	2	121
$Ba_5Ta_4O_{15}$ (3.90)	PC (800°C)	NiO (0.2)	-	7110	3621	122

[a]Calcination temperature. Catalyst: 0.7 g in 350 mL of water. Light source: 400-W high - pressure mercury lamp. PC stands for polymerizable complex while SSR stands for solid state reactions.

Table 6.6 summarizes photocatalytic water splitting properties of various mixed oxides prepared from two different processes, solid-state reaction (SSR) and solution based Pechini-type polymerizable complex routes [56,57,120-122]. $Ba_5Ta_4O_{15}$ prepared by complex polymerization (sol-gel) shows the most remarkable photocatalytic activity of the layered perovskite type mixed oxide photocatalysts. $Ba_5Ta_4O_{15}$ precursor prepared with an excess amount of Ba and Ta, 25 at% above stoichiometric composition, produced 7110 and 3621 μmol/h H_2 and O_2, respectively, a rate nearly 3.5 times greater than that of the same compound fabricated with a precursor of stoichiometric composition (0.2 wt% NiO loading in each case) [122]. The topology of MO_6 (M = Ti, Ta, Nb) bonding in perovskite type compounds and its effect on the electronic structure of these systems is described by Eng and co-workers [110], who predicted that completely isolating the MO_6 octahedra (e.g., the ordered double perovskite structure) dramatically narrows the width of the conduction band leading to a significant increase in the band gap. These tantalates are composed of corner-sharing TaO_6 units. $Sr_4Ta_2O_9$ possesses the structure of the double perovskite and TaO_6 octahedra giving rise to an extremely narrow conduction band that cannot provide efficient carrier migration. The deviation of the M-O-M bond angle from 180° has been found to play a crucial role in photocatalyst performance. Thus despite the different symmetry and slightly different atomic layer arrangement in $Sr_4Ta_2O_9$ (cubic), $Sr_5Ta_4O_{15}$ (hexagonal) and $Sr_2Ta_2O_7$ (hexagonal), the trend in band gap energy should reflect differences in the width of the in conduction band [121], while inductive effects due to the presence of electropositive "spectator" cations (Li^+, Sr^{2+}, Ba^{2+}) tend to be small and can generally be neglected.

Abe and coworkers prepared R_3MO_7 and $R_2Ti_2O_7$ (R = Y, Gd, La; M = Nb, Ta) using the Pechini type complex polymerizable method [123]. These materials are composed of a network of corner-shared octahedral units of metal cations (TaO_6, NbO_6, or TiO_6); materials without such a network remain inactive. It is believed that the octahedral network increases the electron/hole mobility thereby enhancing photocatalytic activity. The photodeposition of gold nanoparticles on these photocatalysts (tantalates, niobates and titanates) containing shared MO_6 octahedra improve the overall water splitting reaction, see **Table 6.4** [124]. It appears that low doping levels of nanoscale gold particles minimizes backward reactions, enhances charge separation, and act as hydrogen evolution sites.

Visible Light Activity [125-133]

Most of the photocatalysts that we have discussed to this point have band gaps (> 3.0 eV) too large for responding to visible light. There are methods by which photocatalysts can be fabricated such that they will absorb visible light, including ion doping of wide band gap materials through hybridization of N 2p and O 2p orbitals as will be discussed in the next section.   Another route is valence band formation using elements other than oxygen, for which single-phase oxide materials are of particular interest due to their intrinsic stability. Such materials include $BiVO_4$ [125], $AgNbO_3$ [126], Ag_3VO_4 [127], $Ca_2Bi_2O_4$ [128], $PbBi_2Nb_2O_9$ [129] and $InMO_4$ (M = Ta^{5+}; Nb^{5+}) [130]. The visible light responses of $BiVO_4$ and $AgNbO_3$ are attributable to a decrease in the band gaps due to valence band formation by not only O_{2p} but also Bi^{3+}_{6s} and Ag^+_{4d} orbitals. Visible light absorption in $PbBi_2Nb_2O_9$ is due to the transition from O_{2p} hybridized with Pb_{6s}/Bi_{6s} to Nb_{4d} and is reported to directly cleave water into H_2 and O_2. $InTaO_4$ and $InNbO_4$, with band gaps in the range $2.8 - 2.4$ eV, are known to split water [130]. There are two kinds of octahedral, InO_6 and NbO_6 (or TaO_6), in a unit cell for both $InNbO_4$ and $InTaO_4$. The difference in unit cell volumes between TaO_6 and NbO_6 leads to a change in the lattice parameters slightly affecting their photocatalytic activity, mainly due to variation in the conduction band levels formed by Ta_{5d} TaO_6 and Nb_{4d} NbO_6.

Tai et al. has reported the preparation of $K_2La_2Ti_3O_{10}$ by a complex polymerization process followed by Au nanoparticle loading, either by impregnation (Au-i) or deposition-precipitation (Au-d) methods [131]. Au-i/$K_2La_2Ti_3O_{10}$ has shown superior water splitting properties in the UV and visible region compared to its counterpart Au-d/$K_2La_2Ti_3O_{10}$. The Au impregnation process resulted in better sample crystallinity than that prepared through Au-deposition-precipitation. Au-i/$K_2La_2Ti_3O_{10}$ is found more active for water splitting than Ni/$K_2La_2Ti_3O_{10}$ [120] under visible light illumination, but less active under UV. The high visible light activity of Au-i/$K_2La_2Ti_3O_{10}$ is attributed to plasma resonance absorption on the surface of the Au nanoparticles. This plasma resonance can polarize the electron distribution on the surface of the Au nanoparticles, which increases the probability for electron transfer from gold to the $K_2La_2Ti_3O_{10}$ conduction band. In the case of Ni there is only absorption from the d–d transition in the visible region, which is not sufficient for transferring an electron from the Ni to the $K_2La_2Ti_3O_{10}$ surface [131].

Intercalation of nanosized Fe_2O_3 and TiO_2 particles within layered compounds such as $HNbWO_6$, $HTaWO_6$, $HTiNbO_5$ and $HTiTaO_5$ have been found to dramatically improve visible light photocatalytic activity [132,133], a behavior ascribed to more effective separation of photogenerated electrons - holes due to their rapid diffusion. Calcined samples of $Fe-HTiNbO_5$ and $Fe-HTiTaO_5$ are yellow in color, showing an absorption band edge at 600 nm. While their band gap energies are nearly 2.0 eV, the conduction band position is not negative enough to reduce water except in the presence of an electron acceptor ($AgNO_3$) [133].

6.5.3 Ion Doped Semiconductors

A common practice in modifying the band gap of a semiconductor is to include a small percentage of foreign elements into the regular crystal lattice. This is especially important in photocatalysis where high band gap oxide semiconductors can only be activated by light energy equal to or greater than their band gap energy, which generally falls in the UV region.

Cation Doping [49,58,62,87,89,134-149]

Doping of transition metal (V, Cr, Fe, Mo, Ru, Os, Re, V, Rh, etc.) and rare earth metal (La, Ce, Er, Pr, Gd, Nd, Sm, etc.) ions have been extensively used as a means of enhancing visible light-induced photocatalytic activities. Several experimental results confirm that metal ion doping expands the photo-response of TiO_2 into the visible region [134,135]. The metal ions incorporated into the TiO_2 lattice create new (impurity) energy levels within the TiO_2 band gap as shown in reactions (6.5.11) and (6.5.12), where M^{n+} represents dopant metal ions.

Charge pair generation

$$M^{n+} + h\nu \rightarrow M^{(n+1)+} + e^-(CB) \qquad (6.5.11)$$
$$M^{n+} + h\nu \rightarrow M^{(n-1)+} + h^+(VB) \qquad (6.5.12)$$

$e^-(CB)$ refers to a conduction band electron, or a trapped electron, and $h^+(VB)$ refers to a valence band hole, or a trapped hole; with respect to charge trapping:

Charge trapping

$$\text{Electron trap: } M^{n+} + e^-(CB) \rightarrow M^{(n-1)+} \qquad (6.5.13)$$
$$\text{Hole trap: } M^{n+} + h^+(VB) \rightarrow M^{(n+1)+} \qquad (6.5.14)$$

As shown in reactions (6.5.13) and (6.5.14), metal ion dopants influence the photoreactivity of metal oxides by acting as electron (or hole) traps thereby altering the e^-/h^+-pair recombination rate. The energy level of $M^{n+}/M^{(n-1)+}$ should be less negative than that of the TiO_2 conduction band (CB) edge, while the energy level of $M^{n+}/M^{(n+1)+}$ should be less positive than that of the TiO_2 valance band (VB) edge. Introduction of such energy levels within the band gap induces a red shift in the band gap transition, with enhanced visible light absorption due to charge transfer between a dopant energy level and the CB (or VB), or a d-d crystal field transition [134]. Since photocatalytic reactions are a surface phenomena, deep doping of metal ions are to be avoided for better charge injection or transfer properties. As we have repeatedly seen in various experiments the concentration of added metal ions affects the photocatalytic activity; above the optimum doping concentration photocatalytic activity decreases due to increased charge recombination [134].

Peng and co-workers [136] investigated the photocatalytic production of hydrogen using Be^{2+} doped TiO_2; the aqueous suspension contained 1.0 ml of ethanol (electron donor) and 0.33 ml of H_2PtCl_6. A 250 W high pressure Hg lamp was used as a light source. Two different methods, coprecipitation (CP) and impregnation (IMP) were chosen to prepare the required photocatalysts. **Figure 6.8(a)** shows the influence of doping method on hydrogen production by water splitting. Compared to the undoped TiO_2 both 1.25 at% Be^{2+}-TiO_2 and Be^{2+}-$P_{25}TiO_2$ (P_{25} is commercially available from Degussa AG, Germany. Surface area = 45 m^2g^{-1}, 70% anatase and 30% rutile, purity >99.5%) prepared by IMP show a 75% increase in hydrogen production. In dramatic contrast, 1.25 at% Be^{2+}-TiO_2 and Be^{2+}-$P_{25}TiO_2$ prepared by coprecipitation (CP) shows a 36% decrease in the photocatalytic hydrogen production compared to pure TiO_2. Low activity in the CP samples can be attributed to the presence of the Be^{2+} ions within the bulk TiO_2 lattice hence the trapped photoexcited electron cannot easily migrate to the surface. The IMP Be^{2+}-TiO_2 doped ions are located on or near the TiO_2 surface, providing ready access of the photoexcited electrons to the reaction interface. As anticipated the amount of dopant further influences the hydrogen production rate, **Fig. 6.8b**. The deposition of Be^{2+} onto TiO_2 particles can produce oxygen vacancies, and lead to formation of a space charge layer facilitating separation of the photo-generated electron-hole pairs. Also Ti-O-Be formed via Be^{2+} entering the TiO_2 surface could promote charge transfer due to the difference in electro-negativity of Ti and Be, resulting in an increase in photocatalytic activity. As anticipated, a continual increase in Be^{2+} doping concentration beyond some optimal point shows degradation in photocatalytic properties. The origin for this behavior can vary with the material system being studied; in the case of Be^{2+} on a TiO_2 surface it appears to increase the number of recombination centers [136]. This observation is similar to one observed by Choi et al with other metal ion dopants [134].

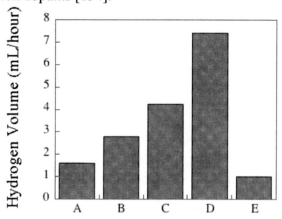

Fig. 6.8a: Photocatalytic hydrogen production from samples doped via coprecipitation (CP) and impregnation (IMP). **(a)** TiO_2; **(b)** 1.25 at% Be^{2+}-TiO_2-IMP; **(c)** P_{25}-TiO_2; **(d)** 1.25 at% Be^{2+}-P_{25}-TiO_2-IMP; **(e)** 1.25 at% Be^{2+}-P_{25}-TiO_2-CP.

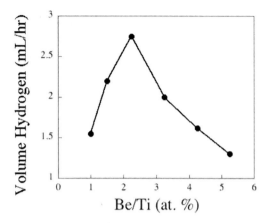

Fig. 6.8b: Photocatalytic hydrogen production as a function of dopant (Be^{2+}) content [136].

Hameed et al. [137] investigated the photolysis properties of transition metal ion doped WO_3 (bandgap = 2.6 eV). As seen in **Fig. 6.3**, the conduction band level of WO_3 is lower than $E°(H_2/H_2O)$ hence WO_3 suspended in water cannot reduce protons to hydrogen molecules. Doping with metal ions such as Fe, Co, Ni, Cu and Zn allows formation of the corresponding oxides on the WO_3 surface. Since the conduction band levels of Fe, Co, Ni and Cu oxides are higher than $E°(H_2/H_2O)$, the conduction band electrons can readily reduce protons [137]. In the case of Mg-doping (10 wt%), the conduction band edge position shifted about 1.82 V higher than $E°(H_2/H_2O)$ thereby thermodynamically allowing H^+ ion to reduce to hydrogen, however the valence band edge shifted about 0.25 V higher than $E°(H_2O/O_2)$ inhibiting water oxidation [49].

Mixed metal oxides having spinel ($MM'O_4$), perovskite ($MM'O_3$) and layered perovskite ($M_2M_2'O_7$) structures, where M and M' represent two different metals, have been doped with various transition metal ions to tailor their band gap towards the visible spectrum. Direct splitting of water under visible light irradiation with Ni doped indium-tantalum-oxide, $In_{(1-x)}Ni_xTaO_4$ (x = 0-0.2) has been reported by Zou and co-workers [138]. This catalyst was prepared as follows: pre-dried solid In_2O_3, Ta_2O_5 and NiO powders were heated at 1100°C to prepare $InNiTaO_4$. 1.0 wt% Ni particles were impregnated on the surface of $InNiTaO_4$ followed by calcination at 350°C and reduction in H_2 atmosphere (200 torr) at 773°C, then treated in O_2 atmosphere (100 torr) at 200°C. The rates of hydrogen and oxygen evolution for $In_{0.90}Ni_{0.10}TaO_4$ are found to be about 16.6 and 8.3 µmol/h ($\lambda > 420$ nm), respectively, and the quantum yield at 402 nm approximately 0.66%. The Ni doping reduced the band gap from 2.6 eV (undoped) to 2.3 eV (doped), attributed to internal d-d transitions in a partly filled d shell.

The introduction of a new valence band or electron donor energy level appears an effective, or at least reasonable, strategy for developing visible light photocatalysts. These photocatalysts and their photocatalytic activities are summarized in **Table 6.7** [139-144]. Rh doped $Pt/SrTiO_3$ shows visible light response ($\lambda > 420$ nm) due to the transition from the electron donor level formed by the Rh^{3+} ion to the conduction band composed of the Ti_{3d} orbitals of $SrTiO_3$ [139]. Similarly, $Pt/SrTiO_3$ codoped with a combination of either antimony, or tantalum and chromium, involves energy gap level transitions from the donor energy level formed by Cr^{3+} to the conduction band of $SrTiO_3$, resulting in visible light absorption. Charge balance is maintained by codoping of Sb^{5+} and Ta^{5+}, resulting in the suppression of the formation of Cr^{6+} ions and oxygen defects within the lattice [140,141].

Table 6.7: Visible-light-driven ($\lambda > 420$ nm) photocatalysts for H_2 and/or O_2 from aqueous solution in the presence of sacrificial reagents.

Semiconductors:doped	Band gap (eV)	Sacrificial reagent	Rate of gas evolution (µmol/h)		Ref.
			H_2	O_2	
$Pt/SrTiO_3$: Rh	2.3	CH_3OH	117	-	139
TiO_2:Cr, Sb	2.5	$AgNO_3$	-	42	140
$Pt/SrTiO_3$:Cr, Sb	2.5	CH_3OH	78	-	140
$Pt/SrTiO_3$:Cr, Ta	2.3	CH_3OH	70	-	141
$Pt/La_2Ti_2O_7$:Cr	2.2	CH_3OH	15	-	142
$Pt/La_2Ti_2O_7$:Fe	2.6	CH_3OH	10	-	142
$InTaO_4$: Ni	2.6	$CH_3OH/AgNO_3$	3.1	1.1	143
TiO_2: Ni, Ta	2.6	$CH_3OH/AgNO_3$	-	7.6	144
$SrTiO_3$:Ni, Ta	2.8	$CH_3OH/AgNO_3$	2.4	0.5	144

Reactions were carried out in presence of sacrificial agents. O_2 evolution: aqueous silver nitrate solution. H_2 evolution: aqueous methanol solution.
Light source: 300-W Xe lamp

Lanthanum doping of NiO-loaded NaTaO₃, NiO(0.2 wt%)/NaTaO₃:La(2%) resulted in a 9x increase in H_2 and O_2 evolution rates to 19.8 and 9.7 mmol/h, respectively, under UV light illumination [145]; the maximum apparent quantum yield reported was 56% at 270 nm. Electron microscope observations revealed the particle size of the NiO/NaTaO₃:La photocatalyst lies in the range of 100-700 nm, smaller the undoped NiO/NaTaO₃ crystals of 2-3 μm. The small particle size, excellent crystallinity, as well as creation of a surface topology comprised of 'nano-steps' due to the La doping appears to help reduce unwanted electron-hole recombination. Loading of ultrafine NiO particles on the step edges helps separate the H_2 and O_2 evolution sites avoiding back reaction during water splitting, **Fig. 6.9**. Evidence suggests that H_2 evolution proceeds on the ultrafine NiO particles, while O_2 evolution occurs at grooves on the step-like structures of the La:NaTaO₃.

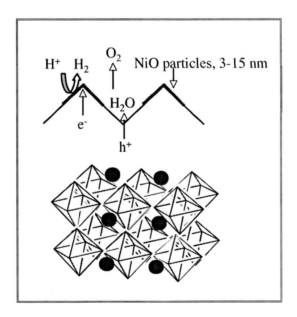

Fig. 6.9: Mechanism of water splitting on surface of NiO/NaTaO₃:La photocatalyst [145].

IrO₂ has also been used as a co-catalyst with NaTaO₃:La [146]; it was found that IrO₂ is involved in the formation of sites active for O_2 evolution. Other metal oxides such as γ-Bi₂O₃ and

Fe_2O_3, doped with various metal ions have been applied to photocatalytic hydrogen production in the presence of methyl viologen (MV^{2+}) that serves as an electron donor [147]. The photocatalytic hydrogen production efficiency of γ-Bi_2O_3 has shown an activity doping dependence of: $Pd^{2+} > Ru^{3+} > Co^{2+} > Ni^{2+} > Fe^{3+} > Cr^{4+} >$ undoped γ-Bi_2O_3. Formation of Schottky-type barrier at the metal dopant - γ-Bi_2O_3 junction leads to more effective charge separation due to the enhanced internal electric field. The smaller the barrier height, given by the difference between the work function of the dopant and electron affinity of γ-Bi_2O_3, the greater the electron flow from γ-Bi_2O_3 to the metal, thus allowing the metal ions to behave as cathodes.

Z-scheme Construction [148-149]

A combined system involving two photoexcitation processes is called a Z-scheme, due to the similarity of the intrinsic process to photosynthesis. For example, a combination of Pt-WO_3 and Pt-$SrTiO_3$ (Cr-Ta doped) were found effective in utilizing visible light for H_2 and O_2 production in a NaI solution [148]. Over an extended duration H_2 and O_2 evolution under monochromatic light (420.7 nm, 57 mW) were measured at 0.21 μmol/h of H_2 and 0.11 μmol/h of O_2; the estimated quantum efficiency was 0.1%. Similarly Fe^{3+}/Fe^{2+} and Ce^{4+}/Ce^{3+} pairs are also effective for hydrogen production by water splitting [149].   Pt/$SrTiO_3$:Rh works as a photocatalyst for H_2 production using Fe^{2+} ions, while $BiVO_4$ acts as a photocatalyst for O_2 production using Fe^{3+} ion, see **Fig. 6.10**.

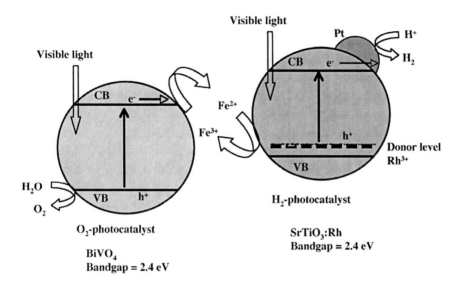

Fig. 6.10: Z-scheme system for water splitting by a two-photon process with visible light response [149].

Anion Doping [29,30,63,90-92,150-159]

Doping of anions such as N, F, C, and S in metal oxide or mixed metal oxides can shift the photoresponse into the visible region [150-159] while showing little tendency to form recombination centers [63, 90-92,155-158]. Asahi and co-workers studied the substitutional doping of C, N, F, P and S for O in anatase TiO_2 [150]. It was found that mixing of N p states with O 2p states could shift the valence band edge upwards, narrowing the band gap of TiO_2. Incorporation of S into the TiO_2 lattice of is a bit difficult due to the large ionic radius, while C, N and P doping has resulted in reduced mobility of the photo-generated charges.

Synthesis of nanoparticulate N-doped TiO_2 can be carried out by a variety of methods, including the heating of titanium hydroxide and urea [151], nitriding of anatase TiO_2 with alkyl ammonium salts [152], and heating TiO_2 powder in presence of NH_3/Ar gas at 550°C [153]. Anion doping results in an upward shift in the valence band energy without affecting the conduction band energy level, resulting in a narrower band gap. Since the valence band level of TiO_2 is far more positive than the oxygen evolution energy level, anion doped TiO_2 suffers from the limited availability of holes for water oxidation. Incorporation of sulpher atoms into the TiO_2 lattice in the form of S^{4+} can be carried out by annealing TiS_2 at 600°C in oxygen [154], and by mixing titanium isopropoxide with thiourea in ethanol [155]. It was found that mixing of S 3p states with the valance band of TiO_2 results in band gap narrowing and thereby improved visible light photocatalytic activities. Co-doped metal oxides such as $Sr_{1-x}La_xTiO_{3-x}N_x$ ($SrTiO_3$:La:N) and rutile TiO_2:La_2O_3:S synthesized by, respectively, sol-gel and hydrothermal processes have been used for photodecomposition of organic compounds under visible light [156,157].

Lee et al. [158] reported the use of N-doped $Sr_2Nb_2O_7$ for water splitting under visible light irradiation. $Sr_2Nb_2O_{7-x}N_x$ (x = 1.47~2.828) catalysts were synthesized from a $Sr_2Nb_2O_7$ precursor by ammonia nitridation at temperatures ranging from 700°C to 1000°C. The photocatalysts prepared at 800°C showed maximum catalytic activity, while above this temperature it gradually decreased. The photocatalyst structure, layered perovskite, appears to be the most important parameter determining the photocatalytic activity. It has been demonstrated that materials of layered-type structure have higher catalytic activity for water splitting than unlayered materials of the same composition. Excess N-doping facilitates the collapse of the layered structure of the parent oxide, $Sr_2Nb_2O_7$, turning it into unlayered $SrNbO_2N$, a behavior that possibly explains the reduction in photocatalytic activity for samples processed at temperatures above 800°C.

As seen from the band model for $Sr_2Nb_2O_7$ (**Fig. 6.11**), the states at the top of the valence band come from O 2p orbitals, while states at the conduction band minimum are predominantly from Nb_{4d}. The main difference observable in the nitrogen-doped oxide is that the major contribution at the valence band comes from N 2p. Density of states (DOS) studies suggest that N substitution doping gives rise to band gap narrowing by mixing of N 2p with O 2p states near the top of the valence band. This band gap narrowing enables the material to shift the optical absorption into visible light range [159].

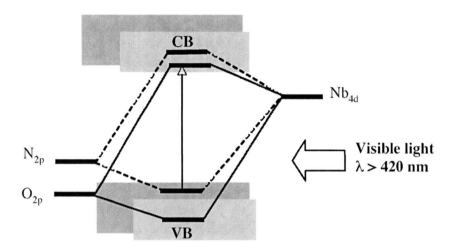

Fig. 6.11: A band model of N-doped $Sr_2Nb_2O_7$ shows mixing of N_{2p} with O_{2p} states near the top of the valence band leading to band gap narrowing and visible light activity.

6.5.4 Dye-Sensitized Oxide Semiconductors [160-165]

Considerable efforts have been made in the area of dye-sensitized photocatalytic hydrogen production [160-165]. In this process dye molecules absorb light with the transfer of an electron from the ground state to an excited state. The excited electron passes from the dye to the conduction band of an appropriate metal oxide (usually TiO_2). The conduction band electron can then be transferred to metal particles loaded on the surface to initiate water reduction. To regenerate the dye, sustaining the reaction, sacrificial reagents such as EDTA can be added to the solution [163,165]. Fast electron injection and slow backward reaction make dye-sensitized semiconductors feasible for hydrogen production [161-163].

Abe and co-workers [160] used xanthene dye on Pt/TiO_2 for hydrogen production from aqueous triethanolamine solution (TEOA) under visible light irradiation. In order to make a stable dye-sensitized photocatalyst a silane-coupling reagent (γ-aminopropyl triethoxysilane) was used to replace the carboxyl group of the xanthene dye with an amino group. Dhanlakshmi et al. [164] carried out a similar experiment using $[Ru(dcpy)_2(dpq)]^{2+}$. With or without Pt loading of TiO_2, no significant changes in the rate of hydrogen production have been seen suggesting that only dye molecules adsorbed on the TiO_2 surface are sufficiently effective in injecting electrons into TiO_2 for water reduction. The effects of various dyes such as eosin blue, rose Bengal, rhodamine B, and $[Ru(bpy)_3]^{2+}$ on photocatalytic hydrogen production by SnO_2, with or without sacrificial agents, have been studied by Gurunathan and co-workers [165]; to date no significant results with reference to hydrogen production have been reported so far.

6.6 Conclusions and Future Prospects

During the past several decades aqueous suspensions of semiconductor particles have been appraised for direct photocatalytic water splitting under UV and visible light irradiation, where each particle behaves as its own electrochemical cell. It has been observed that photogenerated charge separation, prevention of backward reactions, and utilization of a large fraction of the incident solar energy are the essential requirements for achieving high photoconversion efficiencies. The addition of carbonate salts or other electron mediators to the water have shown enhanced hydrogen production by preventing backward reactions. Enhanced hydrogen production has also been shown by the addition of electron donors or hole scavengers that irreversibly react with valence band holes to inhibit charge recombination, while to attain sustainable hydrogen production electron donors must be continuously added.

The most successful of the oxide semiconductors synthesized for aqueous suspension H_2 production by solar light are composed of two or more components. Various syntheses procedures such as loading and/or doping of metal or metal oxide particles on the surface of base photocatalysts have been successfully employed for water photosplitting. Cation doping can lead to an increase in charge carrier recombination centers, introduced by the localized d-states of the dopants deep within the TiO_2 bandgap. Anion doping is found effective in terms of band gap narrowing for harnessing visible light energy; the related impurity energy levels, while near the valence band edge, do not act as charge carriers. Most of the reactions carried out using doped oxide semiconductors are half-reactions that produce either hydrogen or oxygen in the presence of sacrificial agents. These agents scavenge electrons or holes in an irreversible manner making the process unsustainable. Use of a metal and nonmetal co-doped oxide semiconductor system appears a logical extension in the design of a robust photocatalytic system. A variety of perovskite type mixed-oxide nanoparticle photocatalysts have shown significant performances, but stability and phase purity remain a challenge to accomplish overall water splitting.

A concerted effort is needed to screen potentially useful oxide semiconductor systems and find ways to design nanostructure architectures of oxide semiconductors, metals/metal oxides as cocatalysts and/or other light harvesting assemblies. These developments require collaboration with suitable theoretical modeling for a better understanding of the hydrogen production mechanism in order to achieve low-cost and environmental friendly direct water-splitting for hydrogen production.

References

1. Fujishima A, Honda K (1972) Electrochemical photolysis of water at a semiconductor electrode. Nature 238:37–38
2. Schrauzer GN, Guth TD (1977) Photolysis of water and photoreduction of nitrogen on titanium oxide. J Am Chem Soc 99:7189–7193
3. Kawai T, Sakata T (1980) Photocatalytic decomposition of gaseous water. Chem Phys Lett 72:87–89
4. Van Damme H, Hall WK (1979) On the photoassisted decomposition of water at the gas-solid interface on TiO_2. J Am Chem Soc 101:4373–4374
5. Domen K, Naito S, Soma M, Onishi T, Tamaru K (1980) Photocatalytic decomposition of water vapor on a NiO-$SrTiO_3$ catalyst. Chem Commun 543–544
6. Sato S, White JM (1980) Photodecomposition of water over Pt/TiO_2 catalysts. Chem Phys Lett 72:83–86
7. Wagner FT, Somerjai GA (1980) Photocatalytic and Photoelectrochemical hydrogen production on strontium titanate single crystals. J Am Chem Soc 102:5494–5502

8.   Wagner FT, Somerjai GA (1980) Photocatalytic production from water on Pt-free SrTiO$_3$ in aqueous alkaline solution. Nature 285:559–560

9.   Sato S, White JM (1981) Photocatalytic water decomposition and water-gas shift reactions over NaOH coated, platinized TiO$_2$. J Catal 69:128–139

10.   Duonghong D, Borgarello E, Grätzel M (1981) Dynamics of light induced water cleavage in colloidal system. J Am Chem Soc 103:4685–4690

11.   Domen K, Naito S, Onishi T, Tamaru K (1982) Photocatalytic decomposition of liquid water on a NiO-SrTiO$_3$ catalyst. Chem Phys Lett 92:433–434

12.   Lehn JM, Sauvage JP, Ziessel R, Hilaire L (1982) Water photolysis by UV irradiation of rhodium loaded strontium titanate catalysts. Relation between catalytic activity and the nature of the deposit from combined photolysis and ESCA studies. Israel J Chem 22:168–172

13.   Domen K, Kudo A, Onishi T (1986) Mechanism of photocatalytic decomposition of water into H$_2$ and O$_2$ over NiO-SrTiO$_3$. J Catal 102:92–98

14.   Kudo A, Domen K, Maurya K, Aika K, Onishi T (1988) Photocatalytic decomposition of water over NiO-K$_4$Nb$_6$O$_{17}$ catalysts. J Catal 111:67–76

15.   Inoue Y, Kubokawa Y, sato K (1991) Photocatalytic activity of alkali-metal titanates combined with Ru in the decomposition of water. J Phys Chem 95:4059–4063

16.   Inoue Y, Asai Y, sato K (1994) Photocatalysts with tunnel structures for decomposition of water. J Chem Soc Faraday Trans 90:797–802

17.   Sayama K, Arakawa H (1994) Effect of Na$_2$CO$_3$ addition on photocatalytic decomposition of liquid water over various semiconductor catalysts. J Photochem Photobiol A Chem 77:243–247

18.   Sayama K, Arakawa H (2000) Oxide semiconductor materials for solar light energy Utilization. Res Chem Intermed 26:145–152

19.   Baba R, Nakabayashi S, Fujishima A, Honda K (1985) Investigation of the mechanism of hydrogen evolution during photocatalytic water decomposition on metal-loaded semiconductor powders J Phys Chem 89:1902–1905

20.   Aspnes DE, Heller A (1983) Photoelectrochemical hydrogen evolution and water- photolyzing semiconductor suspensions: Properties of platinum group metal catalyst emiconductor contacts in air and in hydrogen. J Phys Chem 87:4919–4929

21.   Bard AJ, Fox MA (1995) Artificial photosynthesis: Solar splitting of water to hydrogen and oxygen. Acc Chem Res 28:141–145

22.   Ashok Kumar M (1998) An overview of semiconductor particulate system for photoproduction of hydrogen. Int J Hydrogen Energy 23:427–438

23.   Kudo A, Kato H, Tsuji I (2004) Strategies for the development of visible-light-driven photocatalysts for water-splitting. Chem Lett 33, 1534-1539

24.   Takata T, Tanaka A, Hara M, Kondo JN, Domen K (1998) Recent progress of photocatalysts for overall water splitting. Catal Today 44:17–26

25.   Kamat PV (1993) Photochemistry on nonreactive and reactive (semiconductor) surfaces. Chem Rev 93: 267–300

26.   Thompson TL, Yates J T (2005) TiO2-based photocatalysis: Surface defects, oxygen and charge transfer. Top Catal 35:197–210

27.   Hagfeldt A, Gratzel M (1995) Light-Induced Redox Reactions in Nanocrystalline Systems Chem Rev 95:49-68

28.   Hara M, Kondo T, komoda M, Ikeda S, Shinohara K, Tanaka A, Kondo JN, Domen K (1998) Cu_2O as a photocatalyst for overall water splitting under visible light irradiation. Chem Commun 357–358

29.   Morikawa T, Asahi R, Ohwaki T, Aoki K, Taga, Y (2001) Bandgap narrowing of titanium dioxide by nitrogen doping. Jpn J Appl Phys Part 2-Lett 40:L561–L563

30.   Wang H, Lewis JP (2006) Second-generation photocatalytic materials: anion-doped TiO_2. J Phys-C, 18:421–434

31.   A. Kudo (2003) Photocatalyst materials for water splitting. Catal Surv Asia 7:31–38

32.   Ni M, Leung MKH, Leung DYC, Sumathy K (2007) A review and recent development in photocatalytic water splitting using TiO_2 for hydrogen production. Renewable and Sustainable Energy Reviews 11:401–425

33.   Kudo A (2006) Development of photocatalyst materials for water splitting. Int J Hygrogen Energy 31:197–202

34.   Murray CB, Kagan CR (2000) Synthesis and Characterization of monodispersed nanocystals and closed-packed nanocrystals assemblies. Annu. Rev Mater Sci 30, 545–610

35.   Cushing BL, Kolesnichenko VL, O'Connor CJ (2004) Recent advances in the liquid-phase syntheses of inorganic nanoparticles. Chem Rev 104:3893–3946.

36.   H. Hahn (1997) Gas phase synthesis of nanocrytalline materials. Nanostuct Mater 9:3–12

37.   Swihart MT (2003) Vapor synthesis of nanoparticles. Curr Opin Colloid Interf Sci 8:127–133

38.   Wang Y, Ma C, Sun X, Li H (2002) Preparation of nanocrystalline metal oxide powders with the surfactant mediated method. Inorg Chem Commun 5:51–755

39.   Zhang Z, Guo L, Wang W (2001) Synthesis and characterization of antimony oxide nanoparticles. J Mater Res 16:803-805

40.   Haruta M, Tsubota S, Kobayashi T, Kageyama H, Genet MJ, Delmon B (1993) Low temperature oxidation of CO over gold supported TiO_2, α-Fe_2O_3 and Co_3O_4. J Catal 144:175–192

41.   Liu ZL, Liu YZ, Yao KL, Ding ZH, Tao J, Wang X (2002) Synthesis and magnetic properties of Fe_3O_4 nanoparticles. J Mater Synth Process 10:83–87

42.   Tang ZX, Sorenson CM, Klabunde KJ, Hajipanayis GC (1991) Size dependent Curie temperature in nanoscale $MnFe_2O_4$. Phy Rev Lett 67:3602–3605

43. You X, Chen, F, Zhang J, Anpo M (2005) A novel deposition-precipitation method for preparation of Ag-loaded TiO$_2$. Catal Lett 102:247–250

44. Zanella R, Giogio S, Shin CH, Henry CR Louis C (2004) Characterization and reactivity in CO oxidation of gold nanoparticles supported on TiO$_2$ prepared by deposition-preparation with NaOH and Urea. J Catal 222:257–267

45. Bamwenda GR, Tsubota S, Nakamura T, Haruta M (1995) Photoassisted hydrogen production from a water-ethanol system: a comparison of activities of Au-TiO$_2$ and Pt-TiO$_2$. J Photochem Photobio A 89:177–189

46. Abe T, Suzuki, E, Nagoshi, K, Miyashita K, Kaneko M (1999) Electron source in photoinduced hydrogen production on Pt supported TiO$_2$ particles. J Phys Chem 103:1119-1123

47. Chan SC, Barteau MA (2005) Preparation of highly uniform Ag/TiO$_2$ and Au/TiO$_2$ supported nanoparticles catalyst by photodeposition. Langmuir 23:5588-5595

48. Borgarello E, Serpone N, Emo G, Harris R, Pelizzetti E, Minero C (1986) Light-induced reduction of rhodium (III) and palladium (II) on titanium dioxide dispersions and the selective photochemical separation and recovery of gold (III), platinum (IV) and rhodium (III) in chloride media. Inorg Chem 25:4499–4503

49. Hwang DW, Kim J, Park TJ, Lee JS (2002) Mg-doped WO$_3$ as a novel photocatayst for visible light-induced water splitting. Catal Lett 80:53-57

50. Galinksa A, Walendziewski J (2005) Photocatalytic water splitting over Pt-TiO$_2$ in the presence of sacrificial reagents. Energy & Fuels 19:1143-1147

51. Cao Y, Hsu JC, Hong ZS, Deng JF, Fan KN (2002) Characterization of high-surface-area zirconia aerogel synthesized from alcohothermal and supercritical fluid drying techniques.Catal Lett 81:107–112

52. Bischoff BL, Anderson MA (1995) Peptization process in the sol-gel preparation of anatase TiO$_2$. Chem Mater 7:1772–1778

53.   Chang YS, Chang YH, Chen IG, Chen GJ, Chai YL (2002) Synthesis and characterization of zinc titanate nano-crystal powders by sol-gel technique. J Cryst Growth 243:319–326

54.   Xuewen W, Zhiyong Z, Shuixian Z (2001) Preparation of nano-crystalline $SrTiO_3$ powder in sol-gel process. Mater Sci Engg B 86:29–33

55.   Kohno M, Ogura S, Inoue Y (1996) Preparation of $BaTi_4O_9$ by a sol-gel method and its photocatalytic activity for water decomposition. J Mater Chem 6:1921–1924

56.   Kim HG, Hwang DW, Bae SB, Jung JH, Lee JS (2003) Photocatalytic water splitting over $La_2Ti_2O_7$ synthesized by the polymerizable complex method. Catal Lett 91:193–198

57.   Yoshino M, Kakihana M, Cho WS, Kato H, Kudo A (2002) Polymerizable complex synthesis of pure $Sr_2Nb_xTa_{2-x}O_7$ solid solutions with high photocatalytic activities for water decomposition into H_2 and O_2. Chem Mater 14:3369–3376

58.   Yener DO, Giesche H (2001) Synthesis or pure and manganese-, nickel- and zinc-doped ferrite particles in water-in-oil microemulsions. J Am Ceram Soc 84:1987–1995

59.   Kapoor MP, Inagaki S, Yoshida H (2005) Novel zirconium-titanium phosphates mesoporous materials for hydrogen production by photoinduced water splitting. 109:9231–9238

60.   Oguri Y, Riman RE, Bowen HK (1988) Processing of anatase prepared from hydrothermally treated alkoxy derived hydrous titania. J Mater Sci 23:2897-2904

61.   Chen H, Ma J, Zhao Z, Qi L (1995) Hydrothermal preparation of uniform nanosize rutile and anatase particle. Chem Mater 7:663–671

62.   Jeon S, Braun PV (2003) Hydrothermal synthesis of Er-doped luminescent TiO_2 nanoparticles. Chem Mater 15:1256–1263

63.   Lou H, Takata T, Lee Y, Zhao J, Domen K, Yan Y (2004) Photocatalytic activity enhancing for titanium dioxide by codoping with chlorine and bromine. Chem Mater 16:846–849

64.   Yan M, Cheng F, Zhang J, Anpo M (2005) Preparation of controllable crystalline titania and study on the photocatalytic properties. J Phys Chem B 109:8673–8678

65.   Ikeda S, Fuboki M, Takahara YK, Matsumura M (2006) Photocatalytic activity of hydrothermally synthesized tantalates pyrochlores for overall water splitting. Appl Catal A 300:186–190

66.   Hulteen JC, Martin CR (1997) A general template base method for the preparation of nanomaterials. J Mater Chem 7:1075-1087

67.   Shchukin DG, Schattka JH, Antonietti M, Curasu RA (2003) Photocatalytic properties of porous metal oxide networks formed by nanoparticles infiltration in a polymer gel template. J Phys Chem B 107:952-957

68.   Lensveld DJ, Mesu JG, van Dillen AJ, de Jong KP (2001) Synthesis and characterization of MCM-41 supported nickel oxide catalysts. Microporous Mesoporous Mat 44-45:401–407

69.   Valdés-Solís T, Marbán G, Fuertes AB (2005) Preparation of nanosized perovskites and spinels through a silica xerogel template route. Chem Mater 17:1919-1922

70.   Nasibulin AG, Richard O, Kauppinen EI, Brown DP, Jokiniemi JK, Altman IS (2002) Nanoparticle synthesis by copper(II) acetylacetonate vapor decomposition in the presence of oxygen. Aerosol Sci Technol 36:899–911

71.   Schmechel R, Kennedy M, von Seggern H, winkler H, Kolbe M, Fischer RA, Xaomao L, Benker A, Winterer M, Hahn H (2001) Luminescence properties of nanocrystalline $Y_2O_3:Eu^{3+}$ in different host materials. J Appl Phys 89: 1679–1686

72.   Ahonen PP, Joutsensaari J, Richard O (2001) Mobility size development and the crystallization path during aerosol decomposition synthesis of TiO_2 particles. J Aerosol Sci 32:615–630

73.   Johannessen T, Jenson JR, Mosleh M, Johansen J, Quaade U, Livbjerg H (2002) Flame synthesis of nanoparticles: Application in catalysis and product/process engineering. Chem Eng Res Des 82:1444-1452

74.   Mädler L, Kammler HK, Mueller R, Pratsinis SE (2002). Controlled synthesis of nanostructured particles by flame spray pyrolysis. J Aerosol Sci 33:369–388

75.   Flagan RC, Lunden MM (2004) Particle structure control in nanoparticle synthesis from the vapor phase. 204:113-124

76.   Ohno T (2002) Morphology of composite nanoparticles of immiscible binary systems prepared by gas-evaporation technique and subsequent vapor condensation. J Nanoparticle Res 4:255–260

77.   Iwasaki M, Iwasaki Y, Tada H, Ito S (2004) One-pot process for anodic oxide films of titanium with high photocatalytic activity. Mater Trans 45: 1607–1612

78.   Harano A, Shimada K, Okubo T, Sadakata M. (2002) Crystal phases of TiO_2 ultrafine particles prepared by laser ablation of solid rods. J Nanoparticle Res 4:215–219

79.   Iwabuchi A, Choo CK, Tanaka K Titania Nanoparticles Prepared with Pulsed Laser Ablation of Rutile Single Crystals in Water J Phys Chem B 108:10863–10871

80.   Guimaraes JL, Abbate M, Betim SB, Alves MCM (2003) Preparation and characterization of TiO_2 and V_2O_5 nanoparticles produced by ball-milling J Alloys & Compounds 352:16–20

81.   Damonte LC, Zelis LAM, Soucase BM, Fenollosa MAH Nanoparticles of ZnO obtained by mechanical milling. Powder Tech 148: 15–19

82.   Shimizu K, Itoh S, Hatamachi T, Kodama T, Sato M, Toda K (2005) Photocatalytic water splitting on Ni-intercalated Ruddlesden-Popper type $H_2La_{2/3}Ta_2O_7$. Chem Mater 17:5161–5166

83.   Ebina Y, Sasaki T, Harada M, Watanabe M (2002) Restacked perovskite nanosheets and their Pt-loaded materials as photocatalysts Chem Mater 14: 4390-4395

84.   Ebina Y, Sakai N, Sasaki T (2005) Photocatalyst of Lamellar Aggregates of RuOx Loaded Perovskite Nanosheets for Overall J Phys Chem B 109:17212-17216

85.   Cheng P, Li W, Zhou T, Jin Y, Gu M (2004) Physical and photocatalytic properties of zinc ferrite doped titania under

visible light irradiation. J Photochem Photobiol A: Chem 168:97–101

86. Sreethawong T, Ngamsinlapasathian S, Suzuki Y, Yoshikawa S (2005) Nanocrystalline mesoporus Ta_2-based photocatalysts prepapred by surfactant-assisted templating sol-gel process photocatalytic H_2 evolution. J Mol Catal A:Chem 235:1–11

87. Wang D, Zou Z, Ye J (2005) Phtotcatalytic water splitting with the Cr-doped $Ba_2In_2O_5/In_2O_3$ composite oxide semiconductors. Chem Mater 17:3255-3261

88. Liu SH, Wang HP (2002) Photocatalytic generation of hydrogen on Zr-MCM-41. Int J Hydrogen Energy 27:859–862

89. Sun B, Reddy EP, Smirniotis PG (2005) Effects of the Cr^{6+} concentration in Cr-incorporated TiO_2-loaded MCM-41 catalysts for visible light photocatalysis. Appl Catal B: Environmental 57:139–149

90. Liu H, Gao L (2004) (Sulfur, Nitrogen)-codoped rutile-titanium oxide as a visible-light-activated photocatalyst. J Am Ceram Soc 87:1582–1584

91. Aita Y, Komatsu M, Yin S, Sato T (2004) Phase-compositional control and visible light photocatalytic activity of nitrogen-doped titania via solvothermal process. J Solid state Chem 177:3235–3238.

92. Yuan J, Chen M, Shi J, Shangguan W (2006) Preparation and photocatalytic hydrogen evolution of N-doped TiO_2 from urea and titanium tetrachloride. Int J Hydrogen Energy 31:1326–1331

93. Senevirathna MKI, Pitigala PKDDP, Tennakone K (2005) Water photoreduction with Cu_2O quantum dots on TiO_2 nano-particles. J Photochem Photobiol A: Chem 171:257–259

94. Tabata S, Nishida H, Masaki Y, Tabata T (1995) Stoichimetetric photochemical decomposition of pure water in Pt/TiO_2 aqueous suspension system. Cat Lett 34:245–249

95. Marathamuthu P, Ashokkumar M (1989) Hydrogen production with visible light using metal loaded WO_3 and

MV^{2+} in aqueous medium. Int J Hydrogen Energy 14:275–277

96.   Mill A, Porter G (1982) Photosensitized dissociation of water using dispersed suspensions of n-type semiconductors. J Chem Soc Faraday Trans.1 78:3659–3669

97.   Sayama K, Arakawa H (1992) Significant effect of carbonate addition on stoichiometric photodecomposition of liquid water into hydrogen and oxygen from platinum-titanium(IV) oxide suspension. J Chem Soc Chem Commun 150–152

98.   Sayama K, Arakawa H (1994) Effect of carbonate addition on the photocatalytic decomposition of liquid water over a ZrO_2 catalyst. J Photochem Photobiol A: Chem 94:67–76

99.   Sayama K, Arakawa H (1997) Effect of carbonate salt addition on the photocatalytic decomposition of liquid water over Pt-TiO_2 catalysts. J Chem Soc Faraday Trans 93:1647–1654

100.   Arakawa H, Sayama K (2000) Solar hydrogen production: significant effect of Na_2CO_3 addition on water splitting using simple oxide semiconductor photocatalysts. Catal Surv Jpn 4:75–80

101.   Abe R, Sayama K, Arakawa H (2003) Significant effect of iodide addition on water splitting into H_2 and O_2 over Pt-loaded TiO_2 photocatalyst: suppression of backward reaction. Chem Phys Lett 371:360–364

102.   Wu NL, Lee MS (2004) Enhanced TiO_2 photocatalysis by Cu in hydrogen production from aqueous methanol solution. Int J Hydrogen Energy 29:1601–1605

103.   Lee SG, Lee SW, Lee HI (2001) Photocatalytic production of hydrogen from aqueous solution containing CN^- as a hole scavengers.Appl Catal A:Gen 207:173–181

104.   Nada AA, Barakat MH, Hameed HA, Mohamad NR, Veziroglu TN (2005) Studies on the photocatalytic hydrogen production using suspended modified TiO_2 photocatalysts. Int J Hydrogen Energy 30:987–991

105.   Abe R, Sayama K, Domen K, Arakawa H (2001) A new type of water splitting system composed of two different TiO_2

photocatalysts (anatase, rutile) and a IO_3^-/I^- shuttle redox mediator. Chem Phys Lett 344:339–344

106. Bamwenda GR, Arakawa H (2001) The photoinduced evolution of O_2 and H_2 from a WO_3 aqueous suspension in presence of Ce^{4+}/Ce^{3+}. Sol Energy Mater Sol Cells 70:1–14

107. Lee K, Nam WS, Han GY (2004) Photocatalytic water splitting in alkaline aqueous solution using redox mediator 1: parameter study. Int J Hydrogen Energy 29:1343–1347

108. Sakata T, Hashimoto K, Kawai T (1984) Catalytic properties of ruthenium oxide on n-type semiconductors under illumination. J Phys Chem 88:5214–5221

109. Srivastava AM, Ackerman JF (1997) On the luminescence of $Ba_5M_4O_{15}$ ($M = Ta^{5+}$ and Nb^{5+}). J Solid State Chem 134, 187–191

110. Eng HW, Barnes PW, Auer BM, Woodword PM (2003) Investigation of the electronic structure of d^0 transition metal oxides belonging to the perovskite family. J Solid State Chem 175:94–109

111. Kato H, Kudo A (1999) Highly efficient decomposition of pure water over $NaTaO_3$ photocatalysts. Catal Lett 58:153–155

112. Kato H, Kudo A (2001) Water splitting into H_2 and O_2 on Alkali tantalate photocatalysts $ATaO_3$. J Phys ChemB 105:4285–4292

113. Kato H, Kudo A (1999) A new tantalate photocatalyst for water decomposition into H_2 and O_2. Chem Phys Lett 295: 487–492

114. Kato H, Kudo A (1999) Photocatalytic decomposition of pure water into H_2 and O_2 over $SrTa_2O_6$ prepared by flux method. Chem Lett 1207–1209

115. Kudo A, Kato H, Nakagawa S (2000) Water splitting into H_2 and O_2 on New $Sr_2M_2O_7$ ($M = Nb$ and Ta) photocatalysts with layered perovskite structures: Factor affecting the photocatalytic activity. J Phys Chem B 104:571–575

116. A. Kudo A, Okutomi H, Kato H (2000)Photocatalytic water splitting into H_2 and O_2 over $K_2LnTa_5O_{15}$. Chem Lett 1212–1213.

117. Yamashita Y, Tada M, Kahihana M, Osada M, Yoshida K (2002) Synthesis of RuO_2-loaded $BaTi_2O_{2n+1}$ (n = 1, 2 and 5) using a polymerizable complex method and its photocatalytic activity for the decomposition of water. J Mater Chem 12:1782–1786

118. Miseki Y, Kato H, Kudo A (2005) Warer splitting into H_2 and O_2 over $Cs_2Nb_4O_{11}$. Chem Lett 54–55

119. Kato H, Kudo A (2003) Photocatalytic water splitting into H_2 and O_2 over various tantalate photocatalysts. Catal Today 78:561–569

120. Ikeda S, Hara M, Kondo JN, Domen K (1998) Preparation of $K_2La_2Ti_3O_{10}$ by polymerized complex method and photocatalytic decomposition of water. Chem Mater 10: 72-77

121. Yoshioka K, Petrykin V, Kakihana M, Kato H, Kudo A (2005) The relationship between photocatalytic activity and crystal structure in Strontium tantalates. J Catal 232:102–107

122. Otsuka H, Kim K, Kouzu A, Takimoto I, Fujimori H, Sakata Y, Imamura H, Matsumoto T, Toda K (2005) Photocatalytic performance of $Ba_5Ta_4O_{15}$ to decomposition of H_2O into H_2 and O_2. Chem Lett 822–823

123. Abe R, Higashi M, Sayama K, Abe Y, Sugihara H (2006) Photocatalytic activity of R_3MO_7 and $R_2Ti_2O_7$ (R = Y, Gd, La; M = Nb, Ta) for water splitting into H_2 and O_2. J Phys Chem B 110:2219–2226

124.   Iwase H, Kato H, Kudo A (2006) Nanosized gold particles as an efficient cocatalyst for photocatalytic overall water splitting. Catal Lett 108:6–9

125. Kudo A, Omri K, Kato H (1999) A novel aqueous process for preparation of crystal form-controlled and highly crystalline $BiVO_4$ powder from layered vanadates at room temperature and its photocatalytic and photophysical properties. J Am Chem Soc 121:11459–11467

126. Kato H, Kobayashi H, Kudo A (2002) Role of Ag^+ in the band structures and photocatalytic properties of $AgMO_3$ (M = Nb and Ta) with the perovskite structure. J Phys Chem B 106:12441–12447

127. Konta R, Kato H, Kobayashi H, Kudo A (2003) Photophysical properties and photocatalytic activities under visible light irradiation of silver vanadates. Phys Chem Chem Phys 5:3061–3065

128. Tang J, Zou J, Ye J (2004) Efficient photocatalytic decomposition of organic contaminants over $Ca_2Bi_2O_4$ under visible-light irradiation. Angew Chem Int Ed 43:4463–4466

129. Kim HG, Hwang DW, Lee JS (2004) An undoped, single-phase oxide photocatalyst working under visible light. J Am Chem Soc 126:8912–8913

130. Zou Z, Ye J, Arakawa H (2003) Photocatalytic water splitting into H_2 and or O_2 under UV and visible light irradiation with a semiconductor catalyst. Int J Hydrogen Energy 28:663–669

131. Tai YW, Chen JS, Yang CC, Wan BJ (2004) Preparation of nano-gold on $K_2La_2Ti_3O_{10}$ for producing hydrogen from photo-catalytic water splitting. Catal Today 97:95–101

132. Wu J, Uchida S, Fujishito Y, Yin S, Sato T (1999) Synthesis and photocatalytic properties of $HNbWO_6/TiO_2$ and $HNbWO_6/Fe_2O_3$ nanocomposite. J Photochem Photobiol A: Chem 128:129–133

133. Jang JS, Kim HG, Reddy VR, Bae SW, Ji SM, Lee JS (2005) Photocatalytic water splitting over iron oxide nanoparticles intercalated in $HTiNb(Ta)O_5$ layered compounds. J Catal 231:213–222

134. Choi WY, Termin A, Hoffman MR (1994) The role of metal ion dopants in quantum-sized TiO_2: Correlation between photoreactivity and charge carrier reaction dynamics. J Phys Chem 98:13669–13679

135. Xu AW, Gao Y, Liu HQ (2002) The preparation, characterization and their photocatalytic activities of rare-earth doped TiO_2 nanoparticles. J Catal 207:151–157

136. Peng S, Li Y, Xiang F, Lu G, Li S (2004) Effects of Be^{2+} doping on its photocatalytic activity. Chem Phys Lett 398:235–239

137. Hameed A, Gondal MA, Yamini ZH (2004) Effects of transition metal doping on photocatalytic activity of WO_3 for

water splitting under laser illumination: role of 3d-orbitals. Catal Commun 5:715–719

138.   Zou Z, Ye J, Sayama K, Arakawa H (2001) Direct splitting of water under visible light irradiation with an oxide semiconductor photocatalyst. Nature 414:625–627

139.   Konta R, Ishii T, Kato H, Kudo A (2004) Photocatalytic activities of noble metal ion doped $SrTiO_3$ under visible light irradiation. J Phys Chem 108:8992–8995

140.   Kato H, Kudo A (2002) Photocatalytic activities of noble metal ion doped $SrTiO_3$ under visible light irradiation. J Phys Chem B 106:5029–5034

141.   Ishii T, Kato H, Kudo A (2004) H_2 evolution from an aqueous methanol solution on $SrTiO_3$ photocatalysts codoped with chromium and tantalum ion under visible light irradiation. J Photochem Photobiol A: Chem 163:181–186

142.   Hwang DW, Kim HG, Lee JS, Kim J, Li W, Oh SH (2005) Photocatalytic hydrogen production from water over M-doped $La_2Ti_2O_7$ (M = Cr, Fe) under visible light irradiation ($\lambda > 420$ nm). J Phys Chem B 109:2093–2102

143.   Zou Z, Ye J, Sayama K, Arakawa H (2002) Photocatalytic hydrogen and oxygen formation under visible light irradiation with M-doped $InTaO_4$ (M = Mn, Fe, Co, Ni, Cu) photocatalysts. J Photochem Photobiol A:Chem 148:65–9

144.   Niishiro R, Kato H, Kudo A (2005) Nickel and either tantalum or niobium co-doped TiO_2 and $SrTiO_3$ photocatalysts with visible-light response for H_2 or O_2 evolution from aqueous solution. Phys Chem Chem Phys 7:2241–2245

145.   Kato H, Asakura K, Kudo A (2003) Highly efficient water splitting into H_2 and O_2 over lanthanum-doped $NaTaO_3$ photocatalysts with high crystallinity and surface nanostructure. J Am Chem Soc 125:3082–3089

146.   Iwase A, Kato H, Kudo A (2005) A novel photodeposition method in the presence of nitrate ions for loading of an iridium oxide cocatalyst for water splitting. Chem Lett 34, 946–947

147. Gurunathan K (2004). Photocatalytic hydrogen production using transition metal ions-doped γ-Bi2O3 semiconductor particles, Int J Hydrogen Energy 29:933–940

148. Sayama K, Masuka K, Abe R, Abe Y, Arakawa H (2001) Stiochiometric water splitting into H_2 and O_2 using a mixture of two different photocatalysts and an IO_3^-/I^- shuttle redox mediator under visible light irradiation. Chem Commun 2416–2417

149. Kato H, Hori M, Konta R, Shimodaira Y, Kudo A (2004) Construction of Z-scheme type heterogeneous photocatalysis systems for water splitting into H_2 and O_2 under visible light irradiation Chem Lett 33:1348–1349

150. Asahi R, Morikawa T, Ohwaki T, Aoki K, Taga Y (2001) Visible light photocatalysis in nitrogen-doped titanium oxide. Science 293:269–271

151. Kobayakawa K, Murakami K, Sato Y (2004) Visible-light active N-doped TiO_2 prepared by heating of titanium hydroxide and urea. J Photochem Photobiol A:Chem 170:177–179

152. Gole JL, Stout JD, Burda C, Lou YB, Chen XB (2004) Highly efficient formation of visible light tunable $TiO_{1-x}N_x$ photocatalysts and their transformation at nanoscale. J Phys Chem B 108:1230–1240

153. Mrowetz M, Balcerski W, Colussi AJ, Hoffmann MR (2004). Oxidative power of N-doped TiO_2 photocatalysts under visible light illumination J Phys Chem B 108:17269–17273

154. Umebayashi T, Yamaki T, Itoh H, Asai K (2002) Band gap narrowing of titanium oxide by sulphur doping. Appl Phys Lett 81:454-456

155. Ohno T, Akiyoshi M, Umebayashi T, Asai K, Mitsui T, Matsumura M (2002) Preparation of S-doped TiO_2 photocatalysts and their photocatalytic activites under visible light. Appl Catal A: Gen 265:115-121

156. Miyauchi M, Takashio M, Tobimatsu H (2004) Photocatalytic activity of $SrTiO_3$ codoped with nitrogen and lanthanum under visible light illumination. Langmuir 20:232–236

157. Liu HY, Gao L (2004) Codoped rutile TiO$_2$ as a new photocatalyst for visible light irradiation. Chem Lett 33:730–731

158. Ji SM, Borse PH, Kim HG, Hwang DW, Jang JS, Bae SW, Lee JS (2005) Photocatalytic hydrogen production from water-methanol mixtures using N-doped Sr$_2$Nb$_2$O$_7$ under visible light irradiation; effect of catalyst structures. Phys Chem Chem Phys 7:1315–1321

159. Lee JS (2006) Photocatalytic water splitting under visible light with particulate semiconductor catalysts. Catal Surv Asia 9:217–227

160. Abe R, Hara K, Sayama K, Domen K, Arakawa H (2000) Steady hydrogen evolution from water on Eosin-Y-fixed TiO$_2$ photocatalyst using a silane coupling reagent under visible light irradiation. J Photochem Photobiol A: Chem 137:63–69

161. Abe R, Sayama K, Arakawa H (2002) Efficient hydrogen evolution from aqueous mixture of I$^-$ and acetonitrile using a merocyanine dye-sensitized Pt/TiO$_2$ photocatalyst under visible light irradiation. Chem Phys Lett 362:441–444

162. Abe R, Sayama K, Arakawa H (2004) Dye-sensitized photocatalysts for efficient hydrogen production from aqueous I$^-$ solution under visible light irradiation. J Photochem Photobiol A: Chem 166:115–122

163. Bae E, Choi W (2006) Effect of the anchoring group (carboxylate vs phosphonate) in Ru-complex-sensitized TiO2 on hydrogen production under visible light. Source. J Phys Chem B 30:14792–14799

164. Dhanlakshmi KB, Latha S, Anandan S, Maruthamuthu P (2001) Dye-sensitized hydrogen evolution from water Int J Hydrogen Energy 26:669–674

165. Gurunathan K, Maruthamuthu P, Shastri VC (1997) Photocatalytic hydrogen production by dye-sensitized Pt/SnO$_2$ and Pt/SnO$_2$/RuO$_2$ in aqueous methyl viologen solution. Int J Hydrogen Energy 22:57–62

Chapter 7

NON-OXIDE SEMICONDUCTOR NANOSTRUCTURES

7.1 General Description of Non-Oxide Semiconductors

This chapter considers photo-electrodes consisting of non-oxide semiconductors, alone and in combination with oxide semi-conductors for water splitting.

7.1.1 Photoelectrodes

Developing a stable, cost-effective, photo-electrochemical based system for splitting water using sunlight as the only energy input remains a formidable task. A metal oxide semiconductor photoanodes typically satisfy three necessary but not sufficient conditions: (1) It must remain stable in aqueous solution during oxygen evolution. (2) Possess suitable overlap of conduction and valence band edges of the semiconductor with water redox potentials. (3) Demonstrate rapid charge transfer from the semiconductor to the water. In general oxide semiconductors such as TiO_2 fulfill these criteria, however as mentioned in the previous chapters most of the stable and corrosion-resistant oxide semiconductors have wide bandgaps and thus are able to utilize only a small fraction of the sunlight energy unless they are catalytically modified. A successful photocatalyst needs to absorb a significant fraction of the incident solar light energy, hence high absorption together with a broad spectral response are critical.

Various inorganic semiconductors (p-type and/or n-type non-oxide semiconducting materials) such as amorphous or crystalline silicon (a-Si or c-Si), gallium arsenide (GaAs), cadmium telluride (CdTe), gallium phosphide (GaP), indium phosphide (InP), copper indium diselenide (CIS), copper indium gallium diselenide (CIGS), and gallium indium phosphide (GaInP$_2$) are known to harness solar light converting it into electricity. There is ongoing interest in using these efficient semiconductors, that have bandgaps appropriate for capturing a large portion of the solar spectrum energy, for conversion of solar energy into chemical energy.

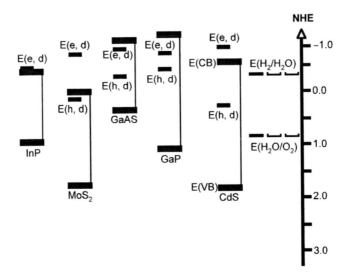

Fig. 7.1: Position of band edges and photodecomposition Fermi energies levels of various non-oxide semiconductors. E(e,d) represents decomposition energy level by electrons, while E(h,d) represents the decomposition energy level for holes *vs* normal hydrogen electrode (NHE). E(VB) denotes the valence band edge, E(CB) denotes the conduction band edge. E(H$_2$/H$_2$O) denotes the reduction potential of water, and (H$_2$O/O$_2$) the oxidation potential of water, both with reference to NHE.

Cadmium sulfide (CdS) is one of the most intriguing non-oxide semiconductors due to its sufficiently negative flat band potential (−0.66 V at pH 7) and optical absorption to 520 nm (bandgap = 2.4 eV), however it suffers from anodic photodecomposition by the photo-generated holes, **Fig. 7.1** [1-4]:

$$CdS + 2h^+ \rightarrow Cd^{2+} + S \qquad (7.1.1)$$

Since the decomposition Fermi level is located significantly above the valence band edge and the Fermi level for water decomposition, the holes are more likely to decompose the semiconductor lattice than water [5]. For aqueous solutions containing n-CdS photoanodes such photoanodic dissolution can be quenched by adding reducing agents such as S^{2-}, SO_3^{2-}, $S_2O_3^{2-}$, $[Fe(CN)_6]^{4-}$, I^-, and $Fe[EDTA]^{2-}$ (EDTA: ethylenediaminetetraacetic acid) [1,4-13]. Frank and Honda observed that coating n-type CdS electrodes with an electrically conducting polymer, e.g. polypyrrole, poly(3-methylthiophene) coupled with the presence of a catalyst, e.g. Pt; Rh; RuO_2, helps prevent photocorrosion thus enabling water oxidation under visible light illumination [14-16]. The conducting polymers play an important role in facilitating interfacial charge transfer thus protecting the n-type semiconductor against photodegradation. Other n-type non-oxide semiconductors such as CdSe (bandgap =1.7 eV), CdTe (bandgap = 1.4 eV), GaP (bandgap = 2.24 eV), GaAs (bandgap =1.35 eV), InP (bandgap =1.35 eV) MoS_2 (bandgap = 1.75 eV) and $MoSe_2$ (bandgap = 1.5 eV) have similar issues with photocorrosion, **Fig. 7.1** [11-13, 16-21].

Several p-type non-oxide semiconducting materials are found stable against cathodic photodecomposition due to migration of excess photoelectrons towards the semiconductor-solution interface [5,20]. In one of the earlier (1976) studies using p-type Si in 0.1 M NaCl solution for water photoelectrolysis, Candea and co-workers [22] suggested that an efficient photoelectrolysis cell could be constructed if the proper non-oxide semiconductor-electrolyte combination is found. Their initial effort [22] focused on applying a bias voltage between a p-Si cathode and Pt anode, potential measured relative to a saturated calomel electrode, from which hydrogen was evolved by water-splitting in a manner analogous to that of the n-TiO_2 electrode as reported by Fujishima and Honda [23]. In 1976, A. J. Nozik [24] reported that the efficiency of photoelectrolysis with solar radiation could be greatly enhanced by using simultaneously illuminated n- and p-type electrodes: homotype n-GaP/p-GaP, and heterotype n-TiO_2/p-GaP [25] which produced H_2 and O_2 without the need of an external anodic bias. In the case of the n-GaP/p-GaP cell, instead of evolving O_2 the anode reaction involves oxidation of P^{3-} in aqueous acid solution by:

$$n\text{-}GaP + 3H_2O + 3H^+ \rightarrow 3H_2 + Ga^{3+} + H_3PO_3 \qquad (7.1.2)$$

These systems are based on immersion of two photoactive electrodes in an electrolyte solution with connection via an external circuit. An overall solar-spectrum hydrogen conversion efficiency of 0.25% was found at zero bias for the n-TiO$_2$/p-GaP cell. Nozik further designed a new type of cell, so-called 'photochemical diodes' that do not require external wires and functions without electrical bias [26]. This device [26], consisting of a small sandwich-like structure, **Fig.7.2**, such as Pt/n-GaP, and n-TiO$_2$/p-GaP connected through ohmic contacts, when suspended in an appropriate electrolyte causes decomposition of water upon exposure to light.

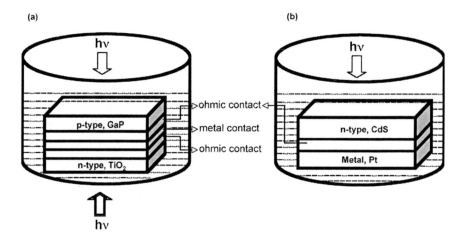

Fig. 7.2: Sandwich (or stacked) configurations placed in an electrolyte solution **(a)** p-n type, **(b)** semiconductor-metal type photochemical diodes. Both p-type and n-type semiconductors are provided with ohmic contacts. In p-n type light is incident from both directions and ohmic contacts are connected through metal contacts.

Incorporating metal catalysts, such as platinum, ruthenium, cadmium, gold, vanadium, etc. on the surface of p-type photocathodes affects semiconductor properties such as barrier height (of the semiconductor-catalyst junction), flat band potential, surface states, the recombination rate of photogenerated electrons at the semiconductor-catalyst interface, and the overpotential for hydrogen production [27-33]. Electrodeposited ruthenium (Ru) on a p-InP photocathode in 1M HCl – 2M KCl solution is the prominent example, with which Heller and Vadimsky have attained 12% efficiency in the photoassisted electrolysis of water under 87 mW/cm^2 sun-light illumination [27]; unfortunately the durability of the p-InP cell is much less than needed for a commercial system.

7.1.2 Photocatalysis in Aqueous Suspensions

Like those of oxide semiconductor particles, see Chapter 6, photocatalytic and photosynthetic reactions of individual non-oxide semiconductor particles suspended in aqueous solutions are one potential route for utilizing solar energy. For example, the use of a platinized suspension of CdS in EDTA as well as in Na_2SO_3 solution was examined, yielding H_2 under solar (visible) light illumination [34-38]. Ideally, the dispersed semiconductor particles should be small, uniform, and stable. Due to their tendency to agglomerate small semiconductor particles, $\approx$ 5-10 nm, are difficult to maintain in solution for long periods of time without a stabilizer. Mau and coworkers fabricated Pt/CdS particles in a polymer (nafion: perfluorosulfonic acid) matrix so that the semiconductor particles did not flocculate or settle [39]. In studies on the use of polymer-stabilized Rh-coated CdS photocatalyst nanoparticles, the presence of benzyl alcohol as a sacrificial electron donor [40] led to a dramatic, 40-fold enhancement of hydrogen production over the use of thiophenol as a sacrificial electron donor [41]. Under visible light irradiation (λ > 420 nm) hydrogen evolution was achieved from an aqueous solution containing sacrificial reagents SO_3^{2-} over $Zn_{1-x}Cu_xS$, which is a solid solution of ZnS and CuS semiconductors, opening a new area of non-oxide semiconductor suspended particle systems [42].

7.1.3 Nano-structured Surfaces and Quantum Size Effects

Nanocrystalline non-oxide semiconductors exhibit a wide range of novel chemical and physical properties [43-63]. For example, the melting temperature of CdS nanocrystals decreases with decreasing particle size [43]. The properties of the nanocrystalline particles can differ significantly from those of bulk materials, with the high surface area-to-volume ratios resulting in almost as many atoms at the surface as in the crystalline lattice. Moreover, the electronic bands are split into discrete energy levels as a result of the three-dimensional confinement of charge carriers, resulting in an increase in the bandgap with decreasing particle size. Advances in chemistry have made it possible to deliberately modify the surface properties of any semiconductor, potentially endowing them with efficient photocatalytic functions. The common ground here for these materials is the synthesis of nanoscale building blocks, or a self-assembly construction process, of different size and shape as well as different composition and surface structures. For example, applying hydrostatic pressure to wurtzite (hexagonal cubic) CdSe nanocrystals yields a less ordered surface but one with a high surface energy rock salt type structure (face centered cubic), a classical example of solid-solid phase transition [44]. In the case of a colloidal dispersion of semiconductor nanoparticles where solvents play a major role in stabilizing the nanocrystals, the nature of the solvent molecules coordinated on the nanoparticle surface strongly influence the nanocrystal surface energies [57-63].

Colloidal CdS particles 2–7 nm in diameter exhibit a blue shift in their absorption and luminescence characteristics due to quantum confinement effects [45,46]. It is known that particle size has a pronounced effect on semiconductor spectral properties when their size becomes comparable with that of an exciton. This so called quantum size effect occurs when $R < a_B$ (R = particle radius, a_B = Bohr radius *see Chapter 4*), coinciding with a gradual change in the energy bands of a semiconductor into a set of discrete electronic levels. The observation of a discrete excitonic transition in the absorption and luminescence spectra of such particles, so called Q-particles, requires samples of very narrow size distribution and well-defined crystal structure [47,48]. Semiconductor nanocrystals, or quantum dots, with dimension(s) smaller than the bulk exciton Bohr radius exhibit unique quantum size effects with strongly size-dependent electronic, optical, electrochemical and photoelectro-chemical properties. Various non-oxide semiconductors and their nanocomposites including CdS [49], ZnS [50], CdSe [51], TiO_2-CdSe [51], TiO_2-CdS [52], CdS-M_xS_y (MxS_y = Bi_2S_3, Sb_2S_3, Cu_2S, Ag_2S) [53], have shown enhanced photocatalytic and photoelectrochemical activities.

Simply reducing the size of the semiconductors to the nanometer-scale, which usually increases the bandgap, is not always helpful for hydrogen production. The design and synthesis of "nano-structured semiconductor systems" [54-56] can be approached by exploiting particle size effects as well as by effectively tailoring the electronic properties of non-oxide semiconductors for hydrogen production, always keeping in mind the challenge of corrosion-stability of such semiconductor nanostructures in water, in the presence of photogenerated holes. For example, theoretical models based on strain-dependent theory suggests core-shell nanorods of CdS-InP and InP-ZnTe of proper dimensions (height, core radius, shell thickness) can be used for solar hydrogen production by exploiting the interplay between quantum confinement and strain [54]. Tetrapyrroles, such as porphyrins and chlorophylls, are often self-organized into nanoscale superstructures that perform light harvesting, and energy- and electron-transfer functions, able to respond to the visible and ultraviolet portions of the solar spectrum. Shelnut [56] suggests nano-tubuler porphyrin, with a gold nanowire inside and platinum nanoparticles outside, as a material architecture for water splitting.

7.2 General Synthesis Techniques of Non-Oxide Semiconductors

Semiconductor nanoparticles have unique size-dependent photoelectrochemical properties. Demand for nanocrystalline semiconductors of uniform size and shape has stimulated research into different synthesis techniques some of which we consider here.

7.2.1 Preparative Methods for Nanoparticulate Materials

Coprecipitation [57-63] and solvothermal/hydrothermal [64-70] appear the most common methods for preparation of non-oxidic nanoparticulate-structured materials. Systematically developed novel strategies using organic chalcogenides, metal molecular carriers, and organic coordinating solvents as the reaction media have resulted in the preparation and characterization of a variety of nanoscale II-VI and III-V semiconductors [57,58]. For the synthesis of nanocrystalline cadmium chalcogenides, (CdE), $Cd(CH_3)_2$ were used as the Cd-source precursor, with the chalcogen source molecules obtained from $[(CH_3)_3Si]_2E$ (E = S, Se), R_3PSe, and R_3PTe (R = C_4-C_8 n-alkyl) [59]. Monodispered size-selective (~1.2-11.5 nm) precipitation occurs upon rapid injection of a room-temperature precursor solution into a preheated (260-300°C) coordinating solvent solution. Control of the solvent temperature allows control of nanocrystal nucleation and growth. The smallest (~1.2 nm) CdS, CdSe and CdTe particles are produced under mild conditions, with injection and growth carried out ~100°C. High temperature baths of solvents such as trioctylphosphine (TOP) and trioctylphosphine oxide (TOPO) also act as a passivating agent and stabilizer. Molecules of TOP and TOPO compounds readily coordinate to metal and chalcogen intermediates, binding to the nanocrystal surfaces thereby reducing their surface energy; nanocrystals grown with their surfaces protected with these capping ligands do not agglomerate.

Alivisatos and co-workers [60,61] have used $(Me_3Si)_3E$ to prepare GaAs, InP, GaAs and InAs nanoparticles.

$$(Me_3Si)_3E + MX_3 \rightarrow ME \qquad (7.2.1)$$
where (M = Ga, In; X = Cl, Br, I; E = P, As, Sb; Me = methyl).

Wells and co-workers [62] have obtained nanocrystalline III-V group semiconductor particles with sizes in the range of 4-35 nm by the reaction

$$(Na/K)_3E + MX_3 \rightarrow ME + 3(Na/K)X \qquad (7.2.2)$$

This reaction is similar to the one used by Alivisatos, except that $(Me_3Si)_3E$ was replaced by $(Na/K)_3E$. Precursors that already have a M-E (M = Ga, In; E = N, P, As) bond can be pyrolyzed to produce ME at relatively low temperatures. The use of N,N,N′,N′-tetramethyl-p-phenylenediamine (TMPD), in comparison to TOP and TOPO, as a capping/stabilizing agent for InP quantum dots has resulted in superior charge transfer properties [63].

Qian [64] first demonstrated a solvothermal synthetic method for the preparation of nanocrystalline III-V semiconductor nanoparticles including InP, GaN, and InAs. As an example, for the GaN particles, prepared in an autoclave of elevated temperature with benzene as the solvent:

$$GaCl_3 + Li_3N \rightarrow GaN + 3LiCl \qquad (7.2.3)$$

The products are usually crystalline, avoiding the need for a high temperature post-treatment [64]. Gautam et al. [65] has introduced an inexpensive, low boiling solvent under solvothermal conditions; in the presence of dodecanethiol as a capping agent, using cadmium stearate as a source monodispersed ~3 nm CdSe particles can be prepared.

A series of sulfides MS_2 (M = Ni, Co, Fe, Ni, Mo), and selenides MSe_2 (M = Ni, Mo) are achieved via low-temperature hydrothermal synthesis using metal chlorides and $Na_2S_2O_3$ or Na_2SeSO_3 as starting materials; and in the case of the molybdenum-containing products Na_2MoO_4 in hydrazine is used in place of the metal chloride [66,67]. Since $Na_2S_2O_3$ is not strong enough to reduce Mo(VI) to Mo(IV) hydrazine monohydride is added to the reaction medium, maintaining a reducing atmosphere in order to assist the formation of uniform MoS_2 [67]; for MoS_2 an average crystallite size of ~4 nm is obtained.

Hydrothermal synthesis of ZnSe and CdSe, using powdered Se with either Zn or Cd heated to 180°C in a water filled autoclave, yields 70-100 nm particles without the need of any stabilizer or capping agent. While most hydrothermal synthesis routes of chalcogenides involve the preparation of binary systems, more complicated ternary compounds can also be prepared. For example, $ZnIn_2S_4$, a ternary chalogenide photocatalyst has been hydrothermally prepared and found to photocatalytically reduce water under visible light irradiation [68]. Other ternary chalcogenide photocatalysts $M^IM^{III}S_2$ e.g.$CuInS_2$ (bandgap = 1.53 eV) and $CuGaS_2$ (bandgap = 2.49 eV) can be hydrothermally prepared using CuCl salt with elemental Ga or In (M^{III}) and excess S in benzene or toluene solution at 200°C [69]. Using the technique of [69], nearly spherical $CuInS_2$ particles are achieved of uniform size, 20-25 nm. Xiao et al. took the process of [69] one step further preparing chalcopyrite quaternary semiconductors, $CuIn(Se_xS_{1-x})_2$, over the full $0 < x < 1$ compositional range [70]. In this case, the starting materials were $InCl_3.4H_2O$, $CuCl_2.2H_2O$, Se, and S, with ethylenediamine the preferred solvent; particle sizes were typically on the order of 15 nm.

7.2.2 Preparative Methods for Nanostructured Photoelectrodes

The primary event for photocurrent generation is photochemical excitation of a material to form an electron–hole pair. However the transfer of conduction-band electrons to the electrode are hindered by competing electron–hole recombination processes, that in turn inhibit photocurrent generation. Simply stated, minimization of electron–hole recombination is essential for enhancing photocurrent. Various approaches have been suggested to maximize charge separation in non-oxide semiconductor nano-architectures, including chemical [51,71-74] and electrochemical deposition [75-78] of nanoparticles on the conducting substrates, as well as subsequent immobilization of surface modified nanoparticles with the use of physical interactions [79,80] and chemical reactions [49, 81-90].

With an aim towards achieving a precisely tailored film structure with enhanced photoelectrochemical properties, photoelectrodes comprised of multiple CdS nanoparticle monolayers have been investigated [87,88]. Electrode preparation involves two steps [87,88]: first, functionalization of a transparent conductive tin oxide electrode using (3-mercaptopropyl) trimethoxysilane in tetrahyrofuran (THF) to yield a surface with pendant thiol groups; second, preparation of 5 nm CdS particles using hexanethiol as a capping agent, see **Fig. 7.3a**. Monodispersed CdS nanoparticles can be immobilized onto a gold electrode substrate coated with a self-assembled monolayer of hexanedithiol ($SH-C_6H_{12}-SH$), 2-aminoethanethiol; dioctyl sulfosuccinate [49,82,84,85]. When surface modified nanoclusters of CdS with 2-mercaptoethane sulfonate are brought into the contact with an allyl ammonium group of poly(allylamine hydrochloride), a multilayer CdS nanocluster assembly is obtained; film thickness is controlled by alternate dip-coating from their respective solutions, see **Fig 7.3b** [89].

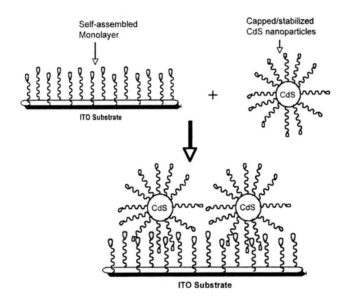

Fig.7.3a: Schematic diagram of thin film of CdS nanoparticles on ITO electrode.

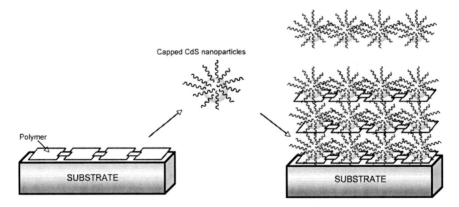

Fig. 7.3b: Schematic diagram of multilayer thin films of CdS nanoparticles on Au electrode in presence of poly allylamine hydrogen chloride.

Physical interactions and chemical reactions are the two most common methods used for immobilizing semiconductor nanoparticles on photoelectrode substrates, by which the size and size distribution of the original nanoparticles are retained with immobilization [79-90]. Prevention of nanoparticle coalescence during immobilization is necessary if the particles are to retain their size-quantization properties, which can be effectively used to promote photoelectrocatalysis. Electrochemical deposition is a convenient method for preparing thin chalogenide films on electrode surfaces [78]. Willner and co-workers [90] prepared a heterostructured photoelectrode using CdS nanoparticles coupled with carbon nanotubes on a gold substrate. The electrodes were prepared as follows: carboxyl functionalized single walled carbon nanotubes (SWCNTs) were covalently linked to the amino group of the cysteamine monolayer-functionalized Au surface in the presence of dicyclohexylcarbodiamide (DCC) and N,N-dimethyl formamide (DMF). 2-thioethanol was added in the mixed monolayer to prevent non-specific adsorption of the SWCNTs onto the electrode surface, and also prevent coupling of the SWCNTs through the sidewall to the surface. A solution of cystamine stabilized CdS nanoparticles was then added to modify the electrode.

It has been found possible to modify the electron flow of many non-oxide semiconductors through their coupling with either oxide and/or non-oxide semiconductors, improving the stability of their performance. For example, improved photocurrent stability is achieved in an electrodeposited CdSe electrode when it is protected from the electrolyte by a thin TiO$_2$ film that retards charge recombination [51]. **Figure 7.4** shows the absorption spectra of CdSe films recorded at different stages of electrochemical deposition. A blue shift in the absorption band edge, attributable to size quantization, is seen for CdSe films less than 7.0 nm thick (spectra a-c). The absorption onset of films above 10 nm (spectra d-f) occurs at a wavelength of about 700 nm, corresponding to the bandgap of the bulk semiconductor (E$_{BG}$ = 1.7 eV). No space charge layer at the electrode/electrolyte interface has been found for films below 20 nm in thickness. Charge separation, unlike that in semiconductor crystals or polycrystalline films, is controlled by the differing rates at which electrons and holes transfer to the electrolyte solution. As seen in the photocurrent action spectra of the TiO_2, CdSe, and CdSe-TiO_2 electrodes, see **Fig. 7.5**, the photocurrent generation onset for the TiO_2 thin film occurs at a wavelength below 350 nm, while the photocurrent response extends to visible region in CdSe and CdSe-TiO_2 electrodes. The 10-20% increase in incident to photon to current efficiency (IPCE) values at excitation wavelengths of 400-600 nm indicates that the TiO_2 film accepts electrons from the photo-excited CdSe layer rapidly transporting them to the electrode surface, thereby suppressing charge recombination and boosting photocurrent amplitudes [51].

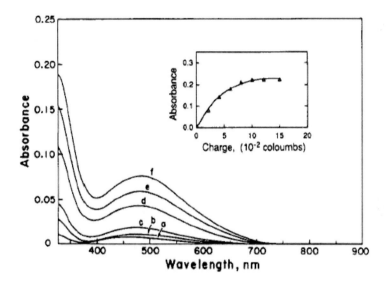

Fig. 7.4: Absorption spectra of CdSe electrodeposited films with various thickness: **(a)** ~20 nm, **(b)** 35 nm, **(c)** 70 nm, **(d)** 155 nm, **(e)** 175 nm, and **(f)** 200 nm. Dependence of absorbance at 500 nm on the cathodic charge during the CdSe deposition is shown in inset. Reproduced with permission from Ref. [51].

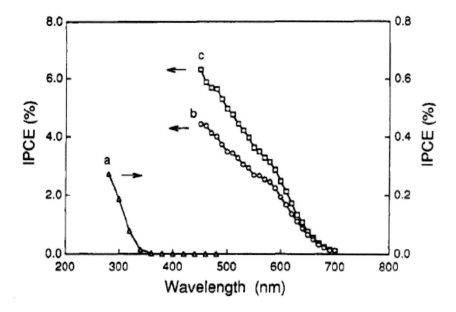

Fig. 7.5: Photocurrent action spectra of **(a)** TiO_2 (thickness = 1 μm, electrolyte = 0.01 M NaOH), **(b)** CdSe, and **(c)** TiO_2-CdSe (thickness = 15 nm, electrolyte = 0.5 M $[Fe(CN)_6]^{4-}$ and 0.1 mM $[Fe(CN)_6]^{3-}$, pH =12) thin film electrodes. Reprinted with permission from Ref. [51].

A systematic study on the electronic structures and wave functions of core-shell nanorods using a strain-dependent theoretical model have been carried out with the specific aim of identifying suitable semiconductor nanostructures as promising photoelectrodes for solar hydrogen production [54]. Core-shell InP-CdS and InP-ZnTe appear to have the most promise for use as water splitting photoelectrodes, see **Fig.7.6.** The smaller bandgap core is protected by the larger bandgap shell from photocorrosion, hence these nanorods should remain stable in aqueous solutions. Moreover, the probability of electron-hole recombination is diminished by localization of energy bands in the different regions. InP-CdS and InP-ZnTe are especially good candidates for photoelectrodes due to small discontinuities in, respectively, their conduction and valence band edges [54]. The appearance of quantum confinement and strain at the layer-to-layer interfaces of the electrode enable efficient charge separation, thereby facilitating water splitting.

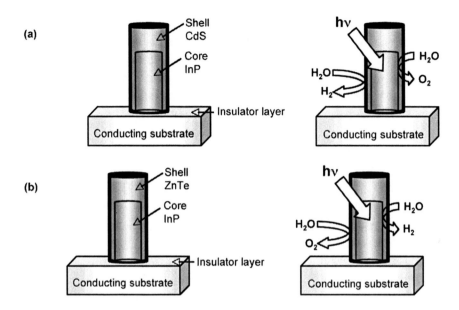

Fig. 7.6: Schematic diagram of core-shell nanorods of **(a)** InP-CdS **(b)** InP-ZnTe and use of nanorods in water photodecomposition. In **(a)**, electrons are holes are localized in the shell and the core, respectively; whereas, in **(b)**, holes and electrons are localized in the shell and the core, respectively.

7.3 Non-Oxide Photoelectrode Systems and Water Photoelectrolysis

The appropriate bandgap energy, band-edge potentials, and chemical stability are acknowledged as the three most essential requirements for a semiconductor that is to be used for water splitting. A semiconductor bandgap of 1.8-2.0 eV can be considered ideal, due to optimum solar light absorption while having enough energy to drive water-splitting reactions. With respect to chemical stability, while electrolyte-soluble semiconductors obviously cannot be used, semiconductors that are insoluble sometimes become soluble upon solar light irradiation. For example,

$$CdS + h\nu \rightarrow Cd^{2+} + S \qquad (7.3.1)$$

There are several non-oxide semiconductor electrodes that have been studied for use in photoelectrochemical cells for water splitting. Materials used as photocathodes include p-Si, p-InP, p-GaAs, and p-CdTe. Materials used as photoanodes include: n-Si, n-InP, n-GaAs, n-CdTe and n-CdSe. Similar to CdS, unfortunately most of these non-oxide semiconductors undergo photocatalytic corrosion under the conditions of oxygen evolution.

7.3.1 n-Si and p-Si [22, 91-99]

For p-type Si, polymer coating [94] and photoelectrochemical platinization (with or without the presence of a polymer) [95] of the Si surface are known treatments by which a significant improvement in photo-assisted hydrogen evolution can be achieved. Different deposition techniques such as electron beam evaporation, magnetron sputtering, and photo-assisted electrochemical deposition have been used to deposit Pt films atop p-Si [96]. Due to the small differences in work functions, $\phi_{Si} = 4.88$ eV and $\phi_{Pt} = 5.28$ eV, Pt ideally forms a relatively small Schottky barrier when in contact with p-Si [96,97]. However each deposition technique comes with its own nuances. Electron beam evaporation results in complete coverage of the p-Si surface with metallic Pt, and residual O and C surface contamination. Sputtering produces a homogeneous $PtSi_2$ layer on the p-Si sample that acts as an ohmic contact, and therefore cannot be used as a photocathode for the reduction of protons to H_2 molecules. XPS analysis of films obtained using photo-assisted electrochemical deposition reveals a platinum island-like topology upon the silicon, due to the coalescence of highly mobile Pt atoms upon the p-Si surface during the initial deposition stages, with 5.0 at% surface coverage (island height 180 nm for average Pt film thickness of 9 nm) [96]. Depositing a sub-monolayer of Pt atop p-Si, with a thickness between 0.05 nm and 0.2 nm, allows its use as an efficient photocathode [96]. Operating in an alkaline medium the photocathodes (island-like Pt film topology) exhibit a photoconversion efficiency of 10%; in acid electrolyte a photoconversion efficiency of 3% [96].

n-type Si can be stabilized against photocorrosion in a variety of ways, including: (a) growing a metal doped SiO_x layer atop the Si [91]; (b) forming metal silicide layer on the n-Si surface [92]; (c) depositing a thin layer of manganese oxide, or analogous oxide, on the surface [93]. Formation of a Pt film, comprised of scattered 5 nm to 50 nm Pt dots, upon n-Si improves the stability of the electrode in aqueous medium, and a significant increase in open-circuit photovoltage (V_{OC}) and quantum conversion efficiency due to enhanced interfacial charge transfer across the electrode-electrolyte interface [98,99]. Minority carrier accumulation is largely controlled by the Si/Pt particles at the interfacial barrier. A V_{OC} of 0.685 V has been obtained from an alkali etched n-Si electrode on which a layer of Pt islands (5-20 nm diameter) was deposited [98].

7.3.2 n-type and p-type CdS, CdSe and CdTe
[11,13-16,81,82,84-90,100-109]

Cadmium sulfide (CdS), 2.4 eV bandgap, is one of the most widely studied non-oxide semiconductors. For a semiconductor irradiated by photons with energy equal to or greater than its bandgap, an electron from the valence band jumps to the conduction band leaving a positively charged hole behind in the valance band; for CdS this is expressed as

$$\text{Semiconductor: CdS} + h\nu \rightarrow e^-(\text{CdS}) + h^+(\text{CdS}) \qquad (7.3.2)$$

$$\text{Water:} \qquad H_2O \rightarrow H^+ + OH^- \qquad (7.3.3)$$

$$2e^-(\text{CdS}) + 2H^+ \rightarrow H_2 \qquad (7.3.4)$$

$$2h^+(\text{CdS}) + 2OH^- \rightarrow 2H^+ + \tfrac{1}{2}O_2 \qquad (7.3.5)$$

Hydrogen is produced by the reduction of H^+ ions by the photogenerated electrons from the semiconductor conduction band. Unfortunately, this semiconductor is not stable in aqueous solution and undergoes self-anodic dissolution under visible light leading to the formation of sulfur as shown in reaction (7.3.6). Such photoanodic dissolution can be quenched by adding reducing agents such as S^{2-}, SO_3^{2-}, $S_2O_3^{2-}$, as shown in reactions (7.3.7) to (7.3.10) [4-13].

$$2h^+(CdS) + CdS \rightarrow Cd^{2+} + S, \tag{7.3.6}$$
$$E° \ (vs \ NHE, \ pH \ 7) = + \ 0.32 \ V$$

$$2S^{2-} + 2h^+ \rightarrow S_2^{2-}, \tag{7.3.7}$$
$$E° \ (vs \ NHE, \ pH \ 14) = - \ 0.52 \ V$$

$$SO_3^{2-} + 2OH^- + 2h^+ \rightarrow SO_4^{2-} + H_2O, \tag{7.3.8}$$
$$E° = - \ 0.92 \ V$$

$$2SO_3^{2-} + 2h^+ \rightarrow S_2O_6^{2-}, \tag{7.3.9}$$
$$E° = - \ 0.25 \ V$$

$$S_2^{2-} + SO_3^{2-} \rightarrow S_2O_3^{2-} + S^{2-} \tag{7.3.10}$$

Photo-induced platinization of CdS, as well as coating with electrically conducting polymers, have been found partially effective in protecting the material against photodecomposition while in an electrolyte. The stabilization mechanism involves, in part, the high conductivity of the polymer, while the electroactive redox species efficiently trap the photogenerated holes before they can react with the semiconductor [14-16]. For CdS in combination with a polymer-catalyst coating (polypyrrole-Pt), nearly 68% of the photogenerated holes are engaged in O_2 production, leaving 32% to produce Cd^{2+} ions that lead to the destruction of semiconductor lattice, **Fig. 7.7** [14]. However in the absence of catalyst or polymer-catalyst films, 99% of all the photogenerated holes go to destruction of the CdS lattice.

Various methods of preparing electrodes modified with an electrochemically active layer of CdS nanoparticles have been reported [80-82,84,85,87,88,100-104]. Spectroscopic investigations suggest that the surface layer acts as an array of independent nanoparticles, with quantum size effects due to exciton confinement observed. A band edge wavelength of 495 nm is observed in the photocurrent spectrum of an electrode modified by 5 nm CdS particles, indicating a bandgap shift of 0.1 eV [87]. Milczarek and coworkers [104] developed a two-compartment photoelectro-chemical cell consisting of a CdS photoanode immersed in an aqueous sulfide solution; a solar hydrogen production efficiency of

12% was obtained under diffuse sun light illumination (20 mW/cm^2). At a given pH, in the compartment containing the sulfide solution immersed CdS photoanode hydrogen evolution depends on the sulfide concentration. Lowering the pH in the compartment containing the Pt electrode further enhances the hydrogen evolution. A Pt-nanoparticles/nafion/Pt cathode design improves hydrogen production by reducing over-voltage values from 0.175 V (for Pt metal electrode) to 0.038 V in 1.0 M H$_2$SO$_4$ solution. The resistance of the membrane leads to a drop in short-circuit current values, hence a another crucial factor in this two-compartment cell is the photoanode-to-membrane (nafion)/photoanode (CdS) surface ratio, which is found optimum as at a ratio of 2.5. While the bandgap of CdS is excellent for capturing solar spectrum energy, the issue of photocorrosion stability remains unsolved limiting its application.

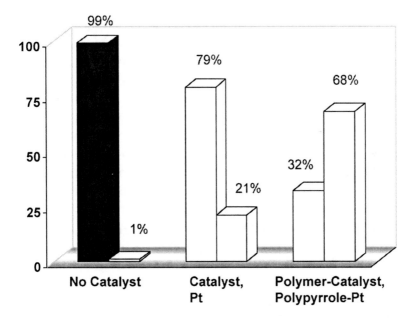

Fig. 7.7: Percentage of photogenerated holes that contribute to anodic decomposition versus O$_2$ evolution from naked (no catalyst), Pt-coated, and polymer-Pt coated n-CdS photoanode in 0.5 M Na$_2$SO$_4$ solution (pH = 8.6) [14].

CdSe photoanodes also suffer from photocorrosion [12,13,106,107]:

$$CdSe + h\nu \rightarrow CdSe + 2e^- + 2h^+ \qquad (7.3.11)$$

$$CdSe + 2h^+ \rightarrow Cd^{2+}(aq.) + Se \qquad (7.3.12)$$

$$4OH^- + 4h^+ \rightarrow O_2 + H_2O \qquad (7.3.13)$$

$$H_2O + 4h^+ \rightarrow O_2 + 4H^+ \qquad (7.3.14)$$

Accumulation of photogenerated holes on the surface leads to dissolution of the CdSe crystal lattice. Both electrodeposited and RF-sputtered CdSe thin films, on stainless steel substrates, have been employed as photoanodes for solar hydrogen production in alkali polysulfide electrolyte solutions with a 150 W halogen lamp light source [106]. While modest hydrogen evolution rates are achieved the system faces twin challenges of stability and practicality of the chosen electrolyte. Modified nanostructured CdSe thin-film electrodes have been reported [51,75,83,105-108]. Perhaps the best of these for solar hydrogen production comes from citrate stabilized CdSe nanoparticles, sensitized with copper phthalocyanine, which exhibit effective charge separation and extended photovoltage response in the range 300-800 nm [108].

Nanocrystalline thin films of CdTe (average grain size 4-7 nm) demonstrating a shift in the absorption spectrum corresponding to a 0.8 eV increase in bandgap have been reported [77]. The films were electrochemically deposited using a pulse-plating technique in a DMSO solution of tri-(n-butyl phosphine telluride and cadmium perchlorate) [77]. In this method removal of excess of Te or Cd from the electrode surface, size distribution and decrease in particle size can be achieved leading to the formation of stoichiometric films. The photocorrosive nature of CdTe limits its use for solar hydrogen production, however solar hydrogen production using surface modified CdTe has been reported [109].

7.3.3 n-Type and p-Type GaN, GaAs and InP
n-GaN and p-GaN [110-116]

GaN, 3.4 eV bandgap, is known to be a material generally resistant to corrosion and hence of interest for application in photoelectrochemical cells. Kocha and co-workers prepared n-GaN using metalorganic vapor phase epitaxy on sapphire substrates, doping with Si to achieve $n = 1.0 \times 10^{19}$ cm^{-3} [111]. The band-edge potential is more negative than the hydrogen production potential, hence properly catalyzed GaN would easily split water. The performance of RuO_2 catalyzed n-GaN photoanodes were investigated in pH 0 and pH 14 cells, demonstrating stability to photocorrosion over time periods of a few minutes (and photocurrent densities less than a mA/cm^2) [111]. Later it was observed that non-catalyzed n-GaN suffers severely from photocorrosion in photoelectrochemical cells containing both acid and alkaline solutions when exposed to above bandgap illumination (3.4 eV) [112,113]. The splitting of aqueous water using n-GaN as the working electrode, Pt counter electrode and Ag/AgCl/NaCl reference electrode (sodium-saturated silver chloride electrode, SSSE) placed in 1 mol/L aqueous KOH solution was carried out by Fujii and co-workers [114]. The conduction and valence band-edge potentials are found at −1.66 and + 1.73 V vs SSSE, respectively. In this experiment, an external bias of 1.0 V at the Pt counter electrode with respect to the working electrode was applied to generate H_2 gas. Evolution of 94% hydrogen and 6% nitrogen at the Pt counter electrode after 5 hours indicates the occurrence of photo-assisted electrolysis as well as n-GaN decomposition due to the following reaction:

$$2GaN + 6h^+ \rightarrow 2Ga^{3+} + N_2(g) \qquad (7.3.15)$$

Beach and co-workers [115] measured the band edge potentials of both p-and n-GaN, relative to SCE, finding the p-GaN band-edge potential appropriate for water splitting. If the reduction potential of water is located below the conduction band edge of p-GaN, photogenerated electrons can be spontaneously injected into the electrolyte solution and H^+ (aq) reduced to H_2. Kobayashi and co-workers [116] prepared p-GaN by Mg doping using a high temperature (1000°C) metal-organic vapor phase epitaxy technique; Hall effect measurements indicate a hole concentration of 7.8×10^{17} cm^{-3}. *I-V* measurements of a p-GaN electrode were carried out using a Pt counter electrode, Ag/AgCl reference electrode, in 1M Na$_2$SO$_4$ electrolyte under 340 nm irradiation [116].   At zero bias a photocurrent of 5 µA was observed, indicating that the energy band bends slightly upward; spontaneous hydrogen evolution is observed indicating the conduction band-edge potential of p-GaN is above the reduction potential of water. The electron traveling from the Pt anode to p-GaN cathode reacts with H$^+$ to form H$_2$ molecules at 70% efficiency, with losses due to charge recombination of trapped photogenerated holes with the accumulated electrons. Although there was no morphological degradation in the sample due to etching, photocurrent degradation and consequently seizure of hydrogen evolution occurred probably due to passivation of Mg impurities in p-GaN by H$_2$ gas and the diffusion of generated H$_2$ into the p-GaN layer [116].

n-GaAs and p-GaAs [18,21,117-131]

With a 1.43 eV bandgap and a flat band potential of nearly −1.58 V (*vs* NHE) in alkaline solution, n-GaAs photoanodes satisfy two of the most desirable criteria for use in water photoelectrolysis [18,21,117,119,123]. The flat band potential in n-GaAs indicates that its conduction band facilitates transfer of electrons with energies suitable for hydrogen reduction at the counter electrode. However, GaAs like most of the non-oxide small bandgap semiconductors, photocorrodes upon illumination in aqueous solution [118-121]. The anodic corrosion of GaAs photoelectrodes in aqueous electrolytes results from either preferential dissolution of the electrode materials [reaction (7.3.16)] or the oxidation of its surface [reaction (7.3.17)]

$$GaAs + 3h^+ \rightarrow Ga^{3+}(aq) + As\ (surf) \hspace{2cm} (7.3.16)$$

$$GaAs + 3H_2O + 6h^+ \rightarrow \tfrac{1}{2}\ Ga_2O_3(surf) + \tfrac{1}{2}\ As_2O_3 + 6H^+ \hspace{0.5cm} (7.3.17)$$

Reaction (7.3.16) enriches the GaAs surface with elemental arsenic, resulting in a negative shift in the onset potential as well as an increase in the photocurrent density, which in turn promotes H_2 photogeneration [118,119]. The stability of elemental arsenic on a GaAs surface depends on the electrolyte pH, as well as the applied potential. At very low pH and a potential of −0.45 V, arsenic is oxidized to stable As_2O_3 [120]:

$$2As \text{ (surf)} + 3H_2O \rightarrow As_2O_3 \text{ (surf)} + 6H^+ + 6e^- \qquad (7.3.18)$$

As reaction (7.3.18) proceeds the result is a decrease in photocurrent amplitude due to oxide layer formation. Oxide growth affects photoactivity in several ways, including: (a) consumption of photo-generated electrons due to introduction of new surface states; (b) less light transmittance through the oxide, due to scattering and absorption, to the underlying semiconductor electrode; (c) anion absorption from the electrolyte which in turn retards movement of the H^+ ion, hence a higher negative potential would be required for the displacement of the adsorbed anions. The existence of the oxide layer shifts the onset potential of the photocurrent towards more negative values [118-120].

Use of non-aqueous organic and inorganic solvents, and concentrated acidic solutions, as electrolytes as well as coating of metals and/or metal oxides are some of the common practices that have been adopted to enable use of GaAs photoelectrodes in photoelectrochemical cells [121-126]. For example, the increase in photogenerated current density from 4 to 18 mA/cm^2 during irradiation at +0.4 V (vs SCE) in 2.5 M H_2SO_4 + 6.0 M LiCl electrolyte solution confirms the suppression GaAs electrode photocorrosion, presumably due to the formation of an arsenic and chlorinated arsenic layer on the electrode surface [121]. The photoelectrochemical dissociation of water at three different pH levels, pH = 4, 10 and 13, has been investigated using a Si-doped n-GaAs photoanode and Pt-counter electrode [124]. Coating of the photoanode with a 5 nm gold layer drastically reduces photodecomposition activity under visible light illumination, with the anodic photocurrent onset shifting toward a cathodic potential with increasing pH [124]. GaAs photoelectrodes display highest photosensitivities in alkaline solutions, with the saturation photocurrent commencing at a potential greater than −0.3 V (*vs* NHE). Effective hydrogen evolution can be achieved because of its more negative flatband potential (−1.58 V) than the standard redox potential of H_2O/H_2 (−0.78 V *vs* NHE) at the same pH; i.e. the minimum energy level of the conduction band is placed at a potential more negative than the H_2O/H_2 half cell potential, indicating thermodynamic feasibility for discharging H_2 at the platinum electrode [124].

p-GaAs electrodes can evolve H_2 at their surface because their flat band potential is more cathodic than the standard H^+/H_2 redox couple potential [123,126-131]. *I-V* characterization of p-GaAs in 0.5 M H_2SO_4 shows the generation of cathodic current at an onset potential of −0.3 V to −1.3 V, corresponding to the evolution of H_2 at the GaAs electrode and O_2 evolution at the Pt counter electrode [127]. Partial modification of the p-GaAs electrode surface due to etching is manifest in a 0.1 V positive shift in potential with a corresponding 15% increase in photocurrent [128]. As shown in reactions (7.3.16) to (7.3.18), the removal of an oxide layer from the GaAs surface and/or the preferential dissolution of one of the elemental (Ga^{3+}) components of GaAs results in an arsenic rich electrode surface that facilitates efficient hydrogen output. Erne [129] elucidated the mechanism of hydrogen gas evolution on a p-GaAs electrode by studying the reaction intermediates using real time, *in-situ* infrared spectroscopy. The spectral studies suggest that hydrogen cathodically absorbs at arsenic sites on the surface, replacing As-OH groups present in the anodic range, which upon reduction yields molecular hydrogen.

The photoelectrochemical behavior of various metal (M = Pt, Ni, Cu, Au, Cd and Pb) loaded n- and p-GaAs electrodes has been investigated [131]. The addition of Pt, Ni and Cu to n-GaAs causes a negative shift in the onset potential of the photocurrent proportional to the work function of the deposited metals. A thin metal film coating on p-GaAs electrodes results in a positive shift of the photocurrent onset potential, with their photoelectrochemical behavior determined by the electrocatalytic activity of the deposited metal thin film. For example, an electrodeposited (by passing current of 40 µA cm^{-2} at 0.2 V *vs* SCE) layer of Pt metal on p-GaAs was found to enhance the *I-V* characteristics by shifting the photocurrent onset potential by ~0.2 V towards more positive values [128].

n-InP and p-InP [19,27-33,132-134]

Like other non-oxidic semiconductors in aqueous solutions, surface oxidized and photocorrosive InP is a poor photoelectrode for water decomposition [19,27,32,33]. To enhance properties several efforts have focused on coupling of the semiconductor with discontinuous noble metal layers of island-like topology. For example, rhodium, ruthenium and platinum thin films, less than 10 nm in thickness, have been electrodeposited onto p-type InP followed by a brief etching treatment to achieve an island-like topology on the surface [27,28]. In combination with a Pt counter electrode, under AM 1.5 illumination of 87 mW/cm^2 the metal (Pt, Rh, Ru) functionalized p-InP photocathodes [27] see a reduction in the threshold voltage for water electrolysis from 1.23 V to 0.64 V, and in aqueous HCl solution a photocurrent density of 24 mA/cm^2 with a photoconversion efficiency of 12% [27].

The semiconductor-metal junction barrier height must be equal to or greater than that of the semiconductor H^+/H_2. The semiconductor-liquid junction barrier height is significantly modified by the metals on the semiconductor surface, with experimental results showing a gain in the onset potential of ~0.7 V in all p-InP(Pt,Rh,Ru)/H^+/H_2 junctions for hydrogen evolution with respect to Pt/H^+/H_2 cathode junction. This lowering of the onset potential is due to a reduction in the metal work function upon hydrogen alloying of the photocathode, which increases the barrier height between the metal catalysts and p-InP [28]. Formation of various metal islands such as Cd, Pb, Co, Au, Ni, V, Mo, and W on p-InP using deposition techniques such as electrochemical [29-32], electron beam [132], laser induced [133], and cathodic decomposition [134], exhibit similar results in different acidic mediums.

7.3.4 Ternary Non-Oxide Semiconductors [135-143]

The 3.4 eV bandgap of GaN is too large for efficient solar water-splitting, but the addition of indium to GaN to form $In_xGa_{1-x}N$ can significantly reduce the bandgap [135]. The bandgap of $In_xGa_{1-x}N$ is governed by its composition in a non-linear manner [136]; the binary endpoint compositions of $In_xGa_{1-x}N$ have bandgaps of 3.4 eV (GaN, $x = 0$) and 1.9 eV (InN, $x = 1$) [137]. Fujii et al. [138] prepared n-type $In_xGa_{1-x}N$ (x = 0.02, 0.09) as a photoanode material using metal-organic vapor phase epitaxy on a sapphire substrate. The bandgap energies for $In_{0.02}Ga_{0.98}N$ and $In_{0.09}Ga_{0.91}N$, obtained from X-ray diffraction and room temperature photoluminescence measurements, are found to be 3.3 eV and 3.0 eV, respectively. At an applied bias of 1.0 V, 150 W Xe lamp light source, gas generation was observed from the photoanodes in both KOH and HCl electrolytes. However in KOH solution $In_xGa_{1-x}N$ undergoes rapid corrosion leading to an almost instantaneous degradation of the photocurrent. In 1 mol/L HCl solution the gas composition generated from the Pt counter electrode was determined to be 90.5% H_2, 9.0% N_2, 0.5% O_2 when using $In_{0.02}Ga_{0.98}N$, and 78.2% H_2, 20.7% N_2, 1.1% O_2 when using the $In_{0.09}Ga_{0.91}N$ photoanode.

Upon irradiation of n-type GaP in alkaline solution, electron flow from GaP (bandgap = 2.24 eV) to the counter electrode leads to photoanodic dissolution by the following reactions:

$$GaP + 4OH^- \rightarrow GaO_2^- + P + 2H_2O + 3e^- \qquad (7.3.19)$$

To overcome GaP photocorrosion suitable redox species such as Se^{2-}/Se_n^{2-} and Te^{2-}/Te_n^{2-} can be added in the electrolyte solution [18]. Alternatively, layers of chemically stable GaPN can be grown or deposited on GaP using, for example, molecular beam epitaxy [139], with a slight lattice mismatch due to the nitrogen. At low nitrogen concentrations GaPN has been found chemically stable in aqueous electrolyte solutions, which coupled with its ≈ 2.0 eV bandgap make this ternary semiconductor appealing for water electrolysis. Both photocurrent and open circuit potential measurements suggest that the flat band potential of GaPN is more positive than that of GaP, with an additional bias of 0.3 V to 0.5 V necessary for water splitting in alkaline 1M KOH solution [139]. GaInP$_2$, a similar ternary compound, has a bandgap energy of 1.82-1.91 eV [140-142]. The band edges of Ga$_{0.5}$In$_{0.5}$P are 0.2 -0.4 V more negative than that required for direct water splitting. Surface modification of p-type GaInP$_2$ by 8-quinoline shifts the flat band potential approximately 0.3 V in the positive direction significantly improving the water splitting energetics [141,142]; use of metal ions (Ru^{3+}, Rh$^+$, Co^{3+}, Os^{3+}, Pt^{4+}) on the p-GaInP$_2$ surface suppresses migration of the bandgap edges, and appears to enhance the P-GAInP$_2$/water interfacial kinetics. Semiconducting Cu(In,Ga)(Se,S)$_2$ thin films prepared by electrodeposition [78] have a bandgap of approximately 2.0 eV, and demonstrate photoelectrochemical hydrogen production in 0.1 M Na$_2$SO$_3$ in pH 10 buffer. Similar work has also been carried out using p-CuIn$_{1-x}$Ga$_x$Se$_2$ thin film photoelectrodes in 0.5 M H$_2$SO$_4$ [143].

7.3.5 Coupled or Composite Semiconductors

The properties and utility of photoanode hetero-systems, for example CdSe-TiO$_2$ [51,144], CdS-TiO$_2$ [52,74,145], CdS-ZnO [144], InP-CdS [54], InP-ZnTe [54] and CdS-ZnS [72,73] can be tailored by means of "semiconductor sensitization". While visible light induced photogenerated holes on the surface of CdS and CdSe leads to dissolution of their crystal lattice, teaming with TiO$_2$ and ZnO provides protection of the small bandgap semiconductors from photocorrosion. Ideally, a semiconductor hetero-system would possess the optimal bandgap of one constituent, e.g. CdS, and the photocorrosion stability of the other, e.g. TiO$_2$. These coupled semiconductors, e.g. CdS-TiO$_2$, CdSe-TiO$_2$ and CdS-ZnO, demonstrate a cathodic shift in their onset potential illustrating n-type behavior [51,144,145]. For CdS-TiO$_2$ crystals in the presence of 0.5 M K$_2$SO$_4$ solution, an anodic potential larger than that required for CdS crystals is necessary to reach photocurrent saturation; under small anodic potentials recombination of the photogenerated charges dominates photocurrent properties [144], which can be improved through addition of a redox couple, such as [Fe(CN)$_6$]$^{4-}$, in aqueous electrolyte solutions [145].

ZnS-CdS (bandgap = 2.3-2.4 eV) composite semiconductor photoelectrodes show a broad spectral response and n-type behavior, with saturation of the anodic photocurrent upon increasing anodic potential making the system suitable for use as a photoelectrochemical cell photoanode [72]. Nanostructured ZnS-CdS thin film electrodes show that anodic photocurrent saturation can be attained with the application of a small, ≈ 0.1 V, bias [73], while hydrogen evolution is observed at the Pt cathode. The performance of the ZnS-CdS photoanodes appear strongly dependent upon the method of film preparation [72,73], with Zn rich films demonstrating superior photocurrent generation, and stability, in comparison to Cd rich films.

7.3.6 p-n Photoelectrolysis Cell

A. J. Nozik [24,26] pioneered the premise of a photoelectrochemical cell where the semiconductor anode and cathode were simultaneously illuminated. Since then various combinations of oxide and non-oxide semiconductors have been used for photoelectrochemical cell construction, with the objective of expanding the spectral response and increasing water-splitting efficiencies. Semiconductor anode-cathode pairs, n-type:p-type, that have been considered include n-TiO$_2$/p-CdTe, n-SrTiO$_3$/p-CdTe, n-GaAs/p-GaAs, and n-GaAs/p-InP [146-148]. The anode and cathode may remain separated, or combined into a monolithic structure as seen in **Fig. 7.2**. Murphy and Bockris [146] constructed a photoelectrochemical cell involving RuO$_2$ coated n,p-GaAs cathode and Pt coated n,p-GaAs anode in an 5M H$_2$SO$_4$ electrolyte solution; a photoelectrochemical solar to hydrogen conversion efficiency of 7.8% was achieved. Similarly, coupled electrodes of Pt functionalized p-InP and MnO$_2$ coated n-GaAs showed a photoconversion efficiency of 8.2% at room temperature and 9.1% at 45°C without external electrical bias [147]. While the photoconversion efficiencies achieved are exciting, issues of limited material availability, high cost, and imperfect photocorrosion stability of the coupled semiconductor anode-cathode geometry remain to be solved.

7.4 Non-oxide Suspended Particle Systems and Direct Water Splitting

In Chapter 6 we were introduced to micro-heterogeneous systems utilizing fine semiconductor particles dispersed throughout an aqueous solution to absorb sunlight and split water simultaneously evolving oxygen and hydrogen. Each suspended particle behaves as its own photoelectrochemical cell, essentially a short-circuited photo-anode and photo-cathode. Various novel material design and synthesis routes of non-oxide semiconductors have been pursued, with an aim of achieving broad spectral response and properties suitable for efficient water splitting. As shown in **Fig. 7.1**, for non-oxidic semiconductors for which the valence band energy is, generally, not sufficiently positive a sacrificial donor is used as an electron source for hydrogen evolution.   Similarly, for those semiconductors for which the conduction band energy is not sufficiently negative, a sacrificial acceptor is used as an electron sink to facilitate oxygen evolution. The complete water splitting reaction, i.e. both H_2 and O_2 generation, by semiconductor particles without the benefit of sacrificial agents has proven elusive.

7.4.1 Non-oxide Semiconductors Mixed with Metals or Metal Oxides [34-41,149-156]

The limiting factors that control photocatalysis efficiency are rapid recombination between photo-generated charge carriers, and the backward reaction leading to recombination of the formed molecular hydrogen and oxygen. To retard these processes efforts have typically focused on surface modification of the semiconductor particles using metals or metal oxides.

Photoplatinization of CdS particles has been found effective in limiting the susceptibility of CdS to photocorrosion [36-39]. Both Reber et al. [38,149] and Matsumara et al. [150] used platinum metal loaded CdS particles, in the presence of sulfite and sulfide ions, for hydrogen production; the amount of metal loading, electrolyte concentration and pH affect the rate of hydrogen production. In sulfide ion containing solutions, the highest hydrogen production rate is obtained near pH = 12.8, while in sulfite ion containing solutions maximum photoreactivity is observed near pH = 8. The hydrogen production efficiency in sulfide ion S^{2-} solutions is low due to the formation of disulfide ions, S_2^{2-}. The presence of a strong reducing agent, such as the sulfite ion SO_3^{2-}, prevents formation of the disulfide ion S_2^{2-} which allows hydrogen to evolve at a high rate [34]. The morphology of CdS particles plays a significant role in determining water photolysis properties. For example, CdS particles of hexagonal crystalline structure have dramatically increased performance over their cubic crystal structured counterparts [149,150]. The activity of these photocatalysts also depends on surface area, which most likely correlates with crystal structure in a manner yet to be eulicidated: only CdS photocatalysts of large ($\geq$100 m^2/g) or low surface areas ($\leq$ 6.7 m^2/g) produce hydrogen at a significant rate [149].

Metal oxide, e.g. RuO$_2$ [35,36] and WO$_3$ [151], loaded CdS colloidal suspensions with and without Pt functionalization were investigated for water decomposition under visible light. WO$_3$-Pt/CdS, in an aqueous solution of methyl viologen (MV^{2+}) which serves as an electron relay, has been found most efficient to date in terms of water splitting, see reactions (7.4.1) to (7.4.8) [151]:

$$CdS + h\nu \rightarrow e^-(CdS) + h^+(CdS) \qquad (7.4.1)$$

$$WO_3 + h\nu \rightarrow e^-(WO_3) + h^+(WO_3) \qquad (7.4.2)$$

$$h^+(CdS) + e^-(WO_3) \rightarrow CdS + WO_3 \qquad (7.4.3)$$

$$e^-(CdS) + Pt \rightarrow CdS + e^-(Pt) \qquad (7.4.4)$$

$$e^-(Pt) + MV^{2+} \rightarrow Pt + MV^{\cdot+} \qquad (7.4.5)$$

$$MV^{\cdot+} + H^+ \rightarrow MV^{2+} + \frac{1}{2}H_2 \qquad (7.4.6)$$

$$h^+(WO_3) + H_2O \rightarrow WO_3 + OH^- + H^+ \qquad (7.4.7)$$

$$2OH^- \rightarrow H_2O + \frac{1}{2}O_2 \qquad (7.4.8)$$

The conduction band electrons of CdS move to Pt that in turns reduces MV^{2+}. The reduced methyl viologen in turn reduces a hydrogen ion to a hydrogen molecule. The valence band holes of WO$_3$ then oxidize water to O$_2$ molecules.

A composite made up of CdS within non-active host materials such as alumina hydrogel or titanosilicate zeolite (ETS-4) have shown photocatalytic activity with no or a little trace of photocorrosion [152-154]. Such embedded systems have better activity than bulk CdS, although the non-active host material generally reduces the amount of light absorption [153]. Sathish and co-workers [154] removed the CdS particles from a host zeolite matrix using a 48% HF solution. The resulting CdS nanoparticles, 6-12 nm in diameter with surface areas of 26-46 m^2/g, have been found an efficient photocatalyst for water decomposition. The photocatalytic properties are further enhanced by loading of the CdS particles with 1.0 wt% Pt, Pd, or Rh (impregnation method), see **Table 7.1**.

Table 7.1: Visible-light-driven (λ > 420 nm) photocatalysts for H$_2$ and/or O$_2$ from aqueous solution in the presence of sacrificial reagents.

Semiconductors	Band gap (eV)	Sacrificial reagent	Rate of gas evolution (μmol/h)		Ref.
			H$_2$	O$_2$	
CdS[a]	2.3	Na$_2$S + Na$_2$SO$_3$	102	-	154
Pt/CdS	-	Na$_2$S + Na$_2$SO$_3$	600	-	154
Pd/CdS	-	Na$_2$S + Na$_2$SO$_3$	144	-	154
Rh/CdS	-	Na$_2$S + Na$_2$SO$_3$	114	-	154
Ru/CdS	-	Na$_2$S + Na$_2$SO$_3$	54	-	154
ZnS: Cu	2.3	K$_2$SO$_3$	450	-	42
ZnS:Ni	2.5	Na$_2$S + Na$_2$SO$_3$	280	-	155
ZnS:Pb, Cl	2.5	Na$_2$S + Na$_2$SO$_3$	40	-	156
Pt/(CuIn)$_{0.09}$Zn$_{1.82}$S$_2$	2.3	Na$_2$S + Na$_2$SO$_3$	1160	-	166
Ru/(CuAg)$_{0.15}$In$_{0.3}$Zn$_{1.4}$S$_2$	2.4	Na$_2$S + Na$_2$SO$_3$	2300	-	167
CdS[b]	2.3	Na$_2$S + Na$_2$SO$_3$	8.75	-	173
LaMnO$_3$/CdS	-	Na$_2$S + Na$_2$SO$_3$	37.5	-	173
Pt-,Ir-, Rh-/TaON	2.4	CH$_3$OH	2-8	-	177
Ru/TaON	2.4	CH$_3$OH	120		178
RuO$_2$/Ge$_3$N$_4$[c]			11200	5600	179

[a] nanoparticle of 8.8 nm in size prepared from H-Y zeolite matrix.
[b] Surfactant (Triton X-100) stabilized
[c] Under UV light irradiation.
Reactions were carried out in presence of sacrificial agents. O$_2$ evolution: aqueous silver nitrate solution. H$_2$ evolution: aqueous methanol solution.
Light source: 300-W Xe lamp

Other non-oxide semiconductors such as ZnS doped with Cu (Zn$_{0.957}$Cu$_{0.043}$S) [42], doped with Ni (Zn$_{0.999}$Ni$_{0.001}$S) [155] and codoped with Pb [156] have shown excellent solar hydrogen production in the presence, respectively, of aqueous potassium sulfite and sodium sulfide, see **Table 7.1.** These photocatalysts are active without noble metal functionalization. The apparent quantum yields for Ni and Cu doped-ZnS are, respectively, reported as 1.3% and 3.7%, indicating that these materials possess active catalytic sites and a sufficiently high conduction band for water reduction to produce H_2.

7.4.2 Mixed Sulfide Semiconductors

ZnS and CdS coprecipitated into nafion film [157], and coprecipitated onto SiO_2 [157,158], showed significant hydrogen evolution from aqueous sulfite suspensions. Mixed system particles prepared by Kakuta et al. [157] via coprecipitation, CdS-rich core with ZnS-rich exterior, provide a route for precise bandgap tuning. A comparison was made of the water splitting properties of silica-supported mixed semiconductor catalyst, ZnS-CdS/SiO_2, and the individual silica supported semiconductors ZnS/SiO_2 and CdS/SiO_2 immersed in aqueous sulfide solutions [158]. A physical mixture of ZnS/SiO_2 and CdS/SiO_2 did not improve water-splitting activity, even when the amounts of ZnS and CdS were the same as those in the ZnS-CdS/SiO_2 catalyst. The results indicate intimate contact between the ZnS and CdS layers is necessary for improved hydrogen photogeneration. Similar observations are reported by Kobayashi et al. [159] for the photogeneration of hydrogen from water over Al_2O_3 supported ZnS-CdS. Other catalyst supports investigated for the ZnS-CdS system include basic oxides [160] and paper [161].

Incorporation of n-Si in the mixed CdS-ZnS system results in electron flow from the n-Si to the CdS conduction band, resulting in greater utilization of solar radiation as well as superior electron-hole separation [162,163]. Under direct sunlight (25 mW/cm^2), a suspension of n-Si/CdS-ZnS yields 15 mL/h/g of hydrogen from an aqueous sulfite/sulfide solution [162]. Hydrogen evolution of 35 mL/h/g has been recorded for CdS-ZnS (2:1 weight ratio) prepared by coprecipitation and then loaded with n-Si (1.4 – 1.6%) [163].

Ternary semiconductors of the I-III-VI groups (I = Cu, Ag; III = Al, Ga, In; VI = S, Se, Te) with a chalcopyrite structure have attracted attention as new functional materials for solar hydrogen production [164-167]. With the aim of capturing a larger fraction of solar light, solid solutions of two or more sulfides were prepared. A solid solution of ZnS with a wide bandgap and AgInS$_2$ with a narrow bandgap to form (AgIn)$_x$Zn$_{2(1-x)}$S$_2$ are reported by Kudo and co-workers [164,165] showing H$_2$ evolution from aqueous solution containing sacrificial agent S^{2-} and SO$_3^{2-}$. Photophysical properties and DFT calculations reveal the band structure of the (AgIn)$_x$Zn$_{2(1-x)}$S$_2$ solid solution: Zn$_{4s4p}$ mixed with In$_{5s5p}$ contribute in forming a conduction band, and thus the conduction band minimum (LUMO) is composed of the hybrid orbital of Zn$_{4s4p}$ + In$_{5s5p}$. Both S$_{3p}$ and Ag$_{4d}$ orbitals contribute to making up the valence band, and valence band maximum (HOMO) is composed of the hybrid orbital of S$_{3p}$ + Ag$_{4d}$. The valence and conduction band levels can be shifted through the ZnS to AgInS$_2$ ratio, as confirmed by the continuous shift in diffuse reflectance and photoluminescence spectra of (AgIn)$_x$Zn$_{2(1-x)}$S$_2$ solid solutions. Pt (3.0 wt%) loaded (AgIn)$_{0.22}$Zn$_{1.56}$S$_2$ with a bandgap of 2.3 eV has shown the highest photocatalytic activity with an apparent quantum yield of 20% under visible light irradiation. H$_2$ evolution under AM 1.5 illumination is shown in **Fig. 7.8** [165]. The selective addition of Pt in the solid solution is claimed to be effective for suppressing recombination, and separating the hydrogen evolution sites from the oxidation reaction sites. The photocatalytic H$_2$ evolution activity of similar solid solutions such as (CuIn)$_{0.09}$Zn$_{1.82}$S$_2$ [166] and (CuAg)$_{0.15}$In$_{0.3}$Zn$_{1.4}$S$_2$ [167] are listed in **Table 7.1**.

7.4.3 Composite Semiconductors

Coupling of non-oxide semiconductors with oxide or non-oxide semiconductors is another approach for achieving efficient charge separation in suspended particle systems, with examples including CdS-TiO$_2$ [168-172], CdS-ZnO [168], and CdS-AgI [169]. As shown in **Fig. 7.9**, in colloidal CdS-TiO$_2$ a photogenerated electron can be transferred from the CdS conduction band to the TiO$_2$ conduction band while the hole remains in the CdS valence band. The difference in the energy level of the two semiconductors plays an important role in achieving such charge separation. For CdS-ZnO, the charge injection efficiency from CdS to ZnO can only be enhanced by decreasing the CdS particle size, allowing size quantization to increase its bandgap [168].

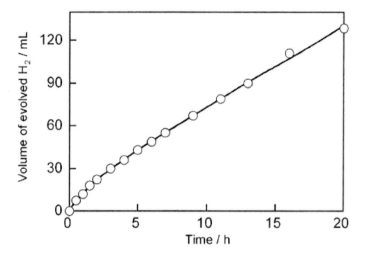

Fig 7.8: Photocatalytic H_2 evolution from an aqueous K_2SO_3 (0.25 mol/L)-Na_2S (0.35 ml/L) solution (150 mL) under a simulated sun light irradiation over Pt (3.0 wt%)-loaded $(AgIn)_{0.22}Zn_{1.56}S_2$ solid solution heat-treated in a quartz ampule tube at 1123 K. Catalyst: 0.3 g. solar simulator: 300 W Xe short arc lamp with AM 1.5 filter. Reaction cell: top window Pyrex cell. Irradiated area: 33 cm^2. Reprinted with permission from Ref. [165].

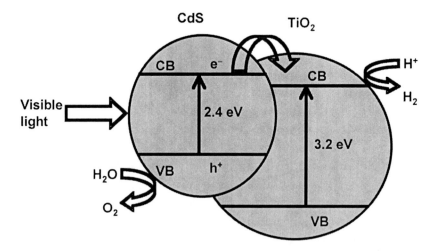

Fig. 7.9: Schematic representation of the energy levels of CdS-TiO_2 coupled semiconductor showing the charge transfer process.

As shown in reaction (7.3.6), trapped holes in CdS result in corrosion. Transient absorption studies indicate increased corrosion in coupled CdS-TiO_2 semiconductors under visible light irradiation due to increased hole trapping in the presence of TiO_2. By using an efficient hole scavenger in the electrolyte such as I^- one can completely suppress the chemical change on the CdS surface and hence anodic corrosion [170]. So and co-workers [171] observed photocatalytic hydrogen production using CdS-TiO_2 nanocomposite particles in an aqueous suspension containing Na_2S to prevent photocorrosion. Under visible light irradiation, composite CdS-TiO_2 particles produced hydrogen at a higher rate than CdS and TiO_2 used separately or in mixed solution. Inter-particle electron transfer between nanoscale clusters of CdS and TiO_2 has been studied [172]; in a non-quantized system, the excited electrons from the CdS particles are quickly transferred to a TiO_2 nanoparticle since the conduction band of bulk CdS is -0.5 V below that of TiO_2, see **Fig. 7.9.** Changing the particle size can, of course, influence the interparticle electron transfer dynamics through size quantization; charge transfer from CdS to TiO_2 can only be observed when the particle size is greater than 1.2 nm [172].

The composite semiconductor CdS/$LaMnO_3$ (10-20 nm diameter), prepared by a microemulsion method (reverse micelle) using triton X-100 as a surfactant and henxan-1-ol as a co-surfactant, has shown improved photocatalytic activities [173]. In contrast to the TiO_2-CdS composite, photogenerated holes in the CdS valence band are transferred to $LaMnO_3$ where they can react with electron donors present in aqueous suspension. The photogenerated electrons in the CdS conduction band, within the $LaMnO_3$-CdS composite, react with water to produce hydrogen. However the photogenerated electrons within $LaMnO_3$ do not contribute to hydrogen evolution since the conduction band edge of $LaMnO_3$ is less negative than that of the H_2/H_2O redox potential, as shown in **Fig. 7.10** [173].

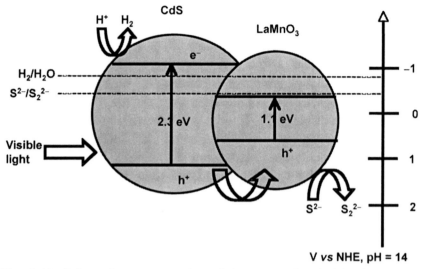

Fig. 7.10: Schematic representation of the energy levels of CdS-LaMnO$_3$ coupled semiconductor showing the charge transfer process.

7.4.4 Nitride and Oxynitride Semiconductors [55,174-182]

In 2002 Hitoki and co-workers [174] reported a transition metal nitride Ta$_3$N$_5$ as a novel visible-light driven catalyst. TaN$_{4.8}$O$_{0.3}$ was prepared from Ta$_2$O$_5$ powder by nitridation under a flow of ammonia gas at 850°C for 15 h. Using this nitridation method, various new materials such as TaON, LaTiO$_2$N and SrNbO$_2$N were synthesized and found to be effective visible-light catalysts [175,176]. These compound catalysts are called oxynitrides. Ultraviolet photoelectron spectroscopic and electrochemical analyses [177] suggest that when N-atoms are partially or fully substituted for O atoms in a metal oxide, the top of the material's valence band shifts to a higher level without affecting the bottom of the conduction band. As a result, the substitution of nitrogen for oxygen in a metal oxide causes the bandgap of the corresponding oxynitride to be small enough to respond to visible light; their bandgaps and photocatalytic activities for H$_2$ production are listed in **Table 7.1**. The photocatalytic activity TaON is enhanced by combination with 2-4 nm Ru nanoparticles [178]; the quantum efficiency for H$_2$ evolution on 0.05 wt%Ru/TaON in aqueous ethanol is 2.1%.

RuO$_2$-loaded β-Ge$_3$N$_4$ as a photocatalyst for water decomposition into H$_2$ and O$_2$ is the first successful example of a non-oxide photocatalyst for overall water splitting under UV light; 11.2 mmol of hydrogen was evolved in an experiment carried out over 1.8 mmol of RuO_2-loaded β-Ge_3N_4 suspended in water for 24 hours [179].

A novel oxynitride compound $(Ga_{1-x}Zn_x)(N_{1-x}O_x)$ [55,180-182], comprised of a solid solution of GaN and ZnO having a wurtzite crystal structure, shows reasonable photocatalytic activity for water splitting under visible light irradiation when modified with appropriate cocatalysts such as RuO_2 and Rh-Cr mixed-oxide $(Rh_{2-y}Cr_yO_3)$ nanoparticles. As shown in **Fig. 7.11**, calcination of $Rh_{2-y}Cr_yO_3/(Ga_{1-x}Zn_x)(N_{1-x}O_x)$ at three different temperatures reveals generation of $Rh_{2-y}Cr_yO_3$ nanoparticles on the $(Ga_{1-x}Zn_x)(N_{1-x}O_x)$ surface, with optimal particle size distribution achieved at 350°C strongly influencing its photocatalytic activity [55].

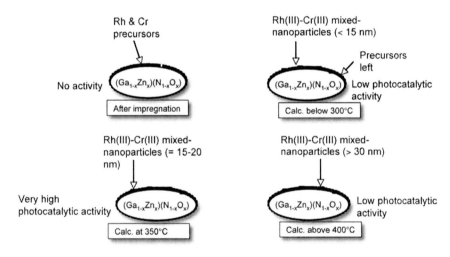

Fig. 7.11: Schematic illustration of structural changes of Rh and Cr species on the $[(Ga_{1-x}Zn_x)(N_{1-x}O_x)]$ surface after various treatments [55].

Loaded with 10-20 nm nanoparticles of rhodium-chromium mixed oxide $(Rh_{2-y}Cr_yO_3)$ as a cocatalyst on the surface of $[(Ga_{1-x}Zn_x)(N_{1-x}O_x)]$, the quantum efficiency of overall water splitting is about 2.5% at 420-440 nm [55,180-182]. Use of $AgNO_3$ as a sacrificial electron acceptor leads to an increase in oxygen evolution. The mixed oxide exhibited high and stable photocatalytic activity in aqueous H_2SO_4 solution at pH 4.5 [182]; both lower and higher pH values saw reduced photocatalytic performance attributable to corrosion of Cr in the cocatalyst, and hydrolysis of the catalyst $(Ga_{1-x}Zn_x)(N_{1-x}O_x)$. A schematic representation of the mechanism of the overall water splitting on $Rh_{2-y}Cr_yO_3/(Ga_{1-x}Zn_x)(N_{1-x}O_x)$ is shown in **Fig. 7.12**. The photogenerated electrons in the conduction band migrate to the $Rh_{2-y}Cr_yO_3$ nanoparticles and reduce H^+ to H_2, whereas photogenerated holes in the valence band oxidize H_2O to O_2 at the surface of $(Ga_{1-x}Zn_x)(N_{1-x}O_x)$. In contrast to the RuO_2-loaded catalyst, the photocatalytic activity of $Rh_{2-y}Cr_yO_3/(Ga_{1-x}Zn_x)(N_{1-x}O_x)$ is largely independent of O_2 gas pressure in the reaction system. It has been observed that 548 μmol/h O_2 evolved under visible light irradiation on $(Ga_{1-x}Zn_x)(N_{1-x}O_x)$ from an aqueous $AgNO_3$ solution without modification, whereas the sample loaded with $Rh_{2-y}Cr_yO_3$ nanoparticles evolved 307 μmol/h of O_2. The rate of O_2 evolution found in the RuO_2 loaded sample, 597 μmol/h, is higher than that of the $(Ga_{1-x}Zn_x)(N_{1-x}O_x)$ samples. These measurements indicate that suppression of O_2 reduction on $Rh_{2-y}Cr_yO_3/(Ga_{1-x}Zn_x)(N_{1-x}O_x)$ at pH 4.5 contributes to efficient photocatalytic water splitting under visible light. Similar to other oxynitride photocatalysts, a negligible amount of N_2 evolution has also been observed in the initial stages of the reaction near the surface of $(Ga_{1-x}Zn_x)(N_{1-x}O_x)$ [55, 181-182].

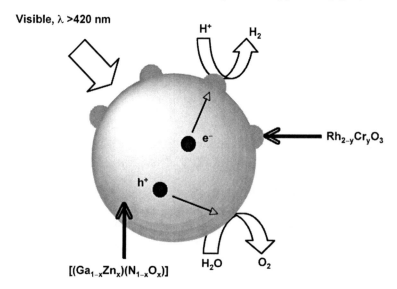

Fig. 7.12: Schematic representation of the overall water splitting mechanism on $Rh_{2-y}Cr_yO_3/(Ga_{1-x}Zn_x)(N_{1-x}O_x)$.

7.5 Concluding Remarks

Save for the rare exception, all non-oxidic semiconductors are prone to photocorrosion in aqueous solution, and thus are not suitable for use in the photoproduction of hydrogen. From the earlier studies on surface metallization to the current exploration of surface nano-architectures, non-oxide based semiconductors have seen a dramatic and continued improvement in terms of photocorrosion stability and water decomposition ability. Still, achieving the complete water splitting reaction, i.e. both H_2 and O_2 generation, by non-oxide semiconductors without benefit of sacrificial agents remains an elusive goal. Composite semiconductors, such as the described oxynitrides and $Rh_{2-y}Cr_yO_3/(Ga_{1-x}Zn_x)(N_{1-x}O_x)$, show excellent traits towards utilizing sunlight and water for hydrogen production however their limited material availability precludes widespread use on an industrial scale.

References

1. Fujishima A, Sugiyama E, Honda K (1971) Photosensitized electrolytic oxidation of iodide ion on cadmium sulfide single crystal electrode. Bull Chem Soc Jpn 44:304
2. Gerischer H (1975) Electrochemical photo and solar cells principles and some experiments. J Electroanal Chem Interfacial Electrochem 58:263–275
3. Williams R (1960). Becquerel photovoltaic effect in binary compound. J Chem Phys 32:1505
4. Ginley DS, Butler MA (1978) Flatband potential of cadmium sulfide (CdS) photoanodes and its dependence on surface ion effects. J Electrochem Soc 125:1968–1974
5. Gerischer H (1979) Solar photoelectrolysis with semiconductor electrodes. In: Seraphin BO, Aranovich JA (Eds.) Solar energy conversion: solid state physics aspects. Springer, Berlin, pp. 114–172
6. Minoura H, Oki T, Tsuiki M (1976) CdS electrochemical photocell with S^{2-} ion-containing electrolyte. Chem Lett 1279–1282

7.   Minoura H, Tsuiki M (1978) Anodic reaction of several reducing agents on illuminated cadmium sulfide electrode. Electrochim Acta 23:1377–1382

8.   Inoue T, Watanabe T, Fujishima A, Honda K, Kohayakawa K (1977) Suppression of Surface Dissolution of CdS Photoanode by Reducing Agents. J Electrochem Soc 124:719–722

9.   Wilson JR, Park SM (1982) Photoanodic dissolution of n-CdS studied by rotating ring-disk electrodes. J Electrochem Soc 129:149–154

10.   Inoue T, Watanabe T, Fujishima A, Honda K (1979) Investigation of CdS photoanode reaction in the electrolyte solution containing S2⁻ ion. Bull Chem Soc Jpn 52:1243–1250

11.   Hodes G, Manassen, Cahen D (1976) Photoelectrochemical energy conversion and storage using polycrystalline chalcogenide electrodes. Nature 261:403–404

12.   Ellis AB, Kaiser SW, Wrighton MS (1976) Optical to electrical energy conversion. Characterization of cadmium sulfide and cadmium selenide based photoelectrochemical cells. J Am Chem Soc 98:6855–6866

13.   Ellis AB, Kaiser SW, Wrighton MS (1977) Study of n-type semiconducting cadmium chalcogenide based photoelectrochemical cells employing poly chalcogenide electrolytes. J Am Chem Soc 99:2839–2847

14.   Frank AJ and Honda K (1982) Visible-light-induced water cleavage and stabilization of n-type CdS to photocorrosion with surface-attached polypyrrole-catalyst coating. J Phys Chem 86:1933–1935

15.   Honda K and Frank AJ (1984) Polymer-catalyst modified cadmium sulfide photochemical diodes in the photolysis of water. J Phys Chem 88:5587–5582

16.   Frank AJ, Glenis S, Nelson AG (1989) Conductive polymer-semiconductor junction: Characterization of poly(3-methylthiophene): cadmium sulfide based photoelectrochemical and photovoltaic cells. J Phys Chem 93:3818–3825

17.   Memming R (1978) The role of energy levels in semiconductor-electrolyte solar cells. J Electrochem Soc 125:117–123

18.   Ellis AB, Bolts JM, Kaiser SW, Wrighton MS (1977) Study of n-type gallium arsenide- and gallium phosphide-based photoelectrochemical cells. Stabilization by kinetic control and conversion of optical energy into electricity.99:2948–2853

19.   Ellis AB, Bolts JM, Wrighton MS (1977) Characterization of n-type semiconducting indium phosphide photoelectrodes stabilization to photoanodic dissolution in aqueous solutions of telluride and ditelluride ions. J Electrochem Soc 124:1603–1607

20.   Tributsch H, Bennett JC (1977) Electrochemistry and photochemistry of MoS_2 layer crystals I. J Electroanal Chem 81:97–111

21.   Tributsch H (1978) Hole reactions from d-energy bands of layer type group VI transition metal dichalcogenides: New perspectives for electrochemical solar energy conversion. J Electrochem Soc 125:1086–1093

22.   Candea RM, Kastner M, Goodman R, Hickok N (1976) Photoelectrolysis of water: Si in salt water. J Appl Phys 47:2724–2726

23.   Fujishima A, Honda K (1972) Electrochemical photolysis of water at a semiconductor electrode. Nature 238:37–38

24.   Nozik AJ (1976) p-n photoelectrolysis cell. Appl Phys Lett 29:150–153

25.   Yoneyama H, Sakamoto H, Tamura H (1975) A photo-electrochemical cell with production of hydrogen and oxygen by a cell reaction. Electrochim Acta 20:341–345

26.   A. J. Nozik (1977) Photochemical diodes. Appl Phys Lett 30:567–569

27.   Heller A and Vadimsky RG (1981) Efficient solar to chemical conversion: 12% efficient photoassisted electrolysis in the [p-type InP(Ru)/HCl-KCl/Pt(Rh)] cell. Phys Rev Lett 46:1153–1156

28.   Heller A, Aharon-Shalom E, Bonner WA, Miller B. (1982) Hydrogen-evolving semiconductor photocathodes. Nature of the junction and function of the platinum group metal catalyst. J Am Chem Soc 104:6942–6948

29.   Szklarczyk M, Bockris JOM (1984) Photoelectrochemical evolution of hydrogen on p-indium phosphide. J Phys Chem 88:5241–5245

30.   Chandra N, Wheeler BL, Bard AJ (1985) Semiconductor electrodes 59. Photocurrent effciences at p-InP electrodes in aqueous solutions. J Phys Chem 89:5037–5040

31.   Kobayashi H, Mizuno F, Nakato Y (1994) Surface states in the band-gap for the Pt-InP photoelectrochemical cells. Appl Surf Sci 81:399–408

32.   Bagdonoff P, Friebe P, Alonso-Vante N (1998) A new inlet system for differential electrochemical mass spectroscopy applied to the photocorrosion of p-InP (111) single crystals. J Electrochem Soc 145:576–582

33.   Schulte KH, Lewerenz HJ (2002) Combined photoelectrochemical conditioning and photoelectron spectroscopy analysis of InP photocathode I. The modification procedures. Electrochim Acta 47:2633–2638

34.   Darwent JR, Porter GJ (1981) Photochemical hydrogen production using cadmium sulfide suspensions in aerated water. J Chem Soc Chem Commun 145–146

35.   Kalyanasundaram K, Borgarello E, Duonghong D, Grätzel M (1981) Cleavage of water by visible-light irradiation of colloidal CdS Solutions; inhibition of photocorrosion by RuO_2. Angew Chem Int Ed 20:987–988

36.   Kalyanasundaram K, Borgarello E, Duonghong D, Grätzel M (1981) Visible light induced water cleavage in CdS dispersion loaded with Pt and RuO_2. Helv Chim Acta 64:362–366

37.   J. R. Harbour, R. Wolkow and M. L. Hair. (1981) Effect of platinization on the photoproperties of cadmium sulfide pigments in dispersion. Determination by hydrogen evolution, oxygen uptake and electron spin resonance spectroscopy. J Phys Chem 85:4026–4029

38. Buhler N, Meier K, Reber JF (1984) Photochemical Hydrogen-Production with Cadmium-Sulfide Suspensions. J Phys Chem 88:3261–3268

39. Mau AWH, Huang CB, Kakuta N, Bard AJ, Campion A, Fox MA, White JM, Webber SE (1984) Hydrogen photoproduction by nafion/cadmium sulfide/ platinum in water/sulfide ion solutions.J Am Chem Soc 106:6537–6542

40. Tricot YM, Emeren A, Fendler JH (1985) In situ Generation of catalyst-coated CdS particles in polymerized and unpolymerized surfactant vesicles and their utilization for efficient visible-light-induced hydrogen-production. J Phys Chem 89:4721–4726

41. Tricot YM, Fendler JH (1984) Colloidal catalyst-coated semiconductor surfactant vesicles. In situ generation of Rh-coated CdS particles in dihexadecylphosphate vesicles and their utilization for photosensitized charge separation and hydrogen generation. J Am Chem Soc 106:7359–7366

42. Kudo A, Sekizawa M (1999) Photocatalytic H_2 evolution under visible light irradiation on $Zn_{1-x}Cu_xS$ solid solution. Catal Lett 58:241–243

43. Goldstein AN, Echer CM, Alivisatos AP (1992) Melting in semiconductor nanocrystals. Science 256:1425–1427

44. Tolbert SH, Alivisatos AP (1994) Size dependence of a first order solid-solid phase transition: the wurtzite to rock salt transformation in CdSe nanocrystals. Science 265:373–376

45. Hayashi S, Nakamori N, Kanamori H, Yodogawa Y, Yamamoto K (1979). Infrared study of surface phonon modes of ZnO, CdS and BeO small crystals. Surf Sci 86:665–671

46. Baral S, Fojtik A, Weller H, Henglein (1986) A Photochemistry and radiation chemistry of colloidal semiconductors 12: intermediates of the oxidation of extremely small particles CdS, ZnS, Cd_3P_2 and size quantization effects (a pulse radiolysis study). J Am Chem Soc 108:375–378

47. Katsikas L, Eychmuller A, Giersig M, Weller H (1990) Discrete excitonic transition in quatum-sized CdS particles. Chem Phys Lett 172:201–204

48.   Wang Y, Herron N (1991) Nanometer-sized semiconductor cluster. Materials synthesis, quantum size effects and photophysical properties. J Phys Chem 99:525–532

49.   Torimoto T, Naohiro T, Nakamura H, Kuwabata S, Sakata T, Mori H, Yoneyama H (2000) Photoelectrochemical properties of size-quantized semiconductor photoelectrodes prepared by two-dimensional cross-linking of monodisperse CdS nanoparticles Electrochim Acta 45:3269–3276

50.   Muller BR, Majoni S, Memming R, Meissner D (1997) Particle size and surface chemistry in photoelectrochemical reactions at semiconductor particles. J Phys Chem B 101:2501–2507

51.   Liu D, Kamat PV (1993) phtotelectrochemical behavior of thin CdSe and coupled TiO_2/CdSe semiconductor films. J Phys Chem 97:10769–10773

52.   Evans JE, Springer KW, Zhang JZ (1994) Femtosecond studies of interparticle electron transfer in coupled CdS-TiO_2 colloidal system. J Chem. Phys 101:6222–6225

53.   Stroyuk AL, Krykov AI, Kuchmii SY, Pokhodenko PD (2005) Quantum size effects in semiconductor photocatalysis. Theoretical and Experimental Chemistry 41:207–228

54.   Yu ZG, Pryor CE, Lau WH, Berding MA, MacQueen DB (2005) Core-shell nanorods for efficient photoelectrochemical hydrogen production. J Phys Chem B 109: 22913–22919

55.   Maeda K, Teramura K, Lu D, Takata T, Saito N, Inoue Y, Domen K (2006) Characterization of Rh-Cr mixed-oxide nanoparticles dispersed on $(Ga_{1-x}Zn_x)(N_{1-x}O_x)$ as a cocatalyst for visible-light-driven overall water splitting. J Phys Chem B 110:13753–13758

56.   Wang Z, Medforth CJ, Shelnutt JA (2004) Self-metallization of photocatalytic porphyrin Nanotubes. J. Am. Chem. Soc. 126, 16720–16721

57.   J. G. Brennan, T. Siegrist, P. J. Carroll, S. M. Stuczynski, P. Reynders, L. E. Brus, and M. L. Steigerwald (1990). Bulk

and Nanostructure Group II-VI Compounds from Molecular Organometallic Precursors. Chem Mater 2:403–429

58.   Stuczynski SM, Opila RL, Marsh P, Brennan JG, Steigerwald ML (1991) Formation of Indium Phosphide from $In(CH_3)_3$ and $P(Si(CH_3)_3)_3$. Chem Mater 3:379–381

59.   Murray CB, Norris DJ, Bawendi MG (1993) Synthesis and characterization of nearly Monodisperse CdE (E = S, Se, Te) Semiconductor Nanocrystallites. J Am Chem Soc 115:8706–8715

60.   M. A. Olshavsky, A. N. Goldstein and A. P. Alivisatos (1990) Organometallic synthesis of GaAs crystallites exhibiting quantum confinement. J Am Chem Soc 112:9438–9449

61.   Guzelian AA, Katari JEB, Kadavanich AV, Banin U, Hamad K, Juban E Alivisatos AP, Wolters RH, Arnold CC, Heath JR (1996) Synthesis of size selected, surface passivated InP nanocrystals. J Phys Chem B, 100:7212–7219

62.   Kher SS, Wells RL (1994) A straight forward, new method for the synthesis of nanocrystalline GaAs and GaP. Chem Mater 6:2056–2662

63.   BlackburnJL, Selmarten DC, Ellingson RJ, Jones M, Micic O, Nojik AJ (2005) Electron and hole transfer in Indium phosphide quantum dots. J Phys Chem B 109:2625–2631

64.   Qian Y (1999) Solvothermal synthesis of III-V nanocrystalline semiconductors. Adv Mater 11:1101–1102

65.   Gautam UK, Seshadri R, Rajamathi M, Meldrum F, Morgan P (2001) A solvothermal route to capped CdSe nanoparticles. Chem Commun 629–630

66.   Chen X, Fan R (2001) Low temperature hydrothermal synthesis of transition metal dichalcogenides. Chem Mater 13:802–805

67.   Peng Q, Dong Y, Deng Z, Sun X, Li Y (2001) Low temperature elemental-direct-reaction route to II-VI semiconductor nanocrystalline ZnSe and CdSe. Inorg Chem 40:3840–3841

68.   Lei Z, You W, Liu M, Zhou G, Takata T, Hara M, Domen K, Li C (2003) Photocatalytic water reduction under visible

light on a novel $ZnIn_2S_4$ catalyst synthesized by hydrothermal method. Chem Commun 2142–2143

69. Lu Q, Hu J, Tang K, Qian Y, Zhou G, Liu X (2000) Synthesis of nanocrystalline $CuMS_2$ (M = In, Ga) through a solvothermal process. Inorg Chem 39:1606–1607

70. Xiao J, Xie Y, Xiong Y, Tang R, Qian Y (2001) A mild solvothermal route for chalcopyrite quarternary semiconductors $CuInSe_xS_{1-x}$ nanocrystallites. J Mater Chem 11:1417–1421

71. Gorer S, Albu-Yaron A, Hodes G (1995) Quantum size effects in chemically deposited, nanocrystalline lead selenide thin film. J Phys Chem 99:16442–16448

72. Arriaga LG, Fernández AM, Solorza O (1998) Preparation and characterization of (Zn,Cd)S photoelectrodes for hydrogen production. Int J Hydrogen Energy 23:995–998

73. Rincon ME, Martinez MW, Harnandez MM (2003) Nanostructured vs polycrystalline CdS/ZnS thim films for photocatalytic application. Thin Solid Films 425:137–144

74. Yamada S, Nosaka AY, Nosaka Y (2005) Fabrication of CdS photoelectrodes coated with titania nanosheets for water splitting with visible light. J Electroanal Chem 585:105–112

75. Golan Y, Margulis L, Rubinstein I, Hodes G (1992) Epitaxial electrodeposition of cadmium selenide nanocrystals on gold. Langmuir 8:749–752

76. Hodes G, howel IDJ, Peter LM (1992) Nanocrystalline photoelectrochemical cells. J Electrochem Soc 139:3136–3140

77. Mastai Y, Hodes G (1997) size quantization in electrodeposited CdTe nanocrystalline films. J Phys Chem B 101:2685–2690

78. Leisch JE, Bhattacharya RN, Teeter G, Turner JA (2004) Preparation and characterization of Cu(In,Ga)(Se,S)(2) thin films from electrodeposited precursors for hydrogen production. Sol Energy Mat Sol Cells 81:249–259

79. Dabbousi BO, Murray CB, Rubner MF, Bawendi MG (1994) Langmuir-Blodgett of size selected CdSe nanocrystalites. Chem Mater 6:216–219

80. Tian Y, Fendler JH (1997) Langmuir-Blodgett film formation from fluorescence-activated, surfactant-capped size selected CdS nanoparticles spread on water surfaces. Chem Mater 8:969–974

81. Zhao XK, McCormick L, Fendler JH (1991) Electrical and photoelectrochemical characterization of CdS nanoparticulate films by scanning electrochemical microscopy, scanning tunneling microscopy, and scanning tunneling spectroscopy. Chem Mater 3: 922–935

82. Ogawa F, Fan FRF, Bard AJ (1995) Scanning tunneling microscopy, tunneling spectroscopy and photoelectrochemistry of Q-CdS particles incorporated in a self-assembled monolayer on a gold surface. J Phys Chem 99:11182–11189

83. Cassagneau T, Mallouk TE, Fendler JH (1998) Layer-by-layer assembly of thin film zener diodes from conducting polymer and CdSe nanoparticles. J Am Chem Soc 120:7848–7879

84. Miyake M, Torimoto T, Sakata T, Mori H, Yoneyama H (1999) Photoelectrochemical characterization of nearly monodispersed CdS nanoparticles-immobilized gold electrode. Langmuir 15:1503–1507

85. Torimoto T, Tsumura N, Miyake M, Nishizawa M, Sakata T, Mori H, Yoneyama H (1999) Preparation and photoelectrochemical properties two dimensionally organized CdS nanoparticle thin films. Langmuir 15:1853–1858

86. Ogawa S, Fu K, Fan FRF, Bard AJ (1997) Photoelectrochemistry of films of quantum sized lead sulfide particles incorporated in self-assembled monolayer of gold. J Phys Chem 101:5707–5711

87. Hickey SG, Riley DJ, Tull EZ (2000) Photoelectrochemical studies of CdS nanoparticle modified electrodes: Absorption and photocurrent investigations. J Phys Chem B 104:7623–7626

88. Hickey SG, Riley DJ (1999) Photoelectrochemical studies of CdS nanoparticle-modified electrodes. J Phys Chem B 103:4599–4602

89.   Uosaki K, Okamura M, Ebina K (2004) Photophysical and photoelectrochemical characteristics of multilayers of CdS nanoclusters. Faraday Discuss 125:39–53

90.   Ichia LAS, Basnar B, Willner I (2005) Efficient generation of photocurrents by using CdS/carbon nanotubes assemblies on electrodes. Angew Chem Int Ed 44:78–83

91.   Contractor AQ, Bockris JOM (1984). Investigation of a protective conductive silica film on n-silicon. Electrochem Acta 29:1427–1434

92.   Nakato Y, Tsumura A, Tsubomura H (1982) Efficient photoelectrochemical conversion of solar energy with n-type silicon semiconductor electrodes surface doped with IIIA elements. Chem Lett 1071–1074

93.   Kainthala RC, Zelenay B, Bockris JOM (1986) Preparation of n-Si photoanode against photocorrosion in photoelectrochemical cell for water electrolysis. J Electrochem Soc 133:248–253

94.   Abruna HD, Bard AJ (1981) Semiconductor electrodes. 40. Photoassisted hydrogen evolution at poly(benzyl viologen)-coated p-type silicon electrodes. J Am Chem Soc 103:6898–6901

95.   Dominey RN, Lewis NS, Bruce JA, Bookbinder DC, Wrighton MS (1982) Improvement of photoelectrochemical hydrogen generation by surface modification of p-type silicon semiconductor photocathodes. J Am Chem Soc 104:467–482

96.   Maier CU, Specht M, Bilger G (1996) Hydrogen evolution on platinum-coated p-silicon photocathodes. Int J Hydrogen Energy 21:859–864

97.   Noda M (1982) Photo-assisted electrolysis of water by Si photoelectrodes. Int J Hydrogen Energy 7:311–320

98.   Nakato Y, Ueda K, Yano H (1988) Effects of microscopic discontinuity of metal over layer on the photovoltages of metal-coated semiconductor-liquid junction photoelectrochemical cells for efficient hydrogen production. J Phys Chem 92:2316–2324

99.   Wunsch F, Nakato Y, Tributsch (2002) Minority carrier accumulation and interfacial kinetics in nano-sized Pt-dotted silicon electrolyte interfaces studied by microwaves techniques. J Phys Chem B 106:11526–11530

100.   Riley DJ, Tull EJ (2001) Potential modulated absorbance spectroscopy: an investigation of the potential distribution at a CdS nanoparticle modified electrode. J Electroanal Chem 504:45–51

101.   Torimoto T, Nagakubo S, Nishizawa M, Yoneyama H (1998) Photoelectrochemical properties of size-quantized CdS thin films prepared by an electrochemical method Langmuir 14:7077–7081

102.   Haram SK, Quinn BM, Bard AJ (2001) Electrochemistry of CdS nanoparticles: A correlation between optical and electrochemical bandgaps. J Am Chem Soc 123:8860–8861

103.   Jia HM, Hu Y, Tang YW, Zhang LZ (2006) Synthesis and photoelectrochemical behavior of nanocrystalline CdS film electrodes. Electrochem Commun 8:1381–1385

104.   Milczarek G, Kasuya A, Mamykin S, Arai T, Shinoda K, Tohji K (2003) Optimization of a two-compartment photoelectrochemical cell for solar hydrogen production. Int J Hydrogen Energy 28:919–926

105.   Morales M, Sebastian PJ, Solorza O (1998) Characterization of screen printed Ti/CdS and Ti/CdSe photoelectrodes for photoelectrochemical hydrogen production. Solar Energy Mater Sol Cells 55:51–55

106.   Bhattacharya RG, Mandal DP, Bera SC, Rohtagi-Mukherjee KK (1996) Photoelectrosynthesis of dihydrogen via water-splitting using S_x^{2-} (x=1,2,3) as an anolyte: A first step for a viable solar rechargeable battery. Int J Hydrogen Energy 21:343–347

107.   Bhattacharya C, Datta J (2005) Studies on anodic corrosion of the electroplated CdSe in aqueous and non-aqueous media for photoelectrochemical cells and characterization of the electrode/electrolyte interface. Mater Chem Phys 89:171–175

108.   Zhang B, Mu J,Gao X, Wang D, Li, Z (2006) Sensitization of CdSe nanostructured electrodes by tetrasulfonated copper phthalocyanine. J Dispersion Sci Technol 27:55–57

109.   Mathew X, Bansal A, Turner JA, Dhere R, Mathews NR, Sebastian PJ (2002) Photoelectrochemical characterization of surface modified CdTe for hydrogen production. J New Mater Electrochem Systems 5:149–157

110.   Pearton J, Abernathy CR, Ren F, Lothian JR, Wisk PW, Katz A (1993) Dry and wet etching characteristics of InN, AlN, and GaN deposited by electron cyclotron resonance metalorganic molecular beam epitaxy. J Vac Sci Technol A 11:1772–1175

111.   Kocha SS, Peterson MW, Arent DJ, Redwing JM, Tischler MA, Turner JA (1995). Electrochemical investigation of the gallium nitride-aqueous electrolyte interface. J Electrochem Soc 142:L238–L240

112.   Youtsey C, Bulman G, Adesida I (1997) Photoelectrochemical etching of GaN. Mat Res Soc Proc 468:349–354

113.   Huygens IM, Strubbe K, Gomes WP (2000) Electrochemistry and photoetching of n-GaN. J Electrochem Soc 147:1797–1802

114.   Fujii K, Karasawa T, Ohkawa K (2005) Hydrogen gas generation by splitting aqueous water using n-type GaN photoelectrode with anodic oxidation. Jpn J Appl Phys 44:L543–L545

115.   Beach JD, Collins RT, Turner JA (2003) Band-edge Potentials of n-type and p-type GaN. J Electrochem Soc 150:A899–A904

116.   Kobayashi N, Narumi T, Morita R (2005) Hydrogen evolution from p-GaN cathode in water under UV light irradiation. Jpn J Appl Phys 44:L784–L786

117.   Rajshwar K, Marz T (1983) The n-GaAs electrolyte interface: evidence for specificity in lattice-ion electrolyte interactions, dependence for potential drops on crystal plane orientation to the electrolyte and implications for solar energy conversion. J Phys Chem 87:742–744

118. Allongue P, Blonkowski S (1991) Corrosion of III-V compounds; a comparative study of GaAs and InP II. Reaction scheme and influence of surface properties. J Electroanal Chem 317:77–99

119. Frese Jr. KW, Madou MJ, Morrison SR (1980) Investigation of photoelectrochemical corrosion of semiconductor. J Phys Chem 84:3174–3178

120. Menezes S, Miller B (1982) Surface and redox reactions at GaAs in various electrolytes J Electrochem Soc 130:517–523

121. Khader MM (1996) Surface arsenic enrichment of n-GaAs photoanodes in concentrated acidic solution. Langmuir 12:1056–1060

122. Ginley DS, Baughman RJ, Butler MA (1983) BP-stabilized n-Si and n-GaAs photoanodes. Electrochem Soc 130:1999–2002

123. Allongue P, Cachet H (1988) Charge transfer and stabilization at illuminated n-GaAs/aqueous electrolyte junction. Electrochim Acta 33:79–87

124. Khader MM, Hannout MM, Eldessouki MS (1991) Photoelectrochemical dissociation of water at silicon doped n-GaAs Electrodes. Int J Hydrogen Energy 16:797–803

125. Miller EA, Richmond GL (1997) Photocorrosion of n-GaAs and passivation by Na$_2$S, Phys Chem B 101:2669–2677

126. Cojucaru A, Simashkevich A, Sherban D, Tiginyahu I, Ursaki V, Tsiulyanu I, Usatyi (2005) Use of porous GaAs electrodes in photoelectrochemical cell. Phys Stat Sol A 202:1678–1682

127. Khader MM, Nasser SA, Hannout MM, El-Dessonki MS (1993) Photoelectrochemical dissociation of water at copper-doped p-GaAs electrodes. Int J Hydrogen Energy 18:921–924

128. Khader MM, Hannout MM, ElDessouki MS (1996) Catalytic effects for hydrogen photogeneration due to metallic deposition on p-GaAs. Int J Hydrogen Energy, 21:547–553

129. Yoon KH, Lee JW, Cho YS, Kang DH (1996) Photoeffects in WO$_3$/GaAs electrode. J Appl Phys 80:6813–6818

130. Erne BH, Ozanam F, Chazalviel JN (1999) The mechanism of hydrogen gas evolution on GaAs cathodes elucidated by in situ infrared spectroscopy. J Phys Chem B 103:2948–2962

131. Khader MM, Saleh MM (1999) Comparative study between the photoelectrochemical behavior of metal loaded n- and p-GaAs. Thin Solid Films 349:165–170

132. Kobayasbi H, Mizuno F, Nakato Y, Tsubomura H (1991) Hydrogen evolution at a Pt-modified InP Photoelectrode: Improvement of current-voltage characterlstics by HCl etching. J Phys Chem 91:819–824

133. Barkschata A, Tributsch H, Dohrmann JK (2003) Imaging of catalytic activity of platinum on p-InP for photocathodical hydrogen evolution. Sol Energy Mater Sol Cells 80:391–403

134. Hassel AW, Aihara M, Seo M (2000) Formation and corrosion of InP:In contacts in hydrochloric acid. Electrochim Acta 45:4673-4682

135. Beach JD, Al-Thani H, McCray S, Collins RT, Turner JA (2002) Bandgap and lattice parameters of 0.9 μm thick $In_{1-x}Ga_xN$ of $0 < x < 0.140$ J Appl Phys 91:5190–5194

136. Wetzel C, Takeuchi T, Yamaguchi S, Katoh H, Amano H, Akasaki I (1998) Optical bandgap in $Ga_{1-x}In_xN$ ($0<x<0.2$) on GaN by photoreflection spectroscopy. Appl Phys Lett 73:1994–1996

137. Parker CA, Roberts JC, Bedair SM, Reed MJ, Liu SX, Masry NA, Robbins LH (1999) Optical bandgap dependence on composition and thickness of $In_xGa_{1-x}N$ grown on GaN. Appl Phys Lett 75:2566–2568

138. Fujii K, Kusakabe K, Ohkawa K (2005) Photoelectrochemical properties of InGaN generation from aqueous water. Jpn J Apl Phys 44:7433–7435

139. Duetsch, TG, Koval CA, Turner JA (2006) III-V nitride epilayer of photoelectrochemical water splitting: GaPN and GaAsPN. J Phys Chem B 110:25297–25307

140. Kocha SS, Turner JA (1994) Study of the Schottky barriers and determination of the energetics of the band edges at the n- and p-type gallium indium phosphide electrode electrolyte interface. J Electroanal Chem 367:27–30

141. Kocha SS, Turner JA (1996) Impedence analysis of surface modified $Ga_{0.5}In_{0.5}P_2$-aqueous electrolyte interface. Electrochim Acta 41:1295–1304

142. Bansal A, Turner JA (2000) Supression of the band edge migration at the p-$GaInP_2$/water under illumination via catalysis. J Phys Chem B, 65:6591–6592

143. Valderrama RC, Sebastian PJ, Enriquez JP, Gamboa SA (2005) Photoelectrochemical characterization of CIGS thin films for hydrogen production. Sol Energy Mat Sol Cells 88:145–155

144. Kohtani S, Kudo A, Sakata T (1993) Spectral sensitization of TiO_2 semiconductor electrode by CdS microcrystals and its photoelectrochemical properties. Chem Phys Lett 206:166–170

145. Suleymanov AS (1991) On the possibility of the transformation of solar energy to chemical energy in the electrochemical cell with photoanode CdSe-TiO_2. Int J Hydrogen Energy 16:741–743

146. Murphy OJ, Bockris JOM (1984) Photovoltaic electrolysis: hydrogen and electricity from water and light. Int J Hydrogen Energy 9:557–561

147. Kainthala RC, Zelenay B, Bockris JOM (1987) Significant efficiency increase in self-driven photoelectrochemical cell for water photoelectrolysis. J Electrochem Soc 134:841–845

148. Kainthala RC, Khan SUM, Bockris JOM (1987) The theory of electrode matching in photoelectrochemical cell for the production of hydrogen. Int J Hydrogen Energy 12:381–392

149. Reber JF, Rusek M (1986) Photochemical hydrogen production with platinized suspension of Cadmium sulfide and cadmium zinc sulfide modified by silver sulfide. J Phys Chem 90:824–834

150. Matsumura M, Saho Y, Tsubomura H (1983) Photocatalytic hydrogen production from solutions of sulfite using platinized cadmium sulfide powder. J Phys Chem 87:3807–3808

151. Ashokkumar M, Maruthamuthu P (1991) Photocatalytic hydrogen production with semiconductor particulate systems

- an effort to enhance the efficiency. Int J Hydrogen Energy 16:591–595

152. Arora MK, Sahu N, Upadhyay SN, Sinha ASK (1999) Activity of cadmium sulfide photocatalysts for hydrogen production from water: Role of support. Ind Eng Chem Res 38:2659–2665

153. Guan G, Kida T, Kusakabe K, Kimura K, Fang X, Ma T, Abe E, Yoshida A (2004) Photocatalytic H_2 evolution under visible light irradiation on CdS/ETS-4 composite. Chem Phys Lett 385:319–322

154. Sathish M, Viswanathan B, Viswanath RP (2006) Alternate synthetic strategy for the preparation of CdS nanoparticles and its exploitation for water splitting. Int J Hydrogen Energy 31:891–898

155. Kudo A, Sekizawa M (2000) Photocatalytic H_2 evolution under visible light irradiation on Ni-doped ZnS photocatalyst. Chem Commun 1371–1372

156. Tsuji I, Kudo A (2003) Hydrogen evolution from aqueous sulfite solution under visible light irradiation over Pb and halogen-codoped ZnS. J Photochem Photobiol A 156:249–252

157. Kakuta N, Park KH, Finlayson MF, Ueno A, Bard AJ, Campion A, Fox MA, Webber SE, White JM (1985) Photoassisted hydrogen production using visible light and coprecipitated ZnS-CdS without a noble metal. J Phys Chem 89:732–735

158. Ueno A, N. Kakuta N, Park KH, Finlayson MF, Bard AJ, Campion A, Fox MA, Webber SE, White JM (1985) Silica supported ZnS-CdS mixed semiconductor catalysts for photogeneration of hydrogen. J Phys Chem 89:3828–3833

159. Kobayashi J, Kitaguchi K, Tanaka H, Tsuiki H, Ueno A (1987) Photogeneration of hydrogen from water over alumina supported ZnS-CdS catalyst. J Chem Soc Faraday Trans 1 83:1395–1404

160. Tambwekar SV, Subrahmanyam M (1998) Enhanced photocatalytic H_2 production over CdS-ZnS supported on super basic oxides. Int J Hydrogen Energy 23:741–744

161. Roy AM, De GC (2003) Immobilisation of CdS, ZnS and mixed ZnS–CdS on filter paper: Effect of hydrogen production from alkaline $Na_2S/Na_2S_2O_3$ solution. J Photochem Photobiol A: Chem 157:87–92

162. De GC, Roy AM, Bhattacharya SS(1996) Effect of n-Si on the photocatalytic production of hydrogen by Pt loaded CdS and ZnS/CdS catalysts. Int J Hydrogen Energy 21:19–23

163. Koca A, Sahin M (2002) Photocatalytic hydrogen production by direct sun light from sulfide/sulfite solution. Int J Hydrogen Energy 27:363–367

164. Kudo A, Tsuji I, Kato H (2002) $AgInZn_7S_9$ solid solution photocatalyst for H_2 evolution from aqueous solutions under visible light irradiation. Chem Commun 1958–1959

165. Tsuji I, Kato H, Kobayashi H, Kudo A (2004) Photocatalytic H_2 evolution reaction from aqueous solutions over band structure-controlled $(AgIn)_xZn_{2(1-x)}S_2$ solid solution photocatalysts with visible-light response and their surface nanostructures. J Am Chem Soc 126:13406–13413

166. Tsuji I, Kato H, Kobayashi H, Kudo A (2005) Photocatalytic H_2 evolution under visible-light irradiation over band-structure-controlled $(CuIn)_xZn_{2(1-x)}S_2$ solid solutions. J Phys Chem B 109:7323–7329

167. Tsuji I, Kato H, Kudo A (2005) Visible-light-induced H_2 evolution from an aqueous solution containing sulfide and sulfite over a ZnS-$CuInS_2$-$AgInS_2$ solid-solution photocatalyst. Angew Chem Int Ed 44:3565–3568

168. Spanhel L, Weller H, Henglein A (1987) Photochemistry of semiconductor colloids 22.Electron injection from illuminated CdS into attached TiO_2 and ZnO particles. J Am Chem Soc 109:6632–6635

169. Gopidas KR, Bohorquez M, Kamat PV (1990) Photophysical and photochemical aspects of coupled semiconductors. Charge-transfer processes in colloidal CdS-TiO_2 and CdS-AgI systems. J Phys Chem 94:6435–6440

170. Kamat PV, Ebbesen TW, Dimitrijevic NM, Nojik AJ (1989) Primary photochemical events in CdS semiconductors

colloids as probed by picosecond laser flash photolysis. Chem Phys Lett 157:384–389

171. So WW, Kim KJ, Moon SJ (2004) Photo-production of hydrogen over the CdS-TiO$_2$ nano-composite particulate films treated with TiCl$_4$. Int J Hydrogen Energy 29:229–234

172. Sant PA, Kamat PV (2002) Interparticle electron between size-quantized CdS and TiO$_2$ semiconductor nanoclusters. Phys Chem Chem Phys 4:198–203

173. Kida T, Guan GQ, Yoshida A (2003) LaMnO3/CdS nanocomposite: a new photocatalyst for hydrogen production from water under visible light irradiation. Chem Phys Lett 371:563–567

174. Hitoki G, Ishikawa A, Takata T, Kondo JN, Hara M, Domen K (2002) Ta$_3$N$_5$ as a novel visible light-driven photocatalyst (lambda < 600 nm) Chem Lett 736–737

175. Hitoki G, Ishikawa A, Takata T, Kondo JN, Hara M, Domen K (2002) An oxynitride, TaON, as an efficient water oxidation photocatalyst under visible light irradiation ($\lambda \leq$ 500 nm) Chem Commun 1698–1699

176. Kasahara A, Nukumizu K, Hitoki G, Takata T, Kondo JN, Hara M, Domen K (2002) Photoreactions on LaTiO$_2$N under visible light irradiation. J Phys Chem A 106:6750–6753

177. Chun WJ, Ishikawa A, Fujisawa H, Takata T, Kondo JN, Hara M, Kawai M, Matsumoto Y, Domen K (2003) Conduction and valence band positions of Ta$_2$O$_5$, TaON, and Ta$_3$N$_5$ by UPS and electrochemical methods. J Phys Chem B 107:1798–1803

178. Hara M, Nunoshige J, Takata T, Kondo JN, Domen K (2003) Unusual enhancement of H$_2$ evolution by Ru on TaON photocatalyst under visible light irradiation Chem Commun 3000–3001

179. Sato J, Saito N, Yamada Y, Maeda K, Takata T, Kondo JN, Hara M, Kobayashi H, Domen K, Inoue Y (2005) RuO$_2$-loaded beta-Ge$_3$N$_4$ as a non-oxide photocatalyst for overall water splitting. J Am Chem Soc 127:4150–4151

180. Maeda K, Takata T, Hara M, Saito N, Inoue Y, Kobayashi H, Domen K (2005) GaN-ZnO solid solution as a

photocatalyst for visible light driven overall water splitting. J Am Chem Soc 127:8286–8287

181.   Maeda K, Teramura K, Lu D, Takata T, Saito N, Inoue Y, Domen K (2006) Photocatalysts releasing hydrogen from water. Nature 440:295

182.   Maeda K, Teramura K, Masuda H, Takata T, Saito N, Inuoe Y, Domen K (2006) Efficient overall water splitting under visible-light irradiation on $(Ga_{1-x}Zn_x)(N_{1-x}O_x)$ dispersed with Rh-Cr mixed-oxide nanoparticles: Effect of reaction condition on photocatalytic activity. J Phys Chem B 110, 13107–13112

Chapter 8

PHOTOVOLTAIC - ELECTROLYSIS CELLS

8.1 Introduction

Steam reforming of methane is currently the most common method of producing hydrogen and, as a by-product, CO_2,

a) $CH_4 + H_2O \rightarrow CO + 3H_2$
b) $CO + H_2O \rightarrow H_2 + CO_2$

which does the world no favors with respect to issues of global warming. As discussed in Chapter 2 water electrolysis is another way to produce hydrogen, however the electricity we use today is predominately obtained by the combustion of the fossil fuels that, like methane, are of finite supply and produce as a combustion product CO_2 as well as other noxious pollutants such as mercury vapor.

Electricity generated from sunlight offers a viable pathway for hydrogen production by water electrolysis [1-34], see **Fig. 8.1**, with photovoltaics (solar cells) converting solar energy into the electric energy needed to achieve water electrolysis resulting in the production of very pure hydrogen. Solar cells are commonly referred to as *photovoltaic* (PV) cells; *Photo* (light) and *Voltaic* (electricity). Several review articles consider the characteristics and performance of such systems, comprised of a photovoltaic converter and electrolysis device [5,6,9,16,17,19,21,29]. The global efforts at developing thin-film solar cell technology over the past decade, moving towards low-cost (plentiful) raw materials for their manufacture coupled with inexpensive (low energy input) mass production techniques, suggest that electricity from PV systems (and therefore PV hydrogen) should become increasingly less costly. The 'less costly' equation becomes particularly meaningful when the true costs of fossil fuels are factored into the equation, e.g. the military costs spent on ensuring oil flow from the Middle East, the associated political bargains necessarily made to keep the oil flowing, pollution and their associated medical affects, and of course the consequences of undesired global climate change.

Solar cells are silent, have no moving parts, cause no environmental pollution in operation, can generate power locally where it is needed, and can operate across extensive temperature regimes. Devices combining photovoltaic solar cells and water electrolysis may be described as integrated, or monolithic, tandem PV-electrolytic devices [35-39]. Multiple bandgap PV devices lead to a more efficient conversion of solar energy, hence a greater production of hydrogen [35-45] but at greater cost.

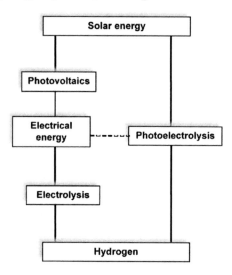

Fig. 8.1: Photovoltaic cell in combination with water electrolysis device for producing solar hydrogen.

8.2 General Description of Solar Cell Technology

8.2.1 From Past to Present

The recorded development of solar cell technology begins with the 1839 research of French experimental physicist Antoine-Edmond Becquerel [46,47]. At the age of nineteen he discovered the photovoltaic effect while experimenting with an electrolytic cell containing two metal electrodes. He found that certain metals and solutions would produce small amounts of electric current when exposed to light. In 1883 Charles Fritts formed photovoltaic junctions by coating selenium with an extremely thin layer of gold [48]. Russell Ohl invented the first silicon solar cell in 1941 [49]. The era of modern solar cell technology began in 1954, when G. L. Pearson, D. Shapin and C. Fuller demonstrated a silicon solar cell capable of 6% energy conversion efficiency with direct sunlight [50]. The first gallium arsenide (GaAs) solar cell was reported in 1956, with a photoconversion efficiency of 6.5% [51]. In 1976 Carlson and Wronksi [52-54] reported solar cells comprised of amorphous silicon. Modern multi-junction solar cells can be viewed as a series of p-n junction photodiodes, each of different bandgap, that commonly include such III-V or II-VI materials as gallium arsenide (GaAs), gallium indium phosphide (GaInP), copper indium diselenide (CIS), copper indium-gallium diselenide (CIGS), and cadmium telluride (CdTe). In 1987, Jerry Olson [55-57] reported a two-junction tandem photovoltaic device consisting of an upper GaInP layer and lower GaAs layer, with a photoconversion efficiency up to 29.5% under concentrated solar light. The addition of a third junction further increases the conversion efficiency, to 34% for a GaInP/GaAs/Ge solar cell [58], to 40% for a GaInP/GaAs/GaInAs cell [59,60]. It is believed that the photoconversion efficiency of multi- (or many) junction solar cells can be increased up to 55% [61]. For example, highly mis-matched alloys such as $Zn_{1-y}Mn_yO_xTe_{1-x}$ have shown utility in the high performance high dollar solar cell markets, such as in space satellites, where dollars are no issue but high photoconversion efficiency is [62]. Unfortunately issues of cost limit application of solar cell technology in the 'real' world. The solar cell market continues to be dominated by silicon, the fabrication of which is energy-intensive requiring a manufacturing energy input equal to several years of energy output of the solar device. Furthermore modest device [63,64] efficiencies correspond to large land area requirements to meet the intrinsic energy demands of modern society. Generally speaking, as of today the cost of energy from solar cells is $\approx$ five times that produced by the burning of fossil fuels. However since fossil fuels are freely provided by nature, and still so cheap as to be commonly treated as free (and not treated as an irreplaceable precious commodity), the factor of five looks pretty modest.

Following M. A. Green [65], 3rd generation photovoltaic solar cells refer to any device that exceeds the efficiency of a single junction solar cell, including the well-known tandem solar cells as well as other approaches that may, for example, circumvent previous assumptions such as one quasi-Fermi level separation existing in a solar cell, one photon generating one electron-hole pair, or a constant temperature existing across the device, etc., giving an approach or pathway to high device efficiencies such as: (A) Multiple absorption path solar cells, in which the absorption process is altered such that either two (low-energy) photons are absorbed to create a single electron-hole pair or alternately one photon creates multiple electron-hole pairs. (B) Multiple spectrum solar cells, where the solar spectrum is changed into a different spectrum with the same energy. (C) Multiple temperature solar cells, which involve the extraction of energy from variations in either carrier or lattice temperature. (D) Multiple energy level solar cells, which have more than a single quasi-Fermi level separation.

8.2.2 Operational Principles

On earth, as opposed to outer space, the most common solar cell material is silicon. An atom of silicon has 14 electrons, arranged in three different shells, $1s^2 2s^2 2p^6 3s^2 3p^2$. The first two shells are full, while the outer shell is half full, hence there are four electrons a silicon atom can and does share with its neighboring atoms. Pure silicon is a poor conductor of electricity since none of its electrons are free to move about.

Intrinsic silicon	n-type silicon	p-type silicon

When a Si atom is replaced with a Column V atom (e.g. P or As), there is an electron more than that which can be bonded through nearest neighbor associations. Relatively little energy is required to manipulate this extra electron; the resulting silicon is called **n-type** (*n* for negative net charge). When silicon is doped with a group III atom (e.g. B or Al), which have only three electrons in their outer shells the result is **p-type** silicon (*p* for positive net charge). The absence of an electron is considered a hole that carries positive charge. When p-type Si is adjacent to n-type Si the concentration gradients result in diffusion of the holes from the p-type side into the n-type side, and diffusion of electrons from the n-side into the p-side. The uncompensated charge result in an electric field across this space charge region that serves to separate any electrons and holes created by photon absorption, see **Fig. 8.2**; the electric field will send the electron to the n-side and the hole to the p-side. This causes further disruption of charge neutrality, and if an external current path is provided electrons will flow through the path to the p-side to unite with the holes the electric field sent there, doing useful work along the way. The electron flow provides the current, and the electric field of the junction provides the voltage; with current and voltage we have power.

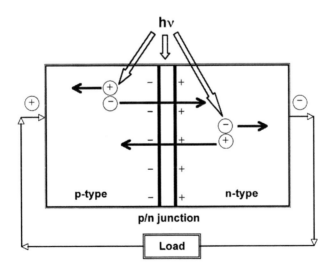

Fig. 8.2: Operation mechanism of single-crystal silicon p/n junction solar cell.

8.2.3 Crystalline Silicon Solar Cells

Since silicon accounts for the large majority of terrestrial solar cells we briefly consider some of their design issues. To minimize series resistance good electrical contacts are of critical importance, commonly achieved by covering the bottom with a metal film contact, and on the top using a transparent conductor such as SnO_2 or metallic contact grid that reduces the distance traveled by the photogenerated electrons while covering only a small part of the cell surface, see **Fig. 8.3**. Antireflection coatings, such as silicon monoxide (SiO_x) or silicon nitride (SiN_x), are used to minimize surface reflections without which silicon will reflect more than 30% of the incident light. Texturing of the surface by selective etching can also be used to reduce reflection. To protect the cell from the environment the cell is covered with a glass plate. PV modules are made by connecting several cells in series or parallel to achieve useful levels of voltage and current, and putting them in a sturdy frame complete with a glass cover and electrical terminals.

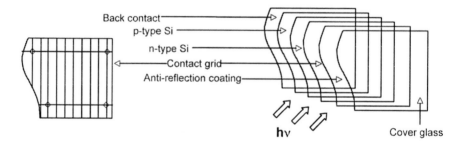

Fig. 8.3: Generic design of single-crystalline solar cell.

8.2.4 Types of Solar Cells

Single crystal silicon (sc-Si), polycrystalline silicon (p-Si), and amorphous silicon (a-Si) can all be used to make solar cells, with fabrication cost and device photoconversion efficiencies decreasing as one moves from single-crystal to amorphous materials. Various properties of these materials are summarized in **Table 8.1**. Other relatively common solar cell materials include gallium arsenide (GaAs), copper indium diselenide (CIS), copper indium-gallium diselenide (CIGS), and cadmium telluride (CdTe). Amorphous silicon is one of the most developed thin film technologies to-date, and offers interesting possibilities in further development through the use of "microcrystalline" silicon which seeks to combine the stable high efficiencies of crystalline Si technology with the simpler and cheaper large area deposition technology of amorphous silicon. However different semiconductor materials, or combinations thereof, are only suited for responding to a specific spectral range. Therefore for a given semiconductor a portion of the radiant energy spectra cannot be used because some of the light quanta will not have enough energy to activate the charge carriers, and some of the light quanta will have surplus amounts of energy that will be transformed into heat rather than electrical energy. There are also optical losses due to reflection, or shadowing of the cell surface. Other losses occur due to the electrical resistances of the semiconductor and connecting cables. Losses intrinsic to a semiconductor photovoltaic include material contamination, surface effects, and crystal defects. Loss mechanisms associated with a bandgap lead to a theoretical maximum efficiency, which in the case of sc-silicon is approximately 33%. The efficiency of a solar cell is calculated by:

$$\eta_{global} = \frac{V_{OC} \times J_{SC} \times FF}{I_S}$$

where V_{oc} represents open circuit voltage, J_{sc} short-circuit current density, FF the fill factor, and I_s the power of the incident light. The fill factor is the ratio of solar cell's actual power to that obtained from its J_{SC} and V_{OC} values.

Table 8.1: Description of silicon based PV solar cells and their efficiencies.

Types	Crystal size range	Deposition method	Laboratory efficiency (%)	Production efficiency (%)
Single-crystal silicon, sc-Si	10-15 cm	Czochralski, Float Zone	~25	21.8
Multi-crystalline silicon, mc-Si	< 10 cm	Sawing (Cast)	~20	18.1
Polycrystalline silicon, p-Si	≈ 1mm	Chemical Vapor Deposition	~18	15.0
Amorphous, a-Si (Microcrystalline silicon)	(< 1mm)	Plasma Deposition	~13	7.0

Solar cells are broadly classified into five types based on their intrinsic structures, that we now briefly consider.

Homojunction

Selective doping alters a single semiconducting material to form within it a p-n junction. Single-crystal or multi-crystalline silicon are the most common examples of this type of cell, with other examples including GaAs and InP, see **Table 8.2**. In order to improve conversion efficiency various design aspects can be considered and varied such as material purity and crystallinity, depth of the p/n junction below the materials surface, and amount and distribution of dopant atoms on either side of the p/n junction.

Heterojunction

Hetero-junction devices are comprised of two different semiconductors. The top, or window, layer uses a material of higher bandgap while the bottom layer uses a material of lower bandgap. Thin-film p-CdTe|n-CdS solar cells are one such promising hetero-junction device, see **Table 8.2**. The 2.4 eV bandgap of CdS makes it transparent down to a wavelength of 515 nm. The p-CdTe layer (bandgap = 1.5 eV) is doped less than that of the n-CdS layer, hence the depletion region is mostly within the p-CdTe layer. Consequently most of the carrier generation and collection occurs within the p-CdTe layer, with ≈ 99% of the incident light absorbed by a 1 μm layer (compared with ≈ 10 μm for Si).

Table 8.2: Various solar cells (modules) and their efficiencies at global AM 1.5 spectrum (1000 W/m²) [10]

Classification	Efficiencies (%)	Classification	Efficiencies (%)
III-V		**Multi-junction (tandem cell)**	
GaAs(Crystalline)	25.1	GaInP/GaAs	30.3
GaAs (thin film)	24.5	GaInP/GaAs/Ge	32.0
GaAs (Multi-crystalline)	18.2	GaAs/CIS	25.8
InP (crystalline)	21.9	a-Si/μC-Si	11.7
Thin film		GaInP/GaInAs/Ge	31.3
CdTe (thin film cell)	16.5	a-Si/a-Si/a-SiGe	12.1
CdTe (thin film submodule)	10.6	AlGaAs/GaAs	28.6
CIGS (thin film cell)	18.4	**Concentrator Cell**	
CIGS (thin film submodule)	16.6	GaInP/GaAs/Ge	34.7
GaAs (thin film cell)	24.5	GaInP/GaInAs/Ge	39.0
		GaAs (single cell)	27.8
		Si (single cell)	26.8
		CIGS (film, single cell)	21.5
		InP (single cell)	25.3

p-i-n Devices

These devices consist of a three-layer stack, with an intrinsic (i-type or undoped) layer between a n-type layer and p-type layer. This geometry results in an electric field between the p- and n-type regions that stretches across the resistive intrinsic region. Each photon absorbed in the intrinsic layer generates an electron-hole pair that is then separated by the electric field. In 1977 Carlson reported an amorphous silicon (a-Si) solar cell with a photoconversion efficiency of 5.5%, produced by glow-discharge decomposition of silane [66]. Twenty years later, Yang and co-workers reported an amorphous silicon p-i-n device with a photoconversion efficiency of 13% [67]. It was first shown by Guha et al. [68] that hydrogen dilution of the active gas mixture during deposition of the amorphous silicon thin films can be used to improve material quality (hydrogenated amorphous silicon, a-Si:H). The promise of a-Si:H solar cells is currently incomplete because of the light induced photo-degradation of its electronic properties known as the Staebler-Wronski effect [69]. As the hydrogen dilution is increased, the transition from amorphous to microcrystalline phase takes place; recent work has shown that the best amorphous silicon alloy is grown at a dilution just below the edge of amorphous to the microcrystalline (μc) transition (μc-Si:H) [70-72]. Materials such as Ge, B, and N have been used to form amorphous silicon alloys to obtain higher open circuit voltages.

Since a-Si:H holes have limited mobility relative to that of electrons, the devices are designed so that light enters through the p-layer to enable efficient hole collection [72], see **Fig. 8.4**. a-Si:H differs from crystalline silicon by the lack of long range order and the high (bonded) hydrogen content (~10% in device quality a-Si:H), with a key feature of this material its relative stability to light induced photodegradation. Although the overall properties of a-Si:H and sc-Si materials are similar due to similarities in short range order, the long range disorder in a-Si:H results in bond length and bond angle distortions, bond defects and microvoids. The major deposition variables in fabrication of a-Si:H p-i-n devices include the hydrogen dilution ratio (H_2/SiH_4), total gas pressure, and substrate temperature which determine the growth regime and resultant surface characteristics of the intrinsic layer [73-75]. Optimum i-layers are obtained at the maximum possible dilution of H_2/SiH_4 without crossing the amorphous to mixed-phase (a+µc)-Si:H boundary [73-75].

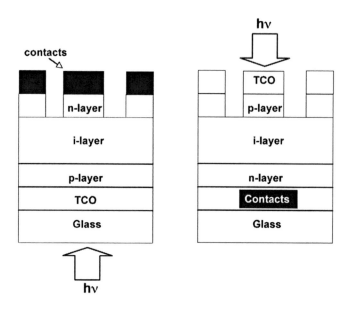

Fig. 8.4: Device structures of Glass/TCO/p-i-n/contacts and glass/contacts/n-i-p/TCO a-Si:H solar cells.

Multijunction

This structure, also called a tandem or stacked cell, can achieve relatively higher total conversion efficiencies by capturing a larger portion of the solar spectrum. Different semiconductor materials, suited for different spectral ranges, are arranged atop each other; the higher bandgap material is on the top surface, absorbing high-energy photons, while allowing lower-energy photons to be absorbed by the lower bandgap materials underneath. These selective absorption processes continue through to the final cell possessing the smallest bandgap. Such cells, see **Fig. 8.5,** have achieved efficiencies of about 40% [76] but are of course more costly to make than single junction devices. **Table 8.2** summarizes the performance of various highly efficient multi-junction or tandem cells.

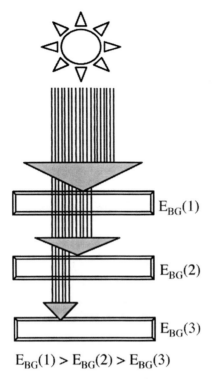

$$E_{BG}(1) > E_{BG}(2) > E_{BG}(3)$$

Fig. 8.5: Light harvesting processes in a multi-junction solar cell. The top cell, of relatively larger bandgap, BG(1), absorbs light of shorter wavelength. Subsequent junctions, of decreasing bandgap, absorb longer wavelength (lower energy) light.

Dye-sensitized Solar Cells

Dye-sensitized solar cells (DSSCs) [77,78] are photoelectrochemical cells that use photo-sensitization of wide-band-gap mesoporous oxide semiconductors. DSSCs are promising, at least in principle, due to their use of low-cost materials and relatively simple apparatus for manufacturing. **Figure 8.6** is a schematic representation of a DSSC comprised of: (1) TCO-glass.   (2) Granular TiO_2 forming a nanoporous structure. (3) A dye (N719), which is a visible light-absorbing substance spread on the TiO_2 surface. (4) A redox couple located in the space between the dye and the cathode. (5) A solvent for the redox couples (I^-/I_3^-) e.g. an organic solvent or Room Temp. Ionic Liquid. (6) Counter electrode.

The current energy conversion efficiencies for a DSSC are 11.1% for an aperture area of 0.219 cm^2, and 6.8% for a larger cell with an aperture area of 101 cm^2 [79]. The energy conversion efficiency of the DSSCs has not yet reached the level of silicon solar cells, however silicon solar cells have a head-start in development of several decades.

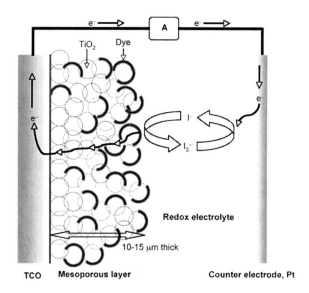

Fig. 8.6a: Schematic representation of a dye-sensitized solar cell (DSSC).

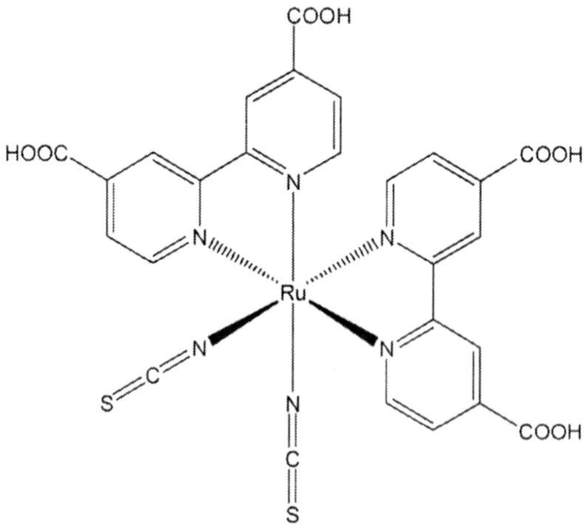

N719

Fig. 8.6b: Molecular structure of N-719 dye, commonly used in dye-sensitized solar cells.

Figure 8.7 illustrates DSSC operation, with the numbered arrows representing process steps as described. Solar light passes through an electrically transparent conductive glass electrode (commonly fluorine doped indium tin oxide) resulting in photoexcitation of dye molecules (Process 1) adsorbed onto the surface of sintered nanocrystalline TiO_2. The excited electron makes a jump from the dye to the TiO_2 conduction band (Process 2); this jump occurs in approximately 10^{-15} s. The electron percolates through the TiO_2 film, reaching the transparent conducting layer upon the glass electrode, goes through external circuit to the counter electrode (Process 3). At the same time dye regenration takes place by receiving one electron from an iodide ion, in turn oxidizing the iodide to triodide (Process 4). The iodide ion regenerates upon receiving one electron from counter electrode, thereby completing the circuit (Process 5).

Primary energy loss pathways include radiative and non-radiative deactivation of the dye sensitizer (Process 6), recombination of the conduction band electrons by the oxidized sensitzer S^+ (Process 7), or recombination of the conduction band electrons by the the oxidized form of the redox system (Process 8).

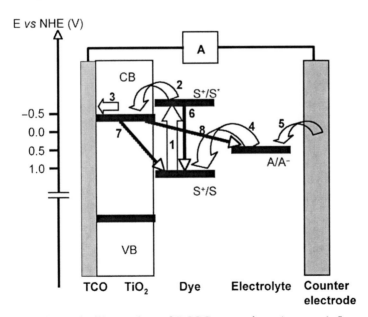

Fig. 8.7: Schematic illustration of DSSC operation. Arrows 1-5 represent the primary pathways for device operation, and arrows 6-8 pathways in which energy can be lost.

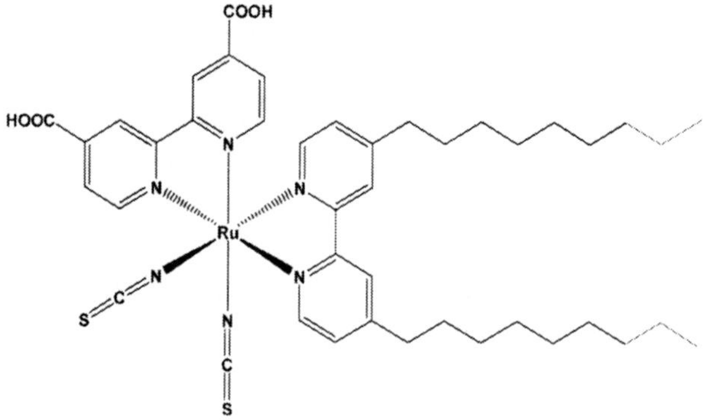

Z-907

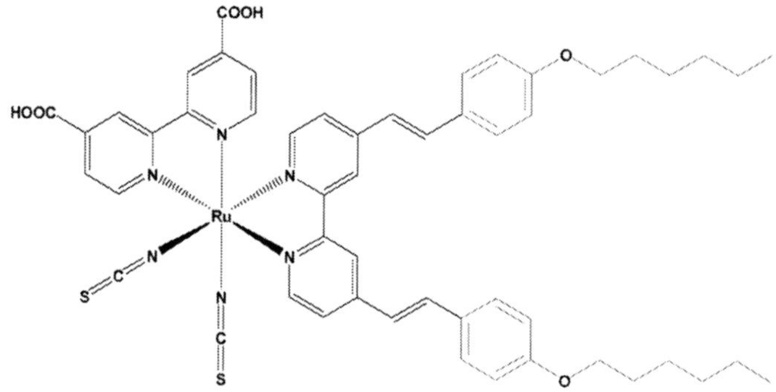

K-19

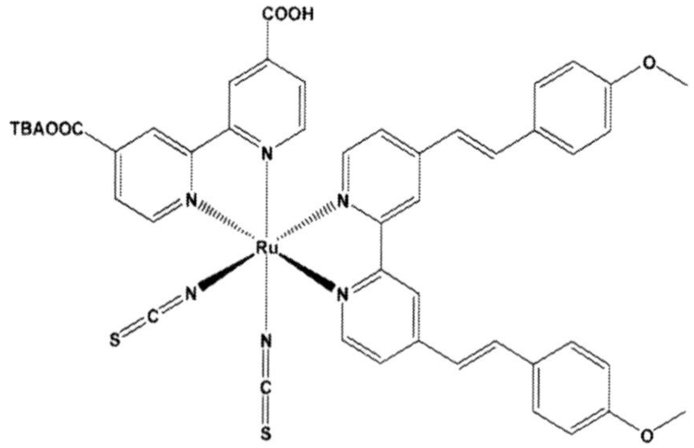

K-73

Fig. 8.8: Molecular structures of: [Z-907] Ru-(4,4′-dicarboxylic acid-2,2′-bipyridine)(4,4′-dinonyl)-2,2′-bipyridine) $(NCS)_2$; [K-19] Ru-(4,4′-dicarboxylic cid-2,2′-bipyridine)(4,4′-bis(p-hexyloxystyryl)-2,2′-bipyridine) $(NCS)_2$; [K-73] Ru-(4,4′-dicarboxylic acid-2,2′-bipyridine)(4,4′-bis(p-methoxystyryl)-2,2′-bipyridine) $(NCS)_2$.

Overall power conversion efficiencies of 6-10% have been achieved by a variety of photoanode nanoporous morphologies [80-82] incorporating hydrophobic tails onto the light sensitive dyes, e.g. K-19 and K-73 (**Fig. 8.8**) [83] and using solvent free electrolytes [84,85]. A DSSC using Z-907 dye (**Fig. 8.8**), underwent a stability test of 1000 hours at 80°C demonstrating a 6% conversion efficiency, losing 4% of its overall performance [86]. This is in contrast to amorphous silicon, which commonly demonstrates over 12,000 hours full-intensity operation with no signs of photo degradation.

8.3 PV-Electrolysis Systems for Hydrogen Production
[1-34,41,43,87-92]

This system employs a PV solar cell to generate electricity that is subsequently passed to a commercial-type water electrolyzer (discussed in Chapter 2), see **Fig. 8.9(a)**. An alternative system involves immersion of the semiconductor photovoltaic cell directly in an aqueous system, see **Fig. 8.9(b)**.

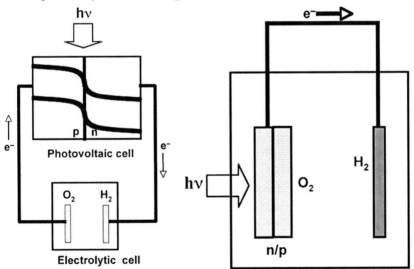

Fig. 8.9: Schematic diagram of PV-electrolysis systems proposed for solar water splitting: **(a)** Electricity generated from photovoltaic cell driving water electrolysis; **(b)** PV assisted cell with immersed semiconductor p/n junction as one electrode.

Commercial single-crystal Si solar cells generally have efficiencies in the 12-16% range, while water electrolysis units have energy efficiencies of about 85%. Thus the efficiency of the combined PV/electrolyzer system using commercially available components is close to 10% [43]. The system shown in **Fig. 8.9(b)** eliminates the costs and mechanical difficulties associated with separate construction and interconnection of solar and electrochemical cells; in these systems the electrodes are composed of single or multiple semiconductor p/n junctions that are irradiated while they are within the cell. This apparatus can be constructed with proper encapsulation of the semiconductors to protect them from the aqueous environment. Since PV-electrolysis units for hydrogen generation are expensive to build it is important to maintain maximum output power independently of variations in solar insolation and temperature [17]. Consequently commercial systems will require the integration of PV arrays with maximum power point tracking (MPPT) devices, see **Fig. 8.10**, which are described by average power conversion efficiency as a function of partial load [17,87].

Germany, Saudi Arabia, Brazil, Spain, Egypt, India, Switzerland and several other countries have chosen PV-electrolysis systems for solar hydrogen production. HYSOLAR (HYdrogen from SOLAR Energy) [17,88,89] is a German-Saudi Arabian cooperative research program that has built a 350 kW photovoltaic-electrolysis hydrogen production plant near Riyadh. The PV-cells are made from 5.7 cm diameter single-crystal silicon wafers mounted behind plastic Fresnel lenses that concentrate sun light by $\approx 33x$. Alkaline water splitting is used for hydrogen production, with a mixture of water and potassium hydroxide (KOH). At start-up this lens-cell combination had a reported conversion efficiency of 13 - 15% depending on the ambient temperature. The electrolytically produced hydrogen contains only small amounts of oxygen, nitrogen and carbon dioxide from the basic water, as well as KOH vapors from the production process [89,90].

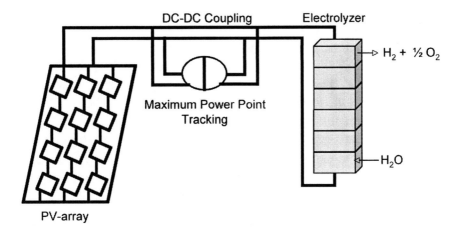

Fig. 8.10: Schematic diagram of PV-electrolysis system pilot plant [88-90].

Ohmari and co-workers [91] used a rf magnetron sputtered p-type c-Si/n-type a-Si:H thin film solar cell for photovoltaic assisted water electrolysis, with a solar to hydrogen conversion efficiency of 3.0%. Currao and co-workers [92] used an amorphous silicon solar cell in combination with a photoelectrochemical cell for water photoelectrolysis, with an AgCl photoanode and Pt cathode; two separate compartments comprising the anode and cathode were connected through a salt bridge. Illumination of both the AgCl photoanode and the amorphous silicon solar cell resulted in photoelectrochemical water splitting, with the voltage generated by the solar cell used to bias the electrodes.

8.4 Multi-junction PV Tandem Cells for Hydrogen Production [35-39,44,45,93-101]

Kocha et al. [93] reported a photovoltaic tandem cell consisting of $GaInP_2$ homojunction grown epitaxially upon a GaAs homojunction, connected through a (transparent) GaAs tunnel diode. This tandem cell is equivalent to two solar cells connected in a series, each utilizing a separate portion of the solar spectra. The $GaInP_2$ p/n junction, bandgap of 1.83 eV, absorbs visible light while the GaAs p/n junction, bandgap of 1.42 eV, absorbs in the near-infrared region. The voltages of the two cells are additive. After modifying the illuminated front surface by a Pt colloid to reduce photocorrosion, photoelectrochemical water decomposition was carried out in 1 M H_2SO_4. The stoichiometry of the simultaneously evolved H_2 and O_2 from the illuminated surface was found to be 2.8:1, after 8 hours O_2 ceased to evolve. Prior to sample decomposition the measured efficiency was found in the range of 4-10% for water splitting.

Khaselev and Turner [35] reported a novel direct water splitting system, shown schematically in **Fig. 8.11**. The integrated, monolithic photovoltaic-photoelectrochemical device consists of a 4.0 µm thick top layer of epitaxially grown p-$Ga_{0.52}In_{0.48}P$, which is connected in a series via a tunnel junction to a GaAs p/n junction bottom cell on a GaAs surface. This device differs from a standard solid-state tandem cell in that a PEC schottky-type junction has replaced the top p/n junction. Under illumination electrons flow toward the illuminated surface and holes flow toward the ohmic contact. For the device to function properly and efficiently the GaAs solar cell must provide sufficient voltage to overcome any energetic mismatch between the band edges of the $GaInP_2$ and the water redox reaction, and must also provide any additional voltage needed to overcome overvoltage losses from the H_2 and O_2 evolution reactions. Hence total photovoltage output must include the thermodynamics of water splitting, polarization losses for anodic and cathodic processes, and the current-resistance potential drop in the bulk of the electrolyte which can be significant when gas evolution occurs. Upon illumination with a 150 W tungsten-halogen lamp, water (freshly prepared 3.0 M H_2SO_4 + 0.01 M *t*-octylphenoxypoly ethoxyethanol used as an electrolyte) splits directly to produce H_2 at the semiconductor electrode and O_2 at the counter electrode in the ratio of 2:1 with a light-to-hydrogen conversion efficiency of 12.4% [35]. Two reasons have been cited for the preferred hydrogen production at semiconductor electrode: one is low overvoltage loss for the H_2 evolution reaction and the other is the cathodically protected semiconductor surface [94].

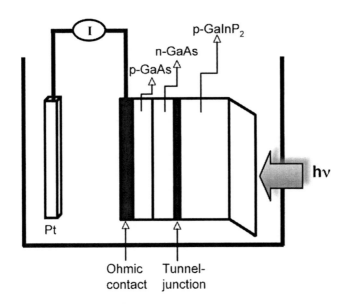

Fig. 8.11: Schematic model of a photoelectrochemical-photvoltaic p-GaAs/n-GaAs/p-Ga$_{0.52}$In$_{0.48}$P device [35].

Several monolithic, multi-junction integrated PV/electrolysis configurations and their efficiencies are summarized in **Table 8.3**. A solar-to-hydrogen conversion efficiency of over 16% has been demonstrated by a tandem configuration of n/p-GaInP$_2$|n/p-GaAs(Pt)|KOH|Pt [95]. Triple junction p-i-n-a-Si(Pt)|KOH|Pt exhibits 7.8% conversion efficiency [95]. The low current density of the a-Si system results in water electrolysis with an equivalent efficiency of 86% (effective electrolysis voltage 1.42 V). Litcht and co-workers [96] designed a novel multi-junction AlGaAs|Si-RuO$_2$|Pt$_{black}$ photoelectrolysis cell demonstrating water splitting at a record solar driven conversion efficiency of 18.3%. The cell structure consists of a thin top layer of sequentially grown p-Al$_{0.15}$Ga$_{0.85}$As (bandgap = 1.6 eV) on a GaAs cap layer atop a p$^+$-Si|n-Si|n$^+$-Si multi-junction. The bipolar configured semiconductors generate an open circuit voltage of 1.30 V and maximum power potential of 1.57 V, well suited to the thermodynamic potential of water splitting.

The principal solar water-splitting models predict dual-band gap photoelectrolysis efficiencies of 16% [40], and 10–18% [41].

These predicted efficiencies are lower than the observed water-splitting efficiency reported by Licht and co-workers [96,97], most likely due to underestimation of the experimental optical energy conversion (η_{photo}) attained by contemporary devices or underestimation of the attainable redox conversion of water to H_2 and O_2 ($\eta_{electrolysis}$). For example, the estimated values of $\eta_{photo} <$ 20% by Bolton and coworkers [41] are inherent to an estimated 10% reflection loss, 10% quantum-yield loss, and 20% absorption loss. However as summarized in **Table 8.3**, demonstrated η_{photo} values are considerably higher than 20%. Furthermore all cells shown in **Table 8.3** exhibit an open circuit photopotential greater than the minimum potential needed to split water, with most of these cells generating a photopotential in excess of 2 V. For example, two GaInP|GaAs cells in series, each cell having a maximum photopotential of 2.0 - 2.1 V and an open-circuit potential of 2.3 V, will drive three series-connected 1.3 - 1.4 V water electrolysis cells. Water electrolysis at 1.36 V yields electrolysis efficiencies of over 90% [97]. Predicted maximum $\eta_{photoelectrolysis}$ values using observed η_{photo} values of various dual bandgap sensitizers are summarized in **Table 8.3**. A challenge remains to select bipolar (multiple) band gaps with a combined maximum power point voltage tuned to the electrolysis potential of water [42,98].

A single chip photovoltaic water electrolysis device with 3% solar to hydrogen conversion efficiency is reported under AM 1.5, 100 mW/cm^2 illumination [99]. The stacked p-i-n a-Si solar cell was deposited on a SnO$_2$ substrate by plasma CVD. Co–Mo and Fe–Ni–O electrodes were prepared on, respectively, a stainless steel and Ni sheet by rf magnetron sputtering. The two electrodes were then adhered to the solar cell with conducting Ag paste, then submerged within a KOH solution for testing. Kelly and Gibson [100] designed a robust photoelectrochemical device using triple junction n-i-p a-Si:H solar cell coated with a fluorine doped tin oxide layer (FTO), that protected it from corrosion, and arranged in such a way that the outer p-type layer remained in contact with KOH electrolyte for water splitting.

Table 8.3: Predicted and measured photoelectrolysis efficiencies derived from equation

Photovoltaics	Light level	η_{photo} (measured)	$\eta_{photoelectrolysis}$ (Predicted maximum)	$\eta_{photoelectrolysis}$ (Experimental)	Ref.
GaInP/GaAs	1 Sun	30.3%	27-29%	-	97
GaInP/GaAs	180 Sun	30.2%	27-29%	-	97
GaAs/Si	350 Sun	29.6%	27-28%	-	97
InP/GaInAs	50 Sun	31.8%	29-30%	-	97
GaInP$_2$/GaAs; p, n/p	11 Sun	-	-	12.4%	35
GaInP$_2$/GaAs; n/p, n/p	1 Sun	28.5%	-	16.5%	95
p-i-n a-Si (triple junction)	1 Sun	9.0%	-	7.8%	95
AlGaAs/Si	1 Sun	21.2%	19-20%	18.3%	96,97
p-i-n a-Si (triple junction)[1]	1 Sun	-	-	2.5%	99
n-i-p a-Si (triple junction)	1 Sun	-	-	5-6%	100
CdTe:CIGS2	1 Sun	16.5%; 18.4%	-	6.77%	102

[1] one-chip PV device dipped into electrolyte.

Multiple Bandgap Tandem Thin-film PV Cells for H_2 Production

This approach is distinct from multi-junction thin-film PV tandem cells where the individual PV cells are grown upon the other [101,102]. Here two PV cells grown on a transparent conducting layer are connected in series. The infrared photons not absorbed by the PV cells are incident on the photoanode to reduce the needed overvoltage required for water splitting. An illustrative PEC set up is shown in **Fig. 8.12** [103], consisting of two illuminated PV cells, with a RuS$_2$ photoanode for oxygen evolution and a platinum foil cathode for hydrogen evolution.

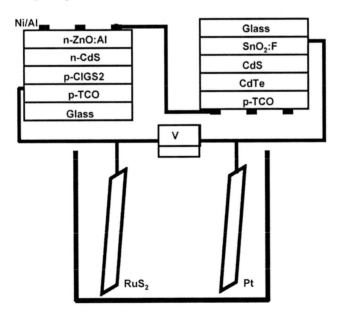

Fig. 8.12: Illustrative photoelectrochemical setup consisting of two thin-film PV cell and RuS$_2$ photoanode for oxygen evolution and Pt cathode for H$_2$ evolution [103].

DSSC-based Tandem Cell for Solar Hydrogen Production

A DSSC-based tandem cell has been described [104] composed of two cells in series. The first, or front cell absorbs the high-energy ultraviolet and blue light in sunlight through nano-crystalline metal oxide (e.g. WO_3) thin films to generate electron-hole pairs, with the valance band holes used to oxidize water. The second cell, a nanocrystalline TiO_2 DSSC cell captures light, passed by the first cell, in the green to red region of the solar spectrum. The electrons photogenerated in the conduction band of the second photosystem are used to generate hydrogen. The two cells operating together in a manner analogous to the Z-scheme of photosynthesis, provide the necessary potential required for water splitting, with an overall solar to hydrogen conversion efficiency of 6% [104], see **Fig.8.13.**

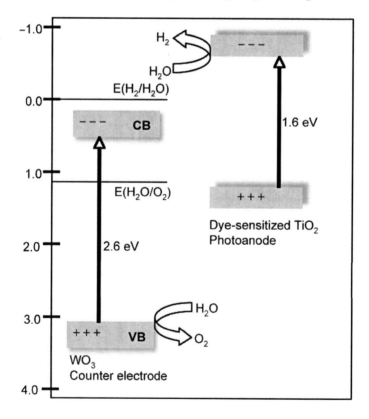

Fig. 8.13:   Z-scheme approach to photocatalytic water splitting using a DSSC based tandem cell.

References

1. Cox KE (1976) Hydrogen from solar energy via water electrolysis. Proc 11th IECEC pp. 926–932

2. Costogue EN, Yasui RK (1977) Performance data for a terrestrial solar photovoltaic/water electrolysis experiment. Sol Energy 19:205-210

3. Esteve D, Ganibal C, Steinmetz D, Vialason A (1980) Performance of a photovoltaic electrolysis system. Proc 3rd word Hydrogen Energy Conference, Tokyo. V. 3, pp.1583-1603

4. Koukouvinos A, Lygerou V, Koumoutsos N (1982) Design of a system for solar energy storage via water electrolysis Int J Hydrogen Energy 7:645-650

5. Carpetis C (1982) A study of water electrolysis with photovoltaic solar energy conversion. Int J Hydrogen Energy 7:287–310

6. Dahlberg R (1982) Replacement of fossil fuels by hydrogen. Int J Hydrogen Energy 7:121-142

7. Estève D, Ganibal C, Steinmetz D, Vialaron A (1982) Performance of a photovoltaic electrolysis system. Int J Hydrogen Energy 7:711-716

8. Dini D (1982) Hydrogen production through solar energy water electrolysis. Int J Hydrogen Energy 8:897–903

9. Carpetis C (1984) An assessment of electrolytic hydrogen production by means of photovoltaic energy conversion. Int J Hydrogen Energy 9:969-991

10. Murphy OJ, Bockris JOM (1984) Photovoltaic electrolysis: Hydrogen and electricity from water and light. Int J Hydrogen Energy 9:557-561

11. Bockris JOM, Dandapani B, Cocke D, Ghoroghchian J (1985) On the splitting of water. Int J Hydrogen Energy 10:179–201

12. Steeb H, Mehrmann A, Seeger W, Schnurnberger W (1985) Solar hydrogen production: Photovoltaic/electrolyzer system with active power conditioning. Int J Hydrogen Energy 10:353-358

13.   Kharkats YI, German ED, Kazarinov VE, Pshenichnikov AG, Pleskov YV.(1985) Hydrogen production by solar energy: Optimization of the plant "solar array + electrolyzer". Int J hydrogen Energy 10:617-621

14.   Delahoy AE, Gao SC, Murphy OJ, Kapur M, Bockris JOM (1985) A one-unit photovoltaic electrolysis system based on a triple stack of amorphous silicon (pin) cells. Int J Hydrogen Energy 10:113-116

15.   Appleby AJ, Delahoy SC, Gau SC, Murphy OJ, Kapur M, Bockris JOM (1985) An amorphous silicon-based one-unit photovoltaic electrolyzer. Int J Hydrogen Energy. 10:871-879

16.   Fischer M (1986) Review of hydrogen production with photovoltaic electrolysis system. Int J Hydrogen Energy 11:495–501

17.   Siegel A, Schott T (1988) Optimization of photovoltaic hydrogen production. Int J Hydrogen Energy 13:659–675

18.   Lin GH, Kapur M, Kainthla RC, Bockris JOM (1989) One step method to produce hydrogen by a triple stack amorphous silicon solar cell. Apl Phys Lett 55:386-387

19.   Ogden JM, Williams RH (1990) Electrolytic hydrogen from thin-film solar cell. Int J Hydrogen Energy 15:155–169

20.   Arashi H, Naito H, Miura H (1991) Hydrogen production from high-temperature steam electrolysis using solar energy. Int J Hydrogen Energy 16:603-608

21.   Bard AJ, Fox MA (1995) Artificial photosynthesis: solar splitting of water to hydrogen and oxygen: Acc Chem Res 28:141–145

22.   Abdel-Aal HK (1992) Storage and transport of solar energy on a massive scale: the hydrogen option. Int J Hydrogen Energy17:875-882

23.   Block DL, Melody I (1992) Efficiency and cost goals for photoenhanced hydrogen production processes. Int J Hydrogen Energy 17:853-861

24.   Barra L, Coiante D (1993) Hydrogen-photovoltaic stand-alone power stations: A sizing method. Int J Hydrogen Energy 18:337-344

25.   Gramaccio CA, Selvagi A,Galluzzi F(1993) Thin-flim multijunction solar cell for photoelectrolysis. Electochim Acta 38:111-113

26.   Kauranen PS, Lund PD, Vanhanen JP (1993) Control of battery backed photovoltaic hydrogen production. Int J Hydrogen Energy 18:383-390

27.   Bolton JR (1996) Solar photoproduction of hydrogen: review. Sol Energy 57:37–50

28.   Shukla PK, Karn RK, Singh AK, Srivastava ON (2002) Studies on PV assisted PEC solar cells for hydrogen production through photoelectrolysis of water. Int J Hydrogen Energy27:135-141

29.   Conibeer GJ, Richards BS (2007) A comparison of hydrogen storage technologies for solar-powered stand-alone power supplies: A photovoltaic system sizing approach. Int J Hydrogen Energy (in press)

30.   Conibeer GJ, Richards BS (2007) A comparison of PV/electrolyser and photoelectrolytic technologies for use in solar to hydrogen energy storage systems. Int J Hydrogen Energy (in press)

31.   Yamaguchi K, Udono H (2007) Novel photosensitive materials for hydrogen generation through photovoltaic electricity. Int J Hydrogen Energy (in press)

32.   Ahmad GE, El Shenawy ET (2006) Optimized photovoltiac system for hydrogen production. Renewable Energy 31:1043-1054

33.   Miri R, Mraoui S (2007) Electrolytic process of hydrogen production by solar energy. Desalination 206:69-77

34.   Rzayeva MP, Salamov OM, Kerimov MK (2001) Modeling to get hydrogen and oxygen by solar water electrolysis. International Journal of Hydrogen Energy 26:195-201

35.   Khaselev O, Turner JA (1998) A monolithic photovoltaic-photoelectrochemical device for hydrogen production via water splitting, Science 280:425–427

36.   Rocheleau RE, Miller EL, Misra A (1998) High efficiency photoelectrochemical hydrogen production using multijunction amorphous photoelectrode. Energy & Fuels 12:3-10

37. Licht S, Ghosh S, Trbutsch, H, Fiecher (2002) High efficiency solar energy water splitting to generate hydrogen fuel: probing RuS_2 enhancement of multiple band electrolysis. Sol Energy Mater Sol Cells. 70:471-480

38. Miller EL, Rocheleau RE, Khan S A (2004) Hybrid multijunction photoelectrode for hydrogen production fabricated with amorphous silicon/germanium and iron oxide thin films Int J Hydrogen Energy 29:907-914

39. Ingler WB, Khan SUM (2006) A self-driven p/n-Fe_2O_3 tandem photoelectrochemical cell for water splitting. 9:G144-G146

40. Weber MF, Dignam MJ (1986) Splitting water with semiconducting photoelectrodes--Efficiency considerations. Int J Hydrogen Energy 11:225–232

41. Bolton JR, Strickler SJ, Connolly JS (1985) Limiting and realizable efficiencies of solar photolysis of water Nature 316:495–500

42. Litcht S (2001) Multiple band gap semiconductor/electrolyte conversion. J Phys Chem B 105:6281

43. Bilgen E (2001) Solar hydrogen from photovoltaic-electrolyzer systems. Energy Conversion and Management 42:1047-1057

44. Litcht S (2005) solar water splitting to generate hydrogen fuel- a photothermal electrochemical analysis. Int J Hydrogen Energy 30:459-470

45. Hanna MC, Nozik AJ (2006) Solar conversion efficiency of photovoltaic and photoelectrolysis cell with carrier multiplication absorbers. J Appl Phys 100:074510-074518

46. Becquerel AE (1839) Recherches sur les effets de la radiation chimique de la lumiere solaire au moyen des courants electriques. Comptes Rendus de L´Academie des Sciences :145-149. *Republished:* Becquerel AE (1841) Annalen der Physick und Chemie 54:8-34

47. Becquerel AE (1839) Memoire sur les effects d´electriques produits sous l´influence des rayons solaires. Comptes Rendus de L´Academie des Sciences 9:561-567. *Republished:* Becquerel AE (1841) Annalen der Physick und Chemie. 54:35-42

48. Fritts CE (1883) On a New Form of Selenium Photocell. Proc American Association for the Advancement of Science. 33:97 and American Journal of Science 26:465

49. RS Ohl (1946) Light sensitive electric device. US Patent US2402662

50. Chapin DM, Fuller CS, Pearson GL (1954) A new silicon p-n junction photocell for converting solar radiation into electrical power. J Appl Phys 25:676–677

51. Jenny DA, Loferski JJ, Rappaport P (1956) Photovoltaic effect in GaAs p-n junctions and solar energy conversion. Phys Rev 101:1208–1209

52. Carlson DE, Wronksi CR (1976) Amorphous silicon solar cell. Appl Phys Lett 28:671-673

53. Carlson DE (1977) Semiconductor device having a body of amorphous silicon. US Patent US4064521

54. Carlson DE (1989) Amorphous silicon solar cell. IEEE Trans Electron Devices 36:2775-2780

55. Olson JM (1987) Current and lattice matched tandem solar cell. US Patent 4667059

56. Olson JM, Kurtz SR (1993) Current-matched high-efficiency, multijunction monolithic solar cell. US patent US 5223043

57. Bertness KA, Kurtz SR, Friedman DJ, Kibbler AE, Crammer C (1994) 29.5% efficient GaInP/GaAs tandem solar cells. Appl Phys Lett 65:989-99

58. King RR, Fetzer CM, Colter PC, Edmondson KM, Ermer JH, Cotal HL, Yoon H, Stavrides AP, Kinsey G, Krut DD, Karam NH (2002) 29th IEEE Photovolyaic Specialist Conference, pp.776-781

59. Wanlass MW, Ahrenkiel SP, Albin DS, Carapella JJ, Duda A, Emery K, Geisz JF, Jones K, Kurtz S, Moriarty T, Romero MJ. GaInP/GaAs/GaInAs Monolithic Tandem Cells for High-Performance Solar Concentrators. Optics & Photonics 2005 San Diego, California, USA

60. King RR, Law DC, Fetzer CM, Sherif RA, Edmondson KM, Kurtz S, Kinsey GS, Cotal HL, Krut DD, Ermer JH, Karam NH (2005) Pathways to 40%-efficient concentrator

photovoltaics. 20th European Photovoltaic Solar Energy Conference and Exhibition, Barcelona, Spain.

61. Bosi M, Pelosi C (2007)The potential of III-V semiconductors as terrestrial photovoltaic devices. Prog Photovolt: Res Appl 15:51-68

62. Yu KM, Walukiewicz W, Wu J, Shan W, Beeman JW, Sarpulla MA, Dubon OD, Becla P (2003) Diluted II-VI oxide semiconductors with multiple band gaps. Phys Rev Lett 91:246403–246405

63. Luque A, Hegedus S (2003). Handbook of Photovoltaic Science and Engineering, John Wiley & Sons New York

64. Green MA (1992) Solar cells-operation principles, technology and system applications, 2nd ed. The University of New South Wales, Kensington, New South Wales, Australia

65. Green MA (2001) Third Generation Photovoltaics: Ultrahigh conversion efficiency at Low Cost. Prog Photovolt Res Appl 9:123–125

66. Carlson DE (1977) Amorphous silicon solar cell. IEEE Trans Electron Devices 24:449-454

67. Yang J, Banerjee A, Guha S (1997) Triple-junction amorphous silicon alloy solar cell with 14.6% initial and 13.0% stable conversion efficiencies. Appl Phys Lett 70:2975 2978

68. Guha S, Narsimhan KL, Pietruszko SM (1981). On light induced effect in amorphous hydrogenated silicon. J Appl Phys 52:859–860

69. Staebler DL, Wronski CR (1977) Reversible conductivity changes in discharge-produced amorphous Si. Appl Phys Lett 31:292-294

70. Tsu DV, Chao BS, Ovshinsky SR, Guha S, Yang J (1997). Effect of hydrogen dilution on the structure of amorphous silicon alloy. Appl Phys Lett 71:1317–1319

71. Guha S, Yang J, Williamson DL, Lubianiker Y, Cohen JD, Mahan AH (1999) Structural, defect, and device behavior of hydrogenated amorphous Si near and above the onset of microcrystallinity. Appl Phys Lett 74:1860-1863

72. Zeman M, Schropp REI (1998) Amorphous and microcrystalline silicon solar cells: Modeling, materials and device technology, Kluwer Academic Publishers, Dordrecht, pp.181-182

73. Koh J, Ferlauto AS, Rovira PI, Wronski CR, Collins RW (1999) Evolutionary phase diagram for plasma enhanced chemical vapor deposition of silicon thin films from hydrogen diluted silane. Appl Phys Lett 75:2286-2289

74. Koh J, Lee Y, Fujiwara H, Wronski CR, Collins RW(1998) Optimization of hydrogenated amorphous silicon p-i-n solar cells with two steps i layers guided by real time ellipsometry. Appl Phys Lett 73:1526 1529

75. Colins RW, Ferlauto AS, Ferreira GM, Chen C, Koh J, Koval RJ, Lee Y, Pearce JM, Wronski CR (2003) Evolution of microstrucutre and phase in amorphous, protocrystalline, and microcrystalline silicon studied by realtime spectroscopic ellipsometry. Sol Energy Mat Sol Cells 78:143–180

76. Green MA, Emery K, King DL, Hishikawa Y, Warta W (2006) Solar efficiency Tables (version 28). Prog Photovolt: Res Appl 14:455–461

77. B.O. Regan BO, Grätzel M (1991) A low-cost high-efficiency solar cell based on dye-sensitized colloidal TiO2 thin film. Nature 353:737–740

78. Nazeeruddin MK, Kay A, Rodocio I, Humphry Baker R, Muller E, Liska P, Vlachopoulos N, Gratzel (1993) Conversion of light to electricity by Cis-X_2Bis(2,2'-bipyridyl-4,4'-dicarboxylate)ruthenium(II) charge-transfer sensitizers(X = Cl⁻, Br⁻, I⁻, CN⁻ and SCN⁻) on nanocrystalline TiO_2 electrodes. J Am Chem Soc 115:6382-6390

79. Han L, Fukui A, Fuke N, Koide N, Yamanaka R (2006) High efficiency of dye-sensitized solar cell and module. Conference Record of the 2006 IEEE 4th World Conference on Photovoltaic Energy Conversion, Hawai, USA pp. 179-182

80. Adachi M, Murata Y, Okada I, Yoshikawa S (2003) Formation of titania nanotube and applications for dye-sensitized solar cells. J Electrochem Soc 150:G488–G493

81. Mor GK, Shankar K, Paulose M, Varghese OK, Grimes CA (2006) Use of highly ordered TiO_2 nanotube arrays in dye-sensitized solar cell. Nano Lett 6:215–218

82. Z. S. Wang, H. Kawauchi, T. Kashima, H. Arakawa. Significance influence of TiO_2 photoelectrode on the energy conversion efficiency N719 dye-sensitized solar cell, Coord Chem Rev 248:1381-89

83. Kuang D, Ito S, Wenger B, Klein C, Moser JE, Baker RH, Zakeeruddin SM, Grätzel M (2006). High molar extinction coefficient heteroleptic ruthenium complxes for thin-film dye sensitized solar cells. J Am Chem Soc 128:4146–4154

84. Wang P, Wenger B, Baker RH, Moser JE, Teuscher J, Kantlehner W, Mezger J, Stoyanov EV, Zakeeruddin SM, Grätzel M (2005) Charge Separation and Efficient Light Energy Conversion in Sensitized Mesoscopic Solar Cells based on binary ionic liquids.. J Am Chem Soc 127:6850–6056

85. Kato T, Okazaki A, Hayase S (2005) Latent gel electrolyte precursors for quasi-solid dye sensitized solar cell. Chem Commun 363–364

86. Wang P, Zakeeruddin SM, Moser JE, Nazeeruddin MK, Sekiguchi T, Gratzel M (2003) A stable-quasi-solid state dye-sensitized solar cell with amphiphilic ruthenium sensitizer and polymer gel electrolyte. Nature Mater 2:402–406

87. Muhida R, Park M, Dakkak M, Matsuura K, Tsuyoshi A, Michira M (2003) A maximum power point tracking for photovoltaic-SPE system using a maximum current controller. Sol Energy Mater Sol Cells 75:697-706

88. Brinner A, Bussmann H, Hug W, Seeger W (1992) Test results of the HYSOLAR 10 kW Int J Hydrogen Energy 17:187–197

89. Brinner A. http://www.hysolar.com; for more details about 350 KW PV-electrolysis plant.

90. Schug CA (1998) Operational characteristics of high-pressure, high-efficiency water-hydrogen-electrolysis. Int J Hydrogen Energy 23:1113-1120

91. Ohmori T, Go H, Yamaguchi N, Nakayama A, Mametsuka H, Suzuki E (2001) Photovoltaic water electrolysis using the sputter-deposited a-Si/c-Si solar cells, Int J Hydrogen Energy 26:661-664

92. Currao A, Reddy VR, van Veen MK, Schropp REI, Calzaferri G (2004) Water splitting with silver chloride photoanode and amorphous silicon solar cell. Photochem Photobio Sci 3:1017–1025

93. Kocha SS, Montgomery D, Peterson MW, Turner JA (1998) Photoelectrochemical decomposition of water utilizing monolithic tandem cells, Sol Energy Mater Sol Cells 52:389–397

94. Gao X, Kocha S, Frank AJ, Turner JA (1999) Photoelectrochemical decomposition of water using modified monolithic tandem cells, Int J Hydrogen Energy 24:319–325

95. Khaselev O, Bansal A, Turner JA (2001) High-efficiency integrated multijunction photovoltaic/electrolysis systems for hydrogen production. Int J Hydrogen Energy 26:127–132

96. Licht S, Wang B, Mukerji S, Soga T, Umeno M, Tributsch H (2000). Efficient solar water splitting, exemplified by RuO2-catalyzed AlGaAs/Si photoelectrolysis. J Phys Chem B 104:8920–8924

97. Licht S, Wang B, Mukerji S, Soga T, Umeno M, Tributsch H (2001) Over 18% solar energy conversion to generation of hydrogen fuel; theory and experiment for efficient solar water splitting, Int J Hydrogen Energy 26:653–659

98. Licht S, Halperin L, Kalina M, Zidman M, Halperin N (2003) Electrochemical potential tuned solar water splitting. Chem Commun 3006-3007

99. Yamada Y, Matsuki N, Ohmori T, Mametsuka H, Kondo M, Matsuda A, Suzuki E (2003) One chip photovoltaic water electrolysis device. Int J Hydrogen Energy 28:1167–1169

100. Kelly NA, Gibson TL (2006) Design and Characterization of a robust photoelectrochemical device to generate hydrogen

using solar water splitting. Int J Hydrogen Energy 31:1658–1673

101. Dheere NG, Jahagirdar AH (2005) Photoelectrochemical water splitting for hydrogen production using combination of CIGS2 solar cell and RuO_2 photocatalyst. Thin Solid Films 480-481:462–465

102. Avachat US, Jahagirdar AH, Dheere NG (2006) Multiple band gap combination of thin film photovoltaic cell and a photoanode for efficient hydrogen and oxygen generation by water splitting. Sol Energy Mat Sol Cells 90:2464–2470

103. Avachat US, Dheere NG (2006) Preparation and characterization of transparent conducting ZnTe:Cu back contact interface layer for CdS/CdTe solar cell for photoelectrochemical application. J Vac Sci Technol A 24:1664–1667

104. Gratzel M (2005) Mesoscopic solar cells for electricity and hydrogen production from sunlight. Chem Lett 34:8–13

Index

Printed in the United States of America.

Printed in the United States
138183LV00001B/40/P

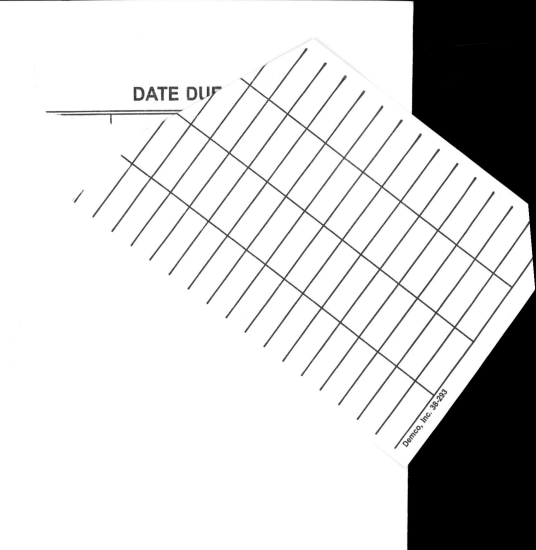

DATE DUE

Demco, Inc. 38-293